Signal Processing in Digital Communications

For a listing of recent titles in the *Artech House Signal Processing Library,* turn to the back of this book.

Signal Processing in Digital Communications

George J. Miao

ARTECH HOUSE
BOSTON | LONDON
artechhouse.com

Library of Congress Cataloging-in-Publication Data
A catalog record for this book is available from the U.S. Library of Congress.

British Library Cataloguing in Publication Data
A catalogue record for this book is available from the British Library.

ISBN: 1-58053-667-0
ISBN 13: 978-1-58053-667-7

Cover design by Yekaterina Ratner

© 2007 ARTECH HOUSE, INC.
685 Canton Street
Norwood, MA 02062

All rights reserved. Printed and bound in the United States of America. No part of this book may be reproduced or utilized in any form or by any means, electronic or mechanical, including photocopying, recording, or by any information storage and retrieval system, without permission in writing from the publisher. All terms mentioned in this book that are known to be trademarks or service marks have been appropriately capitalized. Artech House cannot attest to the accuracy of this information. Use of a term in this book should not be regarded as affecting the validity of any trademark or service mark.

10 9 8 7 6 5 4 3 2 1

To Lisa, Kathleen, and Julia

Contents

Preface   xv

1 Introduction   1
   1.1   A History of Communications Using Electricity   1
   1.2   Digital Communication Systems   3
   1.3   Digital RF Systems   6
       1.3.1   Digital Transceivers   7
       1.3.2   A/D Converter Challenge   9
       1.3.3   Digital Downconversion and Channelization   12
   1.4   Link Budget   17
       1.4.1   Noise Figure   19
       1.4.2   Receiver Sensitivity   19
       1.4.3   Maximum Path Loss   20
   1.5   Summary   21
   References   22

2 Probability, Random Variables, and Stochastic Signal Processing   25
   2.1   Introduction   25
   2.2   Probability   26
       2.2.1   Intuitive Probability   27
       2.2.2   Axiomatic Probability   27
       2.2.3   Conditional Probability   29
       2.2.4   Independence   31
   2.3   Random Variables   31
   2.4   Probability Distributions and Densities   32
       2.4.1   Probability Distributions   32
       2.4.2   Probability Densities   33
       2.4.3   Joint Probability Distributions and Densities   34

		2.4.4 Statistical Averages and Joint Moments	35
		2.4.5 Moment Generation Function	37
	2.5	Functions of Random Variables	39
		2.5.1 Sums of Random Variables	39
		2.5.2 Transformations of Random Variables	43
	2.6	Discrete Distributions and Densities	45
		2.6.1 Discrete Uniform Distribution	45
		2.6.2 Binomial Distribution	46
		2.6.3 Poisson Distribution	48
	2.7	Continuous Distributions and Densities	51
		2.7.1 Gaussian Density Function	51
		2.7.2 Error Function	54
		2.7.3 Q-Function	56
		2.7.4 Multivariate Gaussian Distribution	58
		2.7.5 Uniform Density Function	58
		2.7.6 Chi-Square Distribution	60
		2.7.7 $\mathcal{F}$ Distribution	61
		2.7.8 Rayleigh Distribution	63
		2.7.9 Rice Distribution	64
	2.8	Upper Bounds on the Probability	66
		2.8.1 Chebyshev Inequality	67
		2.8.2 Law of Large Numbers	68
		2.8.3 Central Limit Theorem	69
	2.9	Stochastic Signal Processes	70
		2.9.1 Definition of Discrete-Time Random Process	71
		2.9.2 Stationary Processes	73
		2.9.3 Estimated Functions	73
		2.9.4 Power Spectrum	74
		2.9.5 Stochastic Processes for Linear Systems	76
		2.9.6 Mean Square Estimation	81
	2.10	Detection Theory and Optimum Receivers	83
		2.10.1 Optimality Criterion	83
		2.10.2 Maximum Likelihood Detector	85
		2.10.3 Probability of Error	87
	2.11	Summary	87
	References		89
3	**Sampling Theory**		**91**
	3.1	Introduction	91

	3.2	Discrete-Time Sampled Signals	92
		3.2.1 Instantaneous Sampling	92
		3.2.2 Ideal Sampled Signal	94
	3.3	Nyquist Sampling Theorem	94
		3.3.1 Time-Domain Interpolation Formula	96
		3.3.2 Frequency-Domain Interpolation Formula	98
		3.3.3 Aliasing	99
	3.4	Undersampling	100
		3.4.1 Minimum Sampling Rate	102
		3.4.2 Antialiasing Bandpass Filter	103
	3.5	Stochastic Sampling Theorem	105
	3.6	Summary	106
		References	108

4 Channel Capacity   109

4.1	Introduction	109
4.2	Gaussian Channel Capacity	110
4.3	Bandlimited Channel Capacity	112
4.4	MIMO Channel Capacity	117
4.5	SIMO Channel Capacity	121
4.6	MISO Channel Capacity	121
4.7	Summary	122
	References	123

5 Smart Antenna Systems   125

5.1	Introduction		125
5.2	Smart Antennas and Beamforming Structures		126
	5.2.1	Switched Beamforming	126
	5.2.2	Delay-and-Sum Beamforming	128
	5.2.3	Space-Time Beamforming	129
	5.2.4	Interpolation Beamforming	130
5.3	Beamforming Algorithms		133
	5.3.1	MMSE Beamformer	133
	5.3.2	Maximum SNR of the Beamformer	135
	5.3.3	Minimum Variance Beamformer	137
5.4	Summary		139
	References		140

6 Channel Characterization and Distortion — 141
  6.1 Introduction — 141
  6.2 Wireless Channels — 143
    6.2.1 Free Space Propagation — 143
    6.2.2 Flat Surface Propagation — 146
    6.2.3 Multipath Propagation — 148
    6.2.4 Parameters of Multipath Channels — 150
    6.2.5 Fading Characteristics — 156
    6.2.6 Large-Scale Fading — 157
    6.2.7 Small-Scale Fading — 159
  6.3 Wired Channels — 163
    6.3.1 Transmission Loop — 163
    6.3.2 Crosstalk — 168
    6.3.3 Simulation Loop Model — 169
  6.4 Channel Distortion — 171
    6.4.1 Intersymbol Interference — 172
    6.4.2 Eye Diagrams — 174
    6.4.3 Nyquist Criterion — 178
  6.5 Pulse Shaping — 180
    6.5.1 Raised-Cosine Pulse — 181
    6.5.2 Gaussian Shaping Pulse — 182
  6.6 Matched Filtering — 183
  6.7 Summary — 184
  References — 187

7 Channel Estimation and Blind Identification — 189
  7.1 Introduction — 189
  7.2 Discrete-Time Channel Models — 191
  7.3 Channel Estimators — 194
    7.3.1 Maximum Likelihood Estimator — 194
    7.3.2 Least Squares Estimator — 198
    7.3.3 Generalized Least Squares Estimator — 201
    7.3.4 MMSE Estimator — 202
  7.4 Adaptive Channel Estimation and Algorithms — 208
    7.4.1 The LMS Algorithms — 210
    7.4.2 The LMS Algorithm Convergence — 214
    7.4.3 The LMS EMSE Analysis and Misadjustment — 217
    7.4.4 The RLS Algorithms — 219
    7.4.5 The RLS Algorithm Convergence — 223

        7.4.6   The RLS EMSE Analysis and Misadjustment   224
        7.4.7   Comparison of the Adaptive Algorithms   225
  7.5 Channel Models and HOS Estimations   226
        7.5.1   SISO Channel Model and Estimation   226
        7.5.2   SIMO Channel Model and Estimation   229
        7.5.3   MIMO Channel Model and Estimation   233
  7.6 Blind Channel Identification   238
        7.6.1   Blind Identification for SISO Channel   239
        7.6.2   Subspace Blind Identification for SIMO Channel   242
        7.6.3   Blind Identification for MIMO Channel   246
  7.7 Summary   252
  References   253

8 Adaptive Equalizers in Communication Receivers   257
  8.1 Introduction   257
  8.2 Linear Equalizer   260
        8.2.1   Channel Equalizer   262
        8.2.2   Mean-Square-Error Criterion   266
  8.3 Adaptive Linear Equalizer   272
        8.3.1   Adaptive Algorithms for an Equalizer   274
        8.3.2   Training Methodology   277
        8.3.3   Tap Length of Equalizer Coefficients   278
  8.4 Fractionally Spaced Equalizer   279
        8.4.1   Multirate Communication System Model   280
        8.4.2   Multichannel Model-Based Equalizer   283
        8.4.3   FSE-MMSE Function   287
        8.4.4   FSE Constant Modulus Algorithm   293
        8.4.5   FSE-CM Noisy Cost Function   297
        8.4.6   FSE-CM Performances   298
  8.5 Decision Feedback Equalizer   299
        8.5.1   MMSE for DFE   301
        8.5.2   Predictive DFE   307
        8.5.3   FSE-DFE   310
        8.5.4   Error Propagation   312
  8.6 Space-Time Equalizer   315
        8.6.1   Time-Only Equalizer   316
        8.6.2   Space-Only Equalizer   318
        8.6.3   Space-Time MMSE Equalizer   320

8.7	Diversity Equalizer	323
	8.7.1 Fundamentals of a Rake Receiver	324
	8.7.2 Adaptive Rake Receiver	330
	8.7.3 Equalized Rake Receiver	334
8.8	Summary	338
	References	342

9 Multicarrier Modulation, DMT, and OFDM — 345

9.1	Introduction	345
9.2	Fundamentals of Discrete Multitone Modulation	347
	9.2.1 Multitone Transmission	347
	9.2.2 Geometric SNR	351
	9.2.3 Optimum of Energy Minimum and Bit Loading Maximum	353
9.3	FFT-Based OFDM	359
	9.3.1 OFDM System	359
	9.3.2 OFDM Modulation by IFFT	361
	9.3.3 OFDM Demodulation by FFT	363
	9.3.4 ADC Resolution for the OFDM Modulation	364
	9.3.5 Equalized OFDM	369
9.4	Filter Bank–Based OFDM	373
	9.4.1 Filter Bank Transmultiplexer	373
	9.4.2 The DFT Filter Bank	375
	9.4.3 Polyphase-Based DFT Filter Bank	377
	9.4.4 Maximally Decimated DFT Transmitter Filter Bank	378
	9.4.5 Perfect Reconstruction of the DFT Filter Bank	379
9.5	Summary	382
	References	384

10 Discrete-Time Synchronization — 387

10.1 Introduction	387
10.2 Discrete-Time Phase Locked Loop	389
10.2.1 Discrete-Time Loop Filter	391
10.2.2 Phase Detector	397
10.2.3 Discrete-Time VCO	400
10.3 Timing Recovery	402
10.3.1 Early-Late Gate Synchronizer	402

       10.3.2 Bandedge Timing Recovery   406
       10.3.3 Decision-Directed Timing Recovery   407
       10.3.4 Multirate Timing Recovery   410
       10.3.5 Polyphase Filter Bank Timing Recovery   414
       10.3.6 Multicarrier Modulation Timing Recovery   418
  10.4 Carrier Recovery   424
       10.4.1 Carrier Phase Error   425
       10.4.2 Open-Loop Carrier Recovery   428
       10.4.3 Carrier Recovery for Multiple Phase Signals   430
       10.4.4 Decision-Directed Carrier Recovery   432
  10.5 Summary   435
  References   435

Appendix A: The z-Transform   437

  A.1 Introduction   437
  A.2 The z-Transform   437
      A.2.1 The z-Transform Properties   442
      A.2.2 Common Pairs of the z-Transform   445
      A.2.3 The z-Transform Transfer Function   447
  A.3 The Inverse z-Transform   448
      A.3.1 The Contour Integration   449
      A.3.2 The Examination Approach   449
      A.3.3 The Partial Fraction Expansion   450
  A.4 The z-Transform All-Pass and Phase Systems   452
      A.4.1 All-Pass Systems   452
      A.4.2 Phase Systems   453
      A.4.3 Decomposition of Phase Systems   454
      A.4.4 Compensation Systems   454
      A.4.5 FIR Systems to Minimum-Phase Systems   455
  References   456

Appendix B: Matrix Theory   457

  B.1 Introduction   457
  B.2 Vector Definitions   457
  B.3 Matrix Definitions   459
  B.4 Orthogonal Matrices   461
  B.5 Trace   461
  B.6 Matrix Differentiation   462
  B.7 Determinants   462

B.8 Matrix Inversion   464
B.9 Eigenanalysis   464
B.10 Spectral Decomposition Theorem   465
B.11 Singular Value Decomposition   468
B.12 Quadratic Forms   470
B.13 Maximization and Minimization Analysis   471
References   473

Appendix C: The Discrete Fourier Transform   475
C.1 Introduction   475
C.2 DFT Operation   475
C.3 IDFT Operation   476
C.4 DFT Matrix Operation   477
References   478

Appendix D: The Fast Fourier Transform   479
D.1 Introduction   479
D.2 FFT Methods   479
   D.2.1 Decimation-in-Time FFT Algorithm   480
   D.2.2 Decimation-in-Frequency FFT Algorithm   482
   D.2.3 Computational Complexity   484
D.3 Fixed-Point FFT Algorithm   486
   D.3.1 Quantization   486
   D.3.2 Fixed-Point Overflow, Scaling, and SNR   487
   D.3.3 Quantization Analysis of the FFT Algorithm   489
References   493

Appendix E: Discrete Mathematical Formulas   495
E.1 Complex Exponential Formulas   495
E.2 Discrete Closed-Form Formulas   495
E.3 Approximation Formulas   496
E.4 Logarithmic Formulas   497

About the Author   499

Index   501

Preface

Modern digital communications correspond to a major change in the design paradigm shift from fixed, hardware-intensive to multiband, multimode, and software-intensive for digital communication radios that a large portion of the signal processing functionality is implemented through programmable digital signal processing (DSP) devices. This provides the ability of the digital communication radios to change their operating bandwidths and modes to accommodate new features and capabilities. The digital communication radios not only reduce analog components of radio frequency (RF) but also emphasize DSP to improve overall receiver flexibility and performance for the RF transceiver, while traditional radios still focus on analog component design. This book attempts to present some important and new developments of signal processing technologies and approaches to the digital communications field that are likely to evolve in the coming decade. Signal processing advances will be the key to the future of the digital communication radios.

This book is a complete resource on signal processing for digital communications, including in-depth coverage of theories, algorithms, system design, analysis, and applications. Based on the author's extensive research and industry experience, this authoritative book presents an up-to-date and comprehensive treatment of all aspects, including digital, multirate, adaptive, and statistical signal processing technologies for the digital communication radios. This book provides excellent guidance in overcoming critical challenges in the field involving wireless and wireline channel characterization and distortion, smart antennas, link budget, channel capacities, digital RF transceivers, channel estimation and blind identification, multichannel and multicarrier modulation, discrete multitone (DMT) and orthogonal frequency division multiplexing (OFDM), discrete-time timing and carrier recovery synchronization, and adaptive equalizers at communication receivers.

The book offers a coherent treatment of the fundamentals of cutting-edge technologies and presents efficient algorithms and their implementation methods with detailed examples. Packed with over 1,370 equations and more than 100 illustrations, this book offers a one-stop reference to cover a wide range of key topics, from channel capacity, link budget, digital RF systems, smart antenna systems, probability, random variables and stochastic signal processing, sampling theory, pulse shaping and matched filtering, to channel models, estimation and blind identification, multicarrier, fast Fourier transform (FFT)– and filter bank–based OFDM, discrete-time phase locked loop, fractionally spaced, decision feedback, space-time, and diversity equalizers.

Chapter 1 begins with an introduction of a history of communications using electricity; this chapter also provides an overview of digital communication systems that are intended to present a broad topic of signal processing relative to digital communications. In addition, Chapter 1 addresses basic concepts of digital RF system and link budget.

Chapter 2 reviews fundamental theories of probability, random variable, and stochastic signal processing. This chapter presents probability distribution and density and upper bounds on the probability, and it focuses on stochastic signal processing for linear systems, detection theories, and optimum receivers.

Chapter 3 introduces sampling theory, including instantaneous sampling, Nyquist sampling theorem based on time-domain and frequency-domain interpolation formulas, and aliasing. Undersampling, which is often used for intermediate frequency sampling, is described to sample a bandpass signal at a receiver. In addition, this chapter presents stochastic sampling theorem with applications to bandlimited stochastic processes.

Chapter 4 presents Gaussian and bandlimited channel capacities. This chapter also explains a concept of the channel capacities to single-input multiple-output (SIMO), multiple-input single-output (MISO), and multiple-input multiple-output (MIMO) systems.

Chapter 5 discusses smart antenna systems and focuses on different beamforming structures. In addition, this chapter introduces beamforming algorithms for the smart antenna systems using optimization constraint methods.

The focus of Chapter 6 is channel characterizations and distortions that concentrate on wireless and wireline channels. Pulse shaping with methods of raised-cosine pulse and Gaussian shaping pulse is also addressed. Furthermore, this chapter introduces matched filtering in terms of maximum signal-to-noise ratio.

Chapter 7 considers discrete-time channel models and estimations for SISO, SIMO, and MIMO channels. This chapter discusses four methods of maximum likelihood, least square, generalized least square, and minimum mean-square error (MMSE) estimators for the channels. Moreover, this chapter presents adaptive channel estimations and algorithms and their convergence analysis. Finally, this chapter also introduces the use of blind identifications to estimate the channels in the absence of a training sequence.

Chapter 8 describes a set of equalizers at radio receivers and presents their operation theories, including linear and adaptive linear equalizers, fractional spaced and decision feedback equalizers, and space-time MMSE equalizers. In addition, this chapter introduces diversity equalizers based on adaptive Rake receivers.

Chapter 9 turns our attention to multicarrier modulation, DMT, and OFDM for radio receivers. This chapter begins by introducing fundamentals of DMT modulation and then presents FFT–based and filter bank–based OFDM. In addition, this chapter addresses efficient implementation methods of using polyphase-based and maximally decimated FFT filter banks for designing radio transceivers.

Chapter 10 covers discrete-time synchronizations, which describe discrete-time phase locked loop, timing recovery, and carrier recovery. Various methods of timing and carrier recoveries are introduced. These methods include early-late gate, bandedge, decision-directed, multirate, polyphase filter band, and multicarrier modulation for the discrete-time synchronizations.

I would like to thank many anonymous reviewers for their comments on enhancing the technical presentation of this book. I would especially like to thank the technical reviewers and the copyeditor at Artech House for thoroughly reading all the draft chapters and the final manuscript and providing many detailed and valuable suggestions and comments for the book. I truly appreciate their contributions.

I would like to thank the Artech House staff, including Kevin Danahy, Barbara Lovenvirth, Mark Walsh, and Audrey Anderson, for providing guidance on the writing of the entire manuscript, the publication, and the promotion of this book.

Of course, I would like to thank my family. It's hard to believe that my daughters, Kathleen and Julia, are now in middle school and elementary school, respectively. They are both studying hard at school to enhance their knowledge. They may not be interested in the area of digital signal processing and communications, which I love, but they could learn the fundamentals of this field and use them to explore their own fields of interest when they grow up.

My wife, Lisa, continues to be our cornerstone. Without her love, everything would fall apart. Perhaps there would be no way this book could have been finished without her encouragement.

All things are difficult before they are easy. Without seeking, nothing will be found. Drops of water wear out the stone. Everything is possible to a willing heart.

1

Introduction

1.1 A History of Communications Using Electricity

A history of communications with the first major technical undertaking using electricity is considered starting with the commercial telegraph service by William Cooke and Charles Wheatstone in England in 1839 and by Samuel Morse in the United States in 1844 [1]. The most important element of the telegraph was its instantaneous operation across longer distances. Telegraph was the first technology to the transmission of information data over communication channels. Thus, it led to many fundamental advances in the field of signal processing and communications.

The next major development in communications is the telephone, which was a direct outgrowth of increasing the message-handling capacity of telegraph lines. Joseph Stearns and Thomas Edison in 1870 demonstrated reliable communication systems for the simultaneous transmission of two and four telegraphic signals on a single wire. Meanwhile, in 1875, Alexander Graham Bell and Elisha Gray in the United States both invented practical telephones that could be used to transmit human speech over a single line, thereby leading to telephone for local wireline services. Another advance in communications is radio. Electromagnetic propagation had been discovered in a mathematical formula by James Clerk Maxwell in 1860. In 1888, Heinrich Hertz demonstrated the generation and detection of electromagnetic radiation in numerous experiments. Several years later, Marconi introduced wireless signal instruments to transmit signals over distances of several hundred miles in 1896. This led to widespread advances in wireless communications.

Harry Nyquist was an important contributor in the history of communications and did important work on thermal noise known as the *Johnson-Nyquist noise*, the stability of feedback amplifiers, and information theory. In 1927, Nyquist discovered that the number of independent pulses, which could be put through a telegraph channel per unit time, is limited to twice the bandwidth of the communication channel [2]. His early theoretical work laid the foundations for later advances by Claude Shannon. In 1948, the publications of Claude Shannon [3–5] established the mathematical foundations to reliably transmit the information content of a source over a communication channel with basic limits on the maximum rate. This gave birth to a new field called *information theory*. These results are essentially a combination of what is now known as the *Nyquist-Shannon sampling theorem.*

Information theory provides answers to two fundamental questions in communication theory: (1) What is the ultimate data compression? and (2) What is the ultimate transmission rate of communication in terms of the channel capacity? Information theory aids a basic theory to modem sequence developments in the communications area, including undersea cables for telephony, satellite communications, digital communications, spread spectrum communications, broadcasting, cellular mobile and wireless local area network (WLAN) communications, ultra wideband (UWB) communications (or impulse radio), Internet radio, and software-defined radio (SDR).

There are many textbooks and references that treat various topics on information and communication theories and related areas. For a broad treatment of these subjects, the interested reader may refer to the books by Cover and Thomas [6], Reza [7], Pierce [8], Proakis [9], Haykin [10], Viterbi [11], Rappaport [12], Stüber [13], and Reed [14].

1.2 Digital Communication Systems

In the emergence of new standards and protocols in wireless and wireline communications, one is often faced with the challenge of making applications wireless-friendly. The spread of wireless networks, such as WLAN 802.11 (a, b, and g) and UWB communications, and wireless cellular standards, such as the second generation IS-95 code division multiple access (CDMA), global system mobile (GSM), the third generation wideband CDMA (WCDMA), other future radios, and the new requirements for communication, seamless connectivity is bringing a faster, real-time nature to applications.

The initial deployment of multimode WLAN solutions, including 802.11 (a, b, and g), quickly takes further the discussion of wireless networking standards requiring seamless connectivity. However, these standards can easily coexist in multimode solutions if they occupy different areas of the spectrum. The multimode WLAN solutions deliver the best user experience and performance by providing access across disparate networks through dynamic selection of WLAN standards, depending on system capabilities, channel loads, and type of information. This enables high-speed and high-bandwidth multimedia applications.

Multimode cellular phones, which are able to switch among different wireless cellular standards, are growing much faster than single-mode phones. Technology innovation is accelerating to bring the ability of the multimode cellular phones to interface with other wireless network services. This creates seamless wireless connectivity between the wireless cellular and wireless networks standards, thereby rapidly meeting demand for wireless Internet connectivity.

The expectation of pervasive communication and information access without regard for the means is the fundamental benefit of multimode and multichannel wireless networking and wireless cellular devices. The challenge of creating sophisticated multimode and multichannel communication radios to enable the seamless wireless connectivity is compounded by the desire for next generation communication radios. The next generation communication radios should keep their hardware and software design from becoming obsolete when new standards and new technologies become available. Therefore, we need next generation communication radios to have flexible architecture with reprogrammable or reconfigurable capability to incorporate advanced signal processing techniques and sophisticated algorithms to enhance performance. Hence, we refer to the next generation communication radios as flexible multimode and multichannel-based digital communications systems or software-defined multimode and multichannel-based digital communication systems.

The basic elements of a software-defined multimode and multichannel-based digital communication system are illustrated by the general block diagram, as shown in Figure 1.1. This digital communication system consists of five major blocks, including an antenna, a programmable radio frequency (RF) section, wideband, high-speed analog-to-digital (A/D) and digital-to-analog (D/A) converters, digital down- and up-converters along with a multimode and multichannel, advanced signal processing and algorithms, and a programmable controller. The antenna, which may be an *antenna array* or a *smart antenna*, connects to the programmable RF section followed by a subsystem of the wideband, high-speed A/D and D/A converters. The subsystem of the wideband, high-speed A/D and D/A converters also has an interface with the digital down- and up-converters having multimode and multichannel capability, which is connected with the advanced signal processing and algorithms. The programmable controller is used to control all five of the major blocks.

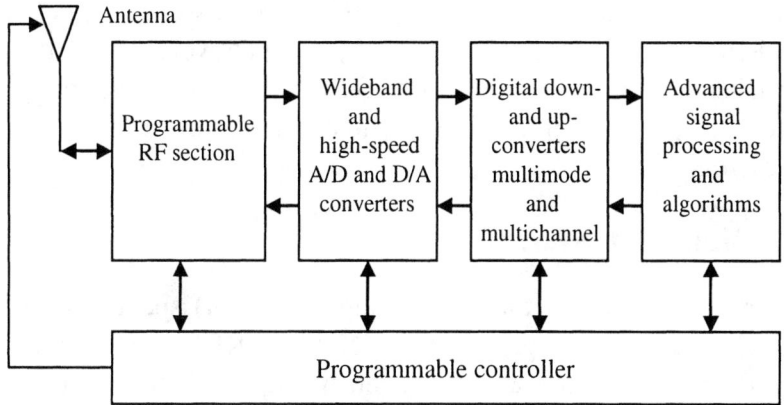

Figure 1.1 A general block diagram of the software-defined multimode and multichannel-based digital communication systems.

The challenge of creating the software-defined multimode and multichannel-based digital communication system is the broad scope of knowledge, including multirate, adaptive, and statistic signal processing and algorithms, multirate A/D and D/A converters, communication concepts and algorithms, RF systems and circuits, digital circuits, and software methodologies. In order to compensate for RF component limitations, understanding the ramifications of selecting RF parameters and the resulting limitations is important so that the appropriate subsequent signal processing can be used. Multirate signal processing offers an efficient way to deal with different sampling rates and can be used to channelize the frequency band into distinct communication channels. Furthermore, it is also a foundation for synchronization at the communication receivers.

There are many excellent books describing techniques in these areas. For multirate and adaptive signal processing, digital filter design, and multirate A/D and D/A converters, the interested reader may refer to Miao and Clements [15]. For multirate systems and filter banks, we recommend Vaidyanathan [16]. For the discrete-time signal processing and adaptive filter theory, we list Oppenheim and Schafer [17] and Haykin [18], respectively. For the RF microelectronics, we suggest Razavi [19]. Finally, we refer the interested reader to Pirsch [20] on the architectures of digital signal processing.

1.3 Digital RF Systems

Any frequency within the electromagnetic (EM) spectrum associated with radio wave propagation is called RF, which stands for radio frequency. When an RF current is supplied to an antenna, it gives rise to an EM field that propagates through space. This field is sometimes referred to as an RF field. In other words, the RF can be referred to the EM spectrum of EM waves that can be generated by alternating current fed to an antenna. Hence, the RF is a terminology that refers to alternating current having characteristics such that the EM field is generated suitable for wireless broadcasting and/or communications if the current is input to an antenna. These radio frequencies cover a significant portion of the EM radiation spectrum, extending from extremely/super low frequency 3–300 Hz (such as communication with submarines) to 1–30 GHz (such as mobile phone, WLAN, UWB, and most modern radars) or even over 300 GHz (such as night vision). Many technologies of wireless communications systems are developed based on RF field propagation.

An RF system design is unique, and it draws upon many disciplines related to RF knowledge, including wireless standards, signal propagation with multiple access, microwave and communication theory, random signal processing, transceiver systems and architectures, and integrated circuits (IC) and their design software tools.

An RF system is traditionally built based on RF electronic components, which have many undesired effects for a communication system. These effects include nonlinearity, harmonic distortion, gain compression, cross modulation, intermodulation, and random noise. Detailed treatments of these for the RF electronics can be found in

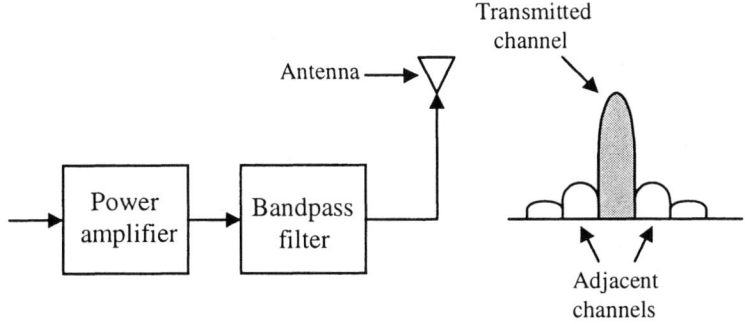

Figure 1.2 A transmitter to avoid leakage to adjacent channels.

Razavi [19]. In this section, we focus on RF theory and system as well as approaches for digital front-end RF radios.

1.3.1 Digital Transceivers

An RF transceiver contains a receiver and a transmitter. The receiver is usually more complex than the transmitter. The goal of the receiver is used to select the desired signal from unwanted interference and noise so that further processing, including demodulation, downconversion, and digital signal processing, can be implemented. In order to reject undesired signals received from an antenna and to provide conditions for further digital signal processing at the receiver, we must first filter it to remove the undesired signals, convert the signal to a center frequency with an amplitude compatible with an A/D converter, and then implement the A/D conversion process to generate a digital signal.

A frequency bandlimited channel for each user impacts the design of the RF transceiver. The transmitter has to consider using bandlimited modulation, amplification, and a bandpass filter (BPF) to avoid leakage to adjacent channels as shown in Figure 1.2. On the other hand, the receiver must have the ability to deal with the desired

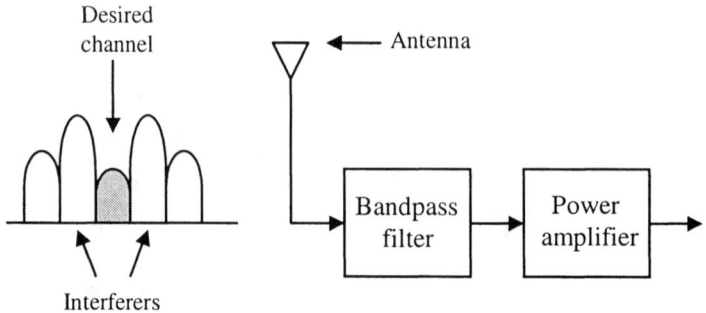

Figure 1.3 A receiver to accept the desired channel and to reject interferers.

channel and sufficiently reject strong neighboring interferences. Figure 1.3 shows a receiver, including an antenna coupled to a BPF followed by a low-noise amplifier (LNA), and a relationship between a desired channel and interferers.

One of the fundamental ideas of digital transceivers is to expand digital signal processing toward the antenna. Of special interest is that analog signal components at the RF front end, which they have dominated so far, are replaced by using digital signal processing, thereby leading to the name of digital transceiver, or digital front end. Thus, the digital transceiver that is derived from the RF front end and digital signal processing is a part of a system to realize front-end functionalities digitally, including downconversion, sample rate conversion, and channelization.

An architecture of the ultimate digital receiver with a minimum of analog components at the RF front end is shown in Figure 1.4. In this system, downconversion and channelization tasks are pushed into the digital signal processing (DSP) for further processing while LNA, bandpass antialiasing filter, and an A/D converter have to process the complete signal bandwidth for which the

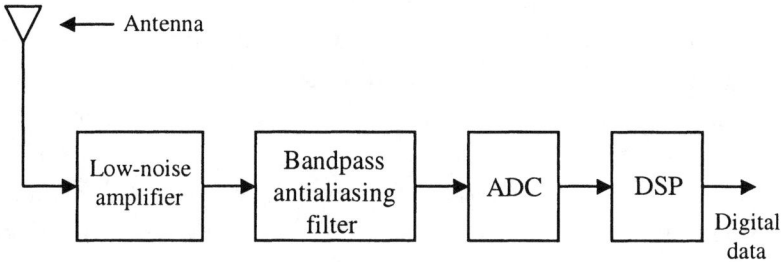

Figure 1.4 An ideal digital receiver with a minimum of analog RF components.

digital receiver is designed. However, the typical characteristics of wireless communications signals, including fading, shadowing, and multipath caused by RF signal propagation and potentially strong blocking and interfering signals due to the coexistence of several transmit signals, require a very high dynamic range. In addition, the digital receiver has to process a large number of channels simultaneously because the downconversion and channelization selection are shifted from the analog domain to the digital domain. Pushing narrowband signals at wideband reception yields a dynamic range far above what conventional receivers have to deal with. Furthermore, extending demodulation to signals of different wireless communications standards, which may appear at the digital receiver simultaneously, increases the dynamic range even more. Thus, the A/D converter is a key component, which has to cover signals of large bandwidth and high dynamic range.

1.3.2 A/D Converter Challenge

Dynamic range of the A/D converter can be increased either by increasing the number of bits or increasing the oversampling ratio.

The dynamic range of the A/D converter is increased 6 dB for every bit added. It is increased 3 dB for every doubling of the oversampling ratio. This is equivalent to improving the resolution of the A/D converter by one-half bit [15]. Benefits of oversampling in the A/D converter are to reduce quantization noise in the frequency band of interest and to decrease the requirement of sharp cutoffs for the anti-aliasing analog filter.

Another type of the A/D converters based on *noise shaping* is referred to as a sigma-delta A/D converter. The first-order sigma-delta A/D converter increases dynamic range 9 dB for every doubling of the oversampling ratio. The 3 dB in this dynamic range increase is due to the reduction in quantization noise power and a further 6 dB is due to the noise-shaping filter. This is equivalent to improving the A/D converter resolution by 1.5 bits. The second-order sigma-delta A/D converter dynamic range increases 15 dB for every doubling of the oversampling ratio. In other words, the resolution of the second-order sigma-delta A/D converter equivalently increases by 2.5 bits. In general, for every doubling of the oversampling ratio, the Nth-order sigma-delta A/D converter can improve $(6.02b + 3.01)$ dB in dynamic range, where b is the number of bits [15]. Equivalently, the dynamic range of the sigma-delta A/D converter can be increased by $(b + 0.5)$ bits approximately.

Recently, a specific architecture of A/D converter achieved a very high speed based on a parallel A/D converter bank by using N low-speed and low-cost A/D subconverters operating at an N-fold lower sampling rate. This type of A/D converter is referred to as the filter bank–based A/D converter. The basic idea of such an A/D converter is to first decompose the analog signal into subband signals by using a set of analog filter banks. The subband signals are then sampled by using a set of A/D subconverters at a sampling rate of $\frac{F_s}{N}$, where F_s is the sampling rate, and converted into digital subband signals, with nonoverlapping frequency bands of bandwidth $2\pi/N$. Such A/D converter architecture has many applications in very high-speed areas, including UWB [21], wireless and wireline communications, SDR, radars, and modern defense applications.

Introduction

In a multicarrier communication system, the synergy of the large number of carriers and the possible large difference in their RF power accounts for steep dynamic range requirements on the A/D converter. To illustrate the dynamic range requirements of A/D converters in a multicarrier communication system, we consider an example with two signals s_d and s_b in the received wideband range, where s_d is the desired signal with the power of P_d, and s_b is the *blocking signal* with the power P_b, and $P_d \ll P_b$. The blocking signal s_b is also assumed to be a Gaussian distribution with a zero mean.

To keep the clipping probability less than a maximum of 5% on the A/D converter with the full-scale range X_{max}, it has been shown [22] that X_{max} is approximately equal to $X_{max} \approx 4\sqrt{P_b}$ by using the properties of the Gaussian distribution. Since $P_d \ll P_b$, the effect of the desired signal s_d is neglected. For a uniform quantization, a step size Δ or a resolution of the quantization of the A/D converter is given by [15]

$$\Delta = \frac{X_{max}}{2^b}, \qquad (1.1)$$

where b is the number of bits, and the variance of the quantization error $e[n]$ is therefore

$$\begin{aligned} \sigma_e^2 &= \frac{\Delta^2}{12} \\ &= \frac{1}{12}\left(\frac{X_{max}^2}{2^{2b}}\right). \end{aligned} \qquad (1.2)$$

If the desired signal s_d has the frequency bandwidth B_d, the power of the quantization noise P_q is then given by

$$\begin{aligned} P_q &= \sigma_e^2 \left(\frac{B_d}{F_s/2}\right) \\ &= \left(\frac{X_{max}^2}{2^{2b}}\right)\left(\frac{B_d}{6F_s}\right) \\ &\approx \frac{8}{3}\left(\frac{P_b B_d}{F_s 2^{2b}}\right), \end{aligned} \qquad (1.3)$$

where F_s is the sampling rate. Let SNR_{min} be the minimum of *signal-to-noise ratio* (SNR) required for the desired signal s_d,

$$\text{SNR}_{min} = \frac{P_d}{P_q}. \tag{1.4}$$

We then obtain the quantization noise power in terms of the desired signal power P_d and SNR_{min} as follows:

$$P_q = \frac{P_d}{\text{SNR}_{min}}. \tag{1.5}$$

Substituting (1.5) into (1.3) yields

$$2^{2b_{min}} = \text{SNR}_{min} \left(\frac{8 B_d P_b}{3 F_s P_d} \right). \tag{1.6}$$

Taking $\log_2$ of both sides of (1.6), we obtain the *minimum number of bits* b_{min} required for the A/D converter resolution as follows:

$$b_{min} = \frac{1}{2} \log_2 \left[\text{SNR}_{min} \left(\frac{8 B_d P_b}{3 F_s P_d} \right) \right]. \tag{1.7}$$

If we now assume that two signals s_d and s_b are between 1.0 MHz and 1.8 MHz, and that the receiver needs to withstand the blocking signal s_b at 80 dB above the desired signal s_d with $\text{SNR}_{min} = 20$ dB, the desired signal bandwidth $B_d = 150$ kHz, overall bandwidth 1.8 MHz, and sampling rate $F_s = 7.2$ MHz. In this case, using (1.7), the minimum number of bits for the A/D converter is 14.52 bits.

1.3.3 Digital Downconversion and Channelization

In the previous sections, we introduced the ideal digital receiver with a requirement of the minimum analog RF components in Figure 1.4. This architecture of the digital receiver is sometimes referred to as *full-band digitization* or *direct conversion*. It needs to have a very high-speed A/D converter with high resolution to cover a very wide frequency bandwidth. Presently, realizing the architecture of

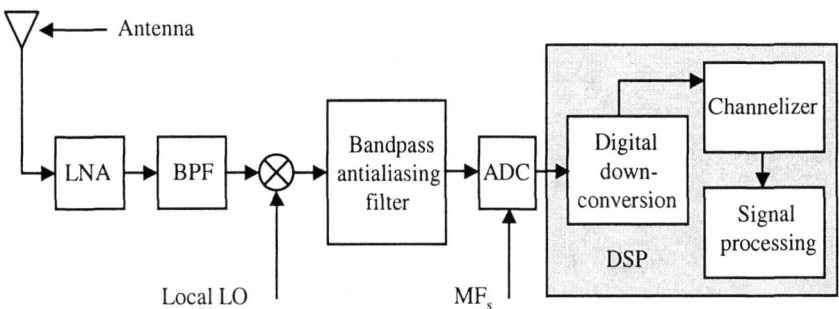

Figure 1.5 A partial-band architecture of a digital receiver using the IF sampling process.

the ideal digital receiver is still a difficult problem and will remain as a challenge in the near future. A more feasible approach is to perform *partial-band digitization*, which is to select a *limited-band digitization* out of the full-band frequency range by using an analog conversion and *intermediate frequency* (IF) filtering [23]. Such an approach leads to having an architecture of the digital receiver employing IF sampling.

Figure 1.5 shows the partial-band architecture of a digital receiver using the IF sampling technique. The BPF is an RF image-reject bandpass filter. The problem of images is a serious issue in the type of architecture designs. This is because each wireless standard imposes constraints upon the signal emissions by its own users based on the RF regulations. It does not have control over the signal in adjacent bands. In some cases, the image power can be much higher than the desired signal power. The most common approach to suppress the images is through the use of an RF image-reject bandpass filter, which is placed before the mixer. Usually, we design the RF image-reject bandpass filter with a relatively small loss in the desired signal band and a large attenuation in the image band. This

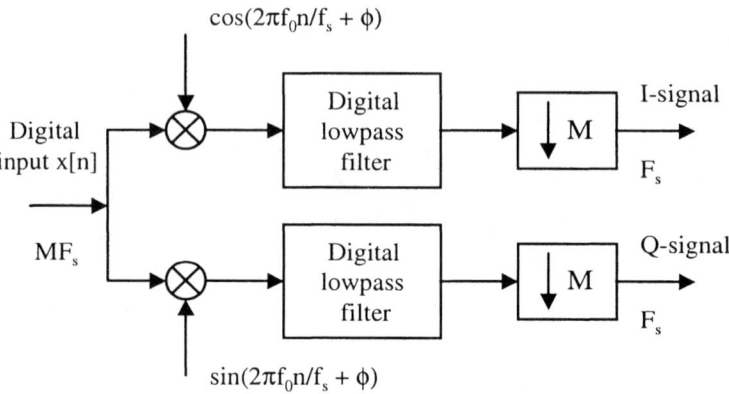

Figure 1.6 A conventional block diagram of a digital downconversion and channelization for the digital receiver.

redundant approach leads to a relatively narrow signal bandwidth for the A/D converter to process. The A/D converter is commonly operated at an oversampling frequency rate of MF_s, where F_s is the sampling rate and M is an integer, due to the higher center frequency of the IF signal. This means that the A/D converter must operate at a higher sampling rate than in the baseband.

A conventional block diagram of a digital downconversion and channelization is shown in Figure 1.6. Here, the digital input signal $x[n]$ with the sampling rate of MF_s is at the IF rather than at baseband. The digital downconversion is performed by using digital multipliers $\cos(2\pi \frac{f_0}{f_s} n + \phi)$ and $\sin(2\pi \frac{f_0}{f_s} n + \phi)$, where f_0 is the IF center frequency, ϕ is the phase, and f_s is the oversampling rate equal to MF_s, to convert the IF signal into the baseband signal. The digital channelization uses digital lowpass filters followed by downsampling by M (or resampling operation) to generate I and Q signals at the sampling rate of F_s because of the higher sampling rate at the input. It is also necessary for channel filtering to extract frequency divided channels. In this case, the digital lowpass filters have to attenuate adjacent channel interferers, meet the blocking characteristics, and reject aliasing after downsampling. Thus, the entire process is a digital translation process in which the digital lowpass filters are designed to have linear phase characteristics. Furthermore, the operation of the digital lowpass filters and downsampling can be implemented in a very efficient way by using polyphase filter structures [15, 24] or a k-stage cascaded integrator comb (CIC) filter capable of larger integer resampling followed by an n-stage half-band filter to finish the compensation of spectral and gain control [25].

A special case of the digital downconversion and channelization is if the center frequency of the digitized channel of interest is equal to a quarter of the sampling rate [23], that is, $f_s = 4f_0$. We also assume that the phase ϕ is equal to zero or can be controlled to zero. In this case, the digital multipliers can be rewritten as follows:

$$\cos\left(2\pi \frac{f_0}{f_s} n + \phi\right) = \cos\left(\frac{\pi n}{2}\right) = \{1, 0, -1, 0, \cdots\} \qquad (1.8)$$

and

$$\sin\left(2\pi \frac{f_0}{f_s} n + \phi\right) = \sin\left(\frac{\pi n}{2}\right) = \{0, -1, 0, 1, \cdots\}. \qquad (1.9)$$

Equations (1.8) and (1.9) indicate that the half of the digital samples in the digital *cosine* multiplier product are set to zero and the complementary set of the digital samples in the digital *sine* multiplier product are also set to zero, respectively. Further note that these zero values cannot contribute to the outputs of the digital lowpass filters operation. Thus, we disregard these zero values since we know the location of these zeros and account for their effect in shifting the nonzero digital samples through the digital lowpass filters.

Let the digital lowpass filters have a finite impulse response (FIR) lowpass filter having the characteristics given by its system

function
$$H(z) = \sum_{n=0}^{N-1} h[n]z^{-n}, \qquad (1.10)$$

where $h[n]$ is the filter impulse response. Because of the downsampling by $M = 4$, it is possible to decompose the filter $H(z)$ into the four-channel filter bank with polyphase form as follows [15]:

$$\begin{aligned} H(z) &= \sum_{k=0}^{3} z^{-k} D_k(z^4) \\ &= D_0(z^4) + z^{-1}D_1(z^4) + z^{-2}D_2(z^4) + z^{-3}D_3(z^4) \end{aligned}$$
(1.11)

where the polyphase filters are

$$D_0(z) = \sum_{n=0}^{q-1} h[4n]z^{-n}, \qquad (1.12)$$

$$D_1(z) = \sum_{n=0}^{q-1} h[4n+1]z^{-n}, \qquad (1.13)$$

$$D_2(z) = \sum_{n=0}^{q-1} h[4n+2]z^{-n}, \qquad (1.14)$$

and

$$D_3(z) = \sum_{n=0}^{q-1} h[4n+3]z^{-n}, \qquad (1.15)$$

where q is the largest integer of $\lfloor \frac{N}{4} \rfloor$. Thus, using digital multipliers in (1.8) and (1.9), and the polyphase filters in (1.12), (1.13), (1.14), and (1.15) for combination of the digital lowpass filters and downsampling, we can further integrate and simplify the entire digital downconversion and channelization shown in Figure 1.6 into a simple and very efficient polyphase filter bank architecture shown in Figure 1.7. This polyphase filter bank architecture of the digital downconversion and channelization can translate the IF frequency

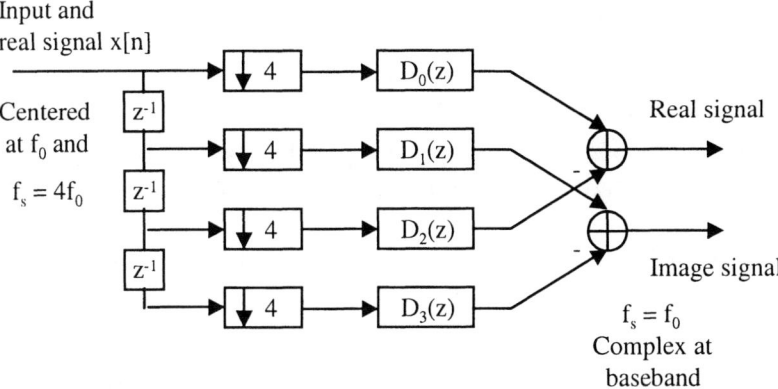

Figure 1.7 An efficient polyphase filter bank structure of the digital downconversion and channelization for the digital receiver.

band signal with quarter-sample rate to the baseband signal with downsampling of 4 and convert a real-input signal $x[n]$ into a complex output signal, real signal $x_I[n]$, and image signal $x_Q[n]$, simultaneously.

1.4 Link Budget

A *link budget* commonly refers to a complete gain and loss equation from a transmitter, through channel mediums (including air, cable, waveguide, fiber, and so on), to a receiver. The calculation of power and noise levels between the transmitter and the receiver by taking account of all gain and loss yields operating values of link margin above threshold in terms of SNR and achieving a minimum bit error rate (BER) requirement.

The effective received carrier signal power is defined by

$$P_c = \frac{P_t G_t G_r}{L_L L_p}, \qquad (1.16)$$

where P_t is the transmitted carrier power, G_t is the transmitted antenna gain, G_r is the received antenna gain, L_L is the receiver implementation loss, L_p is the path loss, and P_c is the received carrier signal power.

The total input noise power at the receiver is given by [7, 13]

$$N = kTBF, \qquad (1.17)$$

where the value of kT at a temperature of $17°$ C (or 290K) is equal to -174 dBm/Hz, B is the noise bandwidth at the receiver, and F is the noise figure (typically 5 to 6 dB). Thus, the received *carrier-to-noise ratio* (CNR) Γ is obtained by

$$\begin{aligned}\Gamma &= \frac{P_c}{N} \\ &= \frac{P_t G_t G_r}{kTBFL_L L_p}. \end{aligned} \qquad (1.18)$$

The modulated symbol *energy-to-noise ratio* (ENR) is defined by $\frac{E_s}{N_0}$, where E_s is the received energy per modulated symbol and N_0 is the white noise power spectral density. There is a relationship between the received CNR and the modulated symbol ENR given by [13]

$$\frac{E_s}{N_0} = \left(\frac{B}{R_s}\right)\Gamma, \qquad (1.19)$$

where R_s is the modulated symbol rate. Substituting (1.18) into (1.19), we can rewrite the link budget of (1.18) into a new form as follows:

$$\frac{E_s}{N_0} = \frac{P_t G_t G_r}{kTR_s FL_L L_p}. \qquad (1.20)$$

Note that there is only one difference between CNR in (1.18) and ENR in (1.20). While CNR uses the noise bandwidth B, ENR uses the modulated symbol rate R_s in the denominator.

1.4.1 Noise Figure

SNR is one of the most important parameters used in many analog circuits, signal processing, and communications. Even though the ultimate goal is to maximize the SNR for the received signal at the receiver, most of the front ends of receivers are characterized in terms of their "noise figure" rather than the input noise. *Noise figure* is normally measured as

$$F = \frac{\text{SNR}_{in}}{\text{SNR}_{out}}, \qquad (1.21)$$

where SNR_{in} and SNR_{out} are the signal-to-noise ratios measured at the input and output of a receiver system.

Equation (1.21) can be expressed in terms of decibels as follows:

$$F = 10\log_{10}\left(\frac{\text{SNR}_{in}}{\text{SNR}_{out}}\right) \quad \text{(dB)}. \qquad (1.22)$$

Understanding the physical meaning of (1.21) or (1.22) is important since the noise figure is a measure of how much the SNR degrades as the signal passes through the receiver system. If the receiver system does not have noise, we then have $\text{SNR}_{out} = \text{SNR}_{in}$, regardless of the gain. This is because using the same factor, without additional noise, attenuates both the input signal and input noise. In this case, the noise factor for a noiseless receiver system is equal to unity or 0 dB. In practice, the receiver system with the finite noise degrades the SNR to yield the noise figure $F > 1$, that is, the noise figure is always greater than 0 dB.

1.4.2 Receiver Sensitivity

Receiver sensitivity of the communication systems is referred to as the ability of the receiver to detect a radio signal in the presence of noise, which can arise from a variety of sources including external and internal to a receiver system. In other words, the receiver sensitivity is also referred to as the minimum signal level in which an RF system can detect with acceptable SNR.

Given (1.20), the receiver sensitivity can be expressed as

$$S_{RS} = kTR_s FL_L \left(\frac{E_s}{N_0}\right), \quad (1.23)$$

where S_{RS} is the receiver sensitivity. Equation (1.23) can be expressed in decibel units as follows:

$$S_{RS} = kT + R_s + F + L_L + \frac{E_s}{N_0}, \quad (1.24)$$

where S_{RS}, kT, R_s, F, L_L, and $\frac{E_s}{N_0}$ are in the units of dBm, dBm/Hz, dBHz, dB, dB, and dB, respectively.

Assume that the receiver implementation loss L_L in (1.24) is equal to 0 dB in an ideal case and $kT = -174$ dBm/Hz at room temperature. Then (1.24) can be further simplied as

$$S_{RS} = -174 + R_s + F + \frac{E_s}{N_0}. \quad (1.25)$$

Note that the sum of the first three terms is the total integrated noise of the receiver system and is sometimes referred to as the *noise floor*. In addition, all the parameters in (1.25) are usually fixed except for $\frac{E_s}{N_0}$. Therefore, in order to determine the minimum of the receiver sensitivity, we first calculate the minimum of $\frac{E_s}{N_0}$ dB, and then substitute it into (1.25). In this case, (1.25) is referred to as the *minimum receiver sensitivity*.

1.4.3 Maximum Path Loss

A path loss can be determined by using the receiver sensitivity in (1.23), the link budget in (1.20), and the minimum receiver sensitivity in (1.25). We can first rewrite (1.20) to obtain the path loss L_P

$$L_P = \frac{P_t G_t G_r}{kTR_s FL_L \left(\frac{E_s}{N_0}\right)}. \quad (1.26)$$

Substituting (1.23) into (1.26), with an ideal case of no implementation loss $L_L = 0$ dB, we obtain the path loss L_P in terms of the receiver sensitivity as follows:

$$L_P = \frac{P_t G_t G_r}{S_{RS}}. \tag{1.27}$$

If S_{RS} is the minimum receiver sensitivity given by (1.25), then (1.27) can be rewritten in terms of decibel units as follows:

$$L_{max} = P_t + G_t + G_r - S_{RS}, \tag{1.28}$$

where S_{RS} is the minimum receiver sensitivity in dBm, P_t is the transmitted carrier power in dBm, G_t is the transmitted antenna gain in dB, G_r is the received antenna gain in dB, and L_{max} is the unit of dB, which is referred to as the *maximum path loss* or the *maximum allowable path loss*.

1.5 Summary

In this chapter, we have first introduced the history of communications from the first major technical milestone of telegraph service in 1893 to Nyquist-Shannon's communication theory in 1948, which ignited subsequent developments in the field of communications. We have then presented a short review of digital communication systems with emphasis on software-defined multimode and multichannel-based digital communication systems. Subsequently, we have addressed designing digital RF systems with emphases on concepts of developing digital transceivers, A/D converter, digital downconversion, and channelization. These discussions have led to a necessary step of understanding the ultimate software-defined radios or digital RF systems that can accept fully programmable operation and control information and support a broad range of frequencies, air interfaces, and applications software in a single communication device with the operation capabilities of multiband and multimode. Furthermore, we have focused on link budgets with respect to noise figure, receiver sensitivity, and maximum path loss, thereby providing a key guideline for designing digital communication systems.

Of particular importance in the evaluation of communication system performance is the Nyquist-Shannon sampling theorem that laid the mathematical foundations for information theory, which has developed rapidly over the last five decades along with the practical applications to digital communications. Digital communication systems and networks that exist today and those currently under development certainly reflect these recent advances in information theory. In fact, this theory is an important mathematical apparatus not only in the statistical modeling of information sources and communication channels but also in the design of modern digital communication systems, including software-defined radios (SDRs), digital RF systems, *multiple-input multiple-output* (MIMO)-based smart antenna systems, orthogonal frequency division multiplexing (OFDM), UWB, and future wireless and wireline communications.

This chapter has attempted to describe some important fundamental theories and new technologies and approaches to the field of wireless communications that are likely to evolve in the coming decades. The advanced development in the field of signal processing will be a key to future digital communication systems with evolving higher data rates and spectral efficiencies. Currently, there are three candidates for providing increased data rates and improved spectral efficiency at the physical layer, including OFDM, UWB transmission, and space-time modulation and coding [26]. Each of these technologies has the potential to greatly increase the data rates and spectral efficiency of the physical layer by using advanced signal processing and will likely find its way into future digital communication systems. Therefore, by overviewing signal processing technologies in this chapter, we have laid technical foundations to introduce later chapters on more advanced developments of signal processing technologies for digital communication systems.

References

[1] IEEE Communications Society, *A Brief History of Communications: A Fifth Year Foundation for the Future*, IEEE Press, Piscataway, New Jersey, 2002.

[2] Nyquist, H., "Certain Factors Affecting Telegraph Speed," *Bell Systems Technical Journal*, Vol. 3, pp. 324–352, April 1924.

[3] Shannon, C. E., "A Mathematical Theory of Communications," *Bell Systems Technology Journal*, Vol. 27, pp. 379–423, July 1948a.

[4] Shannon, C. E., "A Mathematical Theory of Communications," *Bell Systems Technology Journal*, Vol. 27, pp. 623–656, October 1948b.

[5] Shannon, C. E., "Communication in the Presence of Noise," *Proceedings of the IEEE*, Vol. 86, No. 2, pp. 447–457, February 1998 (this paper is reprinted from the *Proceedings of the IRE*, Vol. 37, No. 1, pp. 10–21, January 1949).

[6] Cover, T. M., and J. A. Thomas, *Elements of Information Theory*, John Wiley & Sons, New York, 1991.

[7] Reza, F. M., *An Introduction to Information Theory*, Dover Publications, New York, 1994.

[8] Pierce, J. R., *An Introduction to Information Theory: Symbols, Signals, and Noise*, 2nd ed., Dover Publications, New York, 1980.

[9] Proakis, J. G., *Digital Communications*, 2nd ed., McGraw-Hill Book Company, New York, 1989.

[10] Haykin, S., *Digital Communications*, John Wiley & Sons, New York, 1988.

[11] Viterbi, A. J., *CDMA: Principles of Spread Spectrum Communication*, Addison-Wesley Publishing Company, Reading, Massachusetts, 1995.

[12] Rappaport, T. S., *Wireless Communications: Principles and Practice*, Prentice Hall, Upper Saddle River, New Jersey, 1996.

[13] Stüber, G. L., *Principles of Mobile Communication*, 2nd ed., Kluwer Academic Publishers, Boston, Massachusetts, 2001.

[14] Reed, J. H., *Software Radio: A Modern Approach to Radio Engineering*, Prentice Hall, Upper Saddle River, New Jersey, 2002.

[15] Miao, G. J., and M. A. Clements, *Digital Signal Processing and Statistical Classification*, Artech House, Norwood, Massachusetts, 2002.

[16] Vaidyanathan, P. P., *Multirate Systems and Filter Banks*, Prentice Hall, Englewood Cliffs, New Jersey, 1993.

[17] Oppenheim, A. V., and R. W. Schafer, *Discrete-Time Signal Processing*, Prentice Hall, Englewood Cliffs, New Jersey, 1989.

[18] Haykin, S., *Adaptive Filter Theory*, 3rd ed., Prentice Hall, Upper Saddle River, New Jersey, 1996.

[19] Eazavi, B., *RF Microelectronics*, Prentice Hall, Upper Saddle River, New Jersey, 1998.

[20] Pirsch, P., *Architectures for Digital Signal Processing*, John Wiley & Sons, New York, 1998.

[21] Miao, G. J., "Analog-to-Digital Converter Bank Based Ultra Wideband Communications," U.S. Patent No. 6,744,832, June 2004.

[22] Salkintzis, A. K., "ADC and DSP Challenges in the Development of Software Radio Base Stations," *IEEE Personal Communications*, pp. 47–55, August 1999.

[23] Hentschel, T., M. Henker, and G. Fettweis, "The Digital Front-End of Software Radio Terminals," *IEEE Personal Communications*, pp. 40–46, August 1999.

[24] Vaidyanathan, P. P., "Multirate Digital Filters, Filter Banks, Polyphase Networks, and Applications: A Tutorial," *Proc. of IEEE*, Vol. 78, No. 1, pp. 56–93, 1990.

[25] Harris, F. J., *Multirate Signal Processing for Communication Systems*, Prentice Hall, Upper Saddle River, New Jersey, 2004.

[26] Rappaport, T. S., et al., "Wireless Communications: Past Events and a Future Perspective," *IEEE Communications Magazine*, pp. 148–161, May 2002.

2

Probability, Random Variables, and Stochastic Signal Processing

2.1 Introduction

In many cases, signals that are generated from complex processes cannot be described precisely by a mathematical representation. In such case, the signals are referred to as *random* or *stochastic* signals. A random signal, or random process, is a signal that is not generally repeatable in a predictable manner. For instance, quantization noise, which is generated by using an A/D converter, a fixed-point digital filter, or other fixed-point devices, can be modeled as a random process. In another example, a discrete-time Kalman filter is developed by using an assumption of a discrete-time random process [1]. Thus, in this chapter, we introduce probability, random variables, and stochastic signal processing, which are basic and important concepts to understanding signal processing for digital communications.

A random, discrete-time random, or stochastic signal can be considered a member of an ensemble of signals that is characterized by a set of probabilities. A fundamental mathematical representation or a statistical representation of random or stochastic signals often uses its description in terms of averages, means, and variances. Therefore, in this chapter, we provide the stochastic background that is necessary to understand how a random process can be represented and how its statistical properties are affected by linear shift-invariant systems. We also introduce specific distribution and density functions as well as their means, variances, and moments. Our treatment serves primarily as review, notation definition, and fundamental development. There are many references on this subject, such as Davenport and Root [2], Davenport [3], and Papoulis [4].

This chapter is organized as follows. This section presents a short introduction. In Section 2.2, we begin with a review of probability and random processes, focusing on concepts and definitions, including intuitive, axiomatic, and conditional probabilities and independence. In Section 2.3, we briefly introduce random variables. Probability distribution and density functions are presented in Section 2.4, along with a discussion of joint distribution and density functions, statistical averages, and joint moments. Then, in Section 2.5, we further introduce functions of random variables, including sums of random variables and transformations of random variables, which are useful to derive the joint probability distributions and densities. Specific discrete and continuous distribution and densities are presented in Section 2.6 and Section 2.7, respectively. The upper bound theory on the tail probability is introduced in Section 2.8, where we address the Chebyshev inequality, the law of large numbers, and the central limit theorem. In Section 2.9, stochastic signal processes are discussed, with an emphasis on discrete-time stochastic processes. Then we develop detection theory and optimum receivers in Section 2.10. Finally, a brief summary is given in Section 2.11.

2.2 Probability

In this section, we begin by introducing basic results of probability theory. This brief treatment takes into account the general needs of random variables, subsequent sections, and chapters. Thus, we will present several important concepts, including intuitive, axiomatic, and conditional probabilities and independence.

2.2.1 Intuitive Probability

The probability theory processes the averages of many phenomena occurring sequentially or simultaneously. One purpose of using the probability theory is to describe and predict averages in terms of the probabilities of the events. For instance, in an experiment of coin-tossing, assuming that a fair coin is equally likely to result in heads or tails if the coin is flipped, the outcome would be heads approximately half of the time and tails the other half. Thus, in intuitive probability, if we allow that all possible outcomes of a chance experiment are equally likely, the probability of a particular event, $\mathcal{A}$, is defined as

$$p(\mathcal{A}) = \frac{\mathcal{N}_A}{\mathcal{N}_T}, \qquad (2.1)$$

where N_A and N_T are possible outcomes favoring event $\mathcal{A}$ and total possible outcomes, respectively, and N_T is *sufficiently large*. However, the terminology of "sufficiently large" in (2.1) has no clear meaning. This is an imprecise statement in which cannot be avoided. Thus, intuitive probability has limitations, but still plays an important role in probability theory. In many cases, the simple concept of ratio of possible event outcomes is a useful method for problem solving.

Example 2.1

Assume that if a coin is flipped 5,000 times and heads show 2,565 times, then the probability of heads equals 0.513 and the probability of tails equals 0.487. However, if we assume that the coin is fair, then the probabilities of heads and tails equal 0.5, respectively.

2.2.2 Axiomatic Probability

An axiomatic probability begins with the concept of a *sample space* that is the set of all possible outcomes of the experiment. If the sample space is S and its individual outcomes or elements in the sample space are $\{s_1, s_2, s_3, ...\}$, then the elements in the sample space are *mutually exclusive* or *disjoint*. In other words, there is no overlap of elements in a sample space. The elements are also collectively exhaustive. Every possible outcome is included.

An *event* can be considered as a special subset of the sample space S. Let the events be $\mathcal{A}$, $\mathcal{B}$, $\mathcal{C}$, ..., and so on. Axiomatic probability theory is now stated as follows:

1. The probability of an event $\mathcal{A}$ is a nonnegative number

$$p(\mathcal{A}) \geq 0 \qquad (2.2)$$

2. The probability of a certain event equals 1,

$$p(S) = 1 \qquad (2.3)$$

3. Assuming that $\mathcal{A}$, $\mathcal{B}$, $\mathcal{C}$, ..., are mutually exclusive events, then

$$p(\mathcal{A} + \mathcal{B} + \mathcal{C} + \cdots) = p(\mathcal{A}) + p(\mathcal{B}) + p(\mathcal{C}) + \cdots. \qquad (2.4)$$

These three definitions, along with the traditional axioms of set theory, are fundamental to axiomatic probability theory. These definitions provide the theoretical basis for the formal solution of a wide variety of probability problems. For the probability of an impossible event, a simple consequence is always 0, $p(\emptyset) = 0$.

Furthermore, the operation of intersection is also useful. The intersection of two events $\mathcal{A}$ and $\mathcal{B}$ is the event including elements that are common to both $\mathcal{A}$ and $\mathcal{B}$. Let the intersection of two events $\mathcal{A}$ and $\mathcal{B}$ be $\mathcal{AB}$. For any $\mathcal{A}$ and $\mathcal{B}$ events, their combined probability is given by

$$p(\mathcal{A} + \mathcal{B}) = p(\mathcal{A}) + p(\mathcal{B}) - p(\mathcal{AB}). \qquad (2.5)$$

Note that the subtractive term $p(\mathcal{AB})$ in (2.5) is required. This is because the probabilities in the intersection region have been counted twice in the summation of $p(\mathcal{A})$ and $p(\mathcal{B})$. We refer to the probability $p(\mathcal{AB})$ as the *joint probability* of the events $\mathcal{A}$ and $\mathcal{B}$. The probability $p(\mathcal{AB})$ can be expressed as the probability in which both $\mathcal{A}$ and $\mathcal{B}$ occurred.

2.2.3 Conditional Probability

Consider an experiment that a joint event occurs with the probability $p(\mathcal{AB})$. If the event $\mathcal{B}$ has occurred, we want to determine the probability of the occurrence of the event $\mathcal{A}$. This conditional probability is defined as

$$p(\mathcal{A}|\mathcal{B}) = \frac{p(\mathcal{AB})}{p(\mathcal{B})}, \qquad (2.6)$$

where $p(\mathcal{B}) > 0$.

Let the notation $\subset$ denote the meaning of the subset. If $\mathcal{B} \subset \mathcal{A}$, then $p(\mathcal{A}|\mathcal{B}) = 1$. This can be shown as

$$\begin{aligned} p(\mathcal{A}|\mathcal{B}) &= \frac{p(\mathcal{AB})}{p(\mathcal{B})} \\ &= \frac{p(\mathcal{B})}{p(\mathcal{B})} \\ &= 1. \end{aligned} \qquad (2.7)$$

Similarly, if $\mathcal{A} \subset \mathcal{B}$, then we obtain

$$\begin{aligned} p(\mathcal{A}|\mathcal{B}) &= \frac{p(\mathcal{A})}{p(\mathcal{B})} \\ &\geq p(\mathcal{A}). \end{aligned} \qquad (2.8)$$

Assume that events $\mathcal{A}_i$, $i = 1, 2, \cdots, n$, are mutually exclusive events

$$\bigcup_{i=1}^{n} \mathcal{A}_i = S, \qquad (2.9)$$

and $\mathcal{B}$ is an arbitrary event with $p(\mathcal{B}) > 0$. We then have

$$p(\mathcal{B}\mathcal{A}_i) = p(\mathcal{B}|\mathcal{A}_i)p(\mathcal{A}_i), \qquad (2.10)$$

and

$$p(\mathcal{B}) = p(\mathcal{B}\mathcal{A}_1) + p(\mathcal{B}\mathcal{A}_2) + \cdots + p(\mathcal{B}\mathcal{A}_n). \qquad (2.11)$$

Equation (2.11) is referred to as the *total probability theorem*.

Further note that $p(\mathcal{B}\mathcal{A}_i) = p(\mathcal{A}_i|\mathcal{B})p(\mathcal{B})$; thus, we can rewrite (2.10) as follows:

$$p(\mathcal{A}_i|\mathcal{B}) = p(\mathcal{B}|\mathcal{A}_i)\frac{p(\mathcal{A}_i)}{p(\mathcal{B})}. \qquad (2.12)$$

Substituting (2.11) into (2.12), we obtain

$$p(\mathcal{A}_i|\mathcal{B}) = \frac{p(\mathcal{B}|\mathcal{A}_i)p(\mathcal{A}_i)}{\sum_{j=1}^{n} p(\mathcal{B}|\mathcal{A}_j)p(\mathcal{A}_j)}, \qquad (2.13)$$

where the probabilities $p(\mathcal{A}_i)$ and $p(\mathcal{A}_i|\mathcal{B})$ are referred to as *a priori* and *a posteriori*, respectively. Equation (2.13) is known as *Bayes' theorem*.

Example 2.2

Assume that there are 12 balls in a dish and you want to pick up 2 of them, choosing the balls randomly. You see that 6 of the balls are red and 6 are yellow. You prefer the red ones. What is the probability that you will get 2 red ones?

Let $\mathcal{R}_i$ and $\mathcal{Y}_i$, $i = 1, 2, 3, 4, 5, 6$, denote the numbers of red and yellow balls, respectively. The probability of getting 2 red ones is

$$p(\mathcal{R}_1\mathcal{R}_2) = p(\mathcal{R}_1)p(\mathcal{R}_2|\mathcal{R}_1). \qquad (2.14)$$

Since there are 6 red balls and 12 balls together, then

$$p(\mathcal{R}_1) = \frac{1}{2}. \qquad (2.15)$$

If you get a red one on the first selection, there will be 5 red ones left out of 11 balls in the dish. Thus, you obtain

$$p(\mathcal{R}_2|\mathcal{R}_1) = \frac{5}{11}, \qquad (2.16)$$

and the probability of getting 2 red ones is obtained by

$$\begin{aligned} p(\mathcal{R}_1\mathcal{R}_2) &= p(\mathcal{R}_1)p(\mathcal{R}_2|\mathcal{R}_1) \\ &= \left(\frac{1}{2}\right)\left(\frac{5}{11}\right) = \frac{5}{22}. \end{aligned}$$

Furthermore, if you get 2 red balls, the question is: What is the probability that you will get 2 more red ones? Note that there are 4 red left out of 10 balls because 2 red balls have been removed. Therefore, the probability of getting 2 more red ones is

$$p(\mathcal{R}_3\mathcal{R}_4|\mathcal{R}_1\mathcal{R}_2) = p(\mathcal{R}_3|\mathcal{R}_1\mathcal{R}_2)p(\mathcal{R}_4|\mathcal{R}_1\mathcal{R}_2\mathcal{R}_3)$$
$$= \left(\frac{4}{10}\right)\left(\frac{3}{9}\right) = \frac{2}{15}. \qquad (2.17)$$

2.2.4 Independence

Two events $\mathcal{A}$ and $\mathcal{B}$ are called *independent* if the probability is

$$p(\mathcal{AB}) = p(\mathcal{A})p(\mathcal{B}). \qquad (2.18)$$

The independence of n events can be defined inductively. The events $\mathcal{A}_1, \mathcal{A}_2, \cdots, \mathcal{A}_n$ are said to be independent if the probability is

$$p(\mathcal{A}_1 \cdots \mathcal{A}_n) = p(\mathcal{A}_1) \cdots p(\mathcal{A}_n). \qquad (2.19)$$

Equation (2.19) provides the generalization definition of independence for n events. These above discussions will be useful when we introduce random variables in the next section.

2.3 Random Variables

A *random variable* is a process function, which maps every point in the sample space $\mathcal{S}$ onto a real number. Let a *boldface letter* **x** denote a random variable. The process function must satisfy two conditions as follows:

1. The set $\{\mathbf{x} \leq x\}$ is an event for every real value x.
2. The probabilities of the events $\{\mathbf{x} = \infty\}$ and $\{\mathbf{x} = -\infty\}$ are $p\{\mathbf{x} = \infty\} = 0$ and $p\{\mathbf{x} = -\infty\} = 0$, respectively.

For example, if x_1 and x_2 are real numbers, the notational meaning of $\{x_1 \leq \mathbf{x} \leq x_2\}$ denotes a subset of space including all outcomes ξ such that $x_1 \leq \mathbf{x} \leq x_2$. The notation $\{\mathbf{x} = x\}$ is a subset of space consisting of all outcomes ξ such that $\mathbf{x} = x$, where x is a given number.

A *complex random variable* $\mathbf{z}$ is defined as

$$\mathbf{z} = \mathbf{x} + j\mathbf{y}, \qquad (2.20)$$

where $\mathbf{x}$ and $\mathbf{y}$ are real random variables.

A *random vector* is a vector and defined by

$$\mathbf{X} = [\mathbf{x}_1, \mathbf{x}_2, ..., \mathbf{x}_n], \qquad (2.21)$$

where components $\mathbf{x}_i$, $i = 1, 2, ..., n$, are random variables.

2.4 Probability Distributions and Densities

In this section, we will introduce definitions of distribution and density functions, statistical averages and joint moments, and moment generation functions.

2.4.1 Probability Distributions

The distribution function of the random variable $\mathbf{x}$ is defined by

$$F_x(x) = p\{\mathbf{x} \leq x\}, \qquad (2.22)$$

where $-\infty < x < \infty$. Equation (2.22) is called the *probability distribution function* of the random variable $\mathbf{x}$. It is also known as the *cumulative distribution function* of the random variable $\mathbf{x}$.

The cumulative distribution function of the random variable $\mathbf{x}$ has the properties as follows:

1. Because $F_x(x)$ is a probability, its interval is limited to the range from 0 to 1,

$$0 \leq F_x(x) \leq 1. \qquad (2.23)$$

   In fact, $F_x(\infty) = 1$ and $F_x(-\infty) = 0$.
2. It is a nondecreasing function of x. If $x_1 < x_2$, then

$$F_x(x_1) \leq F_x(x_2). \qquad (2.24)$$

3. The probability $p\{\mathbf{x} > x\}$ of the event $\{\mathbf{x} > x\}$ is obtained by
$$p\{\mathbf{x} > x\} = 1 - F_x(x). \qquad (2.25)$$

4. The probability $p\{x_1 < \mathbf{x} \leq x_2\}$ of the event $\{x_1 < \mathbf{x} \leq x_2\}$ is given by
$$p\{x_1 < \mathbf{x} \leq x_2\} = F_x(x_2) - F_x(x_1). \qquad (2.26)$$

2.4.2 Probability Densities

The derivative of the cumulative distribution function of the random variable $\mathbf{x}$ is called the *density function*:
$$f_x(x) = \frac{dF_x(x)}{dx}, \quad -\infty < x < \infty. \qquad (2.27)$$

Equation (2.27) is also referred to as the *probability density function*. This is equivalent to
$$F_x(x) = \int_{-\infty}^{x} f_x(u) du, \quad -\infty < x < \infty. \qquad (2.28)$$

The probability density function has the properties as follows:

1. Because $F_x(x)$ is monotonic, it follows that
$$f_x(x) \geq 0. \qquad (2.29)$$

2. Since $F_x(\infty) = 1$, (2.28) yields
$$\int_{-\infty}^{\infty} f_x(x) dx = 1. \qquad (2.30)$$

3. From (2.28), we obtain
$$F_x(x_1) - F_x(x_2) = \int_{x_1}^{x_2} f_x(x) dx. \qquad (2.31)$$

2.4.3 Joint Probability Distributions and Densities

The joint cumulative distribution function of two random variables $\mathbf{x}_1$ and $\mathbf{x}_2$ is defined as

$$\begin{aligned}F_{x_1x_2}(x_1, x_2) &= p\{\mathbf{x}_1 \leq x_1, \mathbf{x}_2 \leq x_2\} \\ &= \int_{-\infty}^{x_2}\int_{-\infty}^{x_1} f_{x_1x_2}(u_1, u_2)du_1 du_2,\end{aligned} \quad (2.32)$$

where $f_{x_1x_2}(x_1, x_2)$ is the joint density function. The joint density function of random variables $\mathbf{x}_1$ and $\mathbf{x}_2$ is given by

$$f_{x_1x_2}(x_1, x_2) = \frac{\partial^2 F_{x_1x_2}(x_1, x_2)}{\partial x_1 \partial x_2}. \quad (2.33)$$

Given two random variables $\mathbf{x}_1$ and $\mathbf{x}_2$, the marginal distribution is defined by

$$F_{x_1}(x_1) = F_{x_1x_2}(x_1, \infty) \quad (2.34)$$
$$F_{x_2}(x_2) = F_{x_1x_2}(\infty, x_2) \quad (2.35)$$

and the marginal density is defined by

$$f_{x_1}(x_1) = \int_{-\infty}^{\infty} f_{x_1x_2}(x_1, x_2)dx_2 \quad (2.36)$$
$$f_{x_2}(x_2) = \int_{-\infty}^{\infty} f_{x_1x_2}(x_1, x_2)dx_1. \quad (2.37)$$

If joint density function $f_{x_1x_2}(x_1, x_2)$ is integrated over both variables x_1 and x_2, we then obtain

$$\int_{-\infty}^{\infty}\int_{-\infty}^{\infty} f_{x_1x_2}(x_1, x_2)dx_1 dx_2 = F_{x_1x_2}(\infty, \infty) = 1. \quad (2.38)$$

Moreover, we also have the results as follows:

$$F_{x_1x_2}(-\infty, -\infty) = F_{x_1x_2}(-\infty, x_2) = F_{x_1x_2}(x_1, -\infty) = 0. \quad (2.39)$$

The generalization of the joint cumulative distribution function for multidimensional random variables is straightforward. Given

n random variables, $\mathbf{x}_1, \mathbf{x}_2, ..., \mathbf{x}_n$, the joint cumulative distribution function is obtained by

$$\begin{aligned} F_{x_1...x_n}(x_1, ..., x_n) &= p(\mathbf{x}_1 \leq x_1, ..., \mathbf{x}_n \leq x_n) \\ &= \int_{-\infty}^{x_1} \cdots \int_{-\infty}^{x_n} f_{x_1...x_n}(u_1, ..., u_n) \\ &\quad \cdot du_1 \cdots du_n, \end{aligned} \quad (2.40)$$

where $f_{x_1...x_n}(x_1, ..., x_n)$ is the joint density function. On the other hand, the joint density function is obtained by using (2.40)

$$f_{x_1...x_n}(x_1, ..., x_n) = \frac{\partial^n F_{x_1...x_n}(x_1, ..., x_n)}{\partial x_1 \cdots \partial x_n}. \quad (2.41)$$

The n random variables, $\mathbf{x}_1, \mathbf{x}_2, ..., \mathbf{x}_n$, are said to be statistically independent if and only if the following condition is satisfied,

$$F_{x_1...x_n}(x_1, ..., x_n) = F_{x_1}(x_1) \cdots F_{x_n}(x_n), \quad (2.42)$$

or

$$f_{x_1...x_n}(x_1, ..., x_n) = f_{x_1}(x_1) \cdots f_{x_n}(x_n). \quad (2.43)$$

2.4.4 Statistical Averages and Joint Moments

In this section, we introduce statistical averages and joint moments that have an important role in the characterization of the outcomes of experiments. Of great importance are the first and second moments of a random variable $\mathbf{x}$. In addition, we define joint moments, including the correlation and covariance, which are between pairs of random variables within a set of n random variables.

Given a density function $f_x(x)$, the *mean* or *expected value* of random variable $\mathbf{x}$ is defined by

$$m_x = E\{\mathbf{x}\} = \int_{-\infty}^{\infty} x f_x(x) dx, \quad (2.44)$$

where $E\{\}$ denotes expectation or statistical average.

The *variance* of random variable **x** is defined by

$$\sigma_x^2 = \int_{-\infty}^{\infty} (x - m_x)^2 f_x(x) dx, \qquad (2.45)$$

where $m_x = E\{\mathbf{x}\}$ and σ_x is called the *standard deviation* of random variable **x**. The variance σ_x^2 (2.45) can be considered to be the mean of the random variable $(\mathbf{x} - m_x)^2$. Thus, the relationship between the variance and mean is obtained by

$$\begin{aligned}
\sigma_x^2 &= E\{(\mathbf{x} - m_x)^2\} \\
&= E\{\mathbf{x}^2 - 2\mathbf{x}m_x + m_x^2\} \\
&= E\{\mathbf{x}^2\} - 2m_x E\{\mathbf{x}\} + m_x^2 \\
&= E\{\mathbf{x}^2\} - E^2\{\mathbf{x}\}. \qquad (2.46)
\end{aligned}$$

If a new random variable is given by $\mathbf{y} = (\mathbf{x} - m_x)^n$, where m_x is the mean value of the random variable **x**, the mean of the random variable **y** is then obtained by

$$\begin{aligned}
E\{\mathbf{y}\} &= E\{(\mathbf{x} - m_x)^n\} \\
&= \int_{-\infty}^{\infty} (\mathbf{x} - m_x)^n f_x(x) dx. \qquad (2.47)
\end{aligned}$$

Equation (2.47) is known as the nth *central moment* of the random variable **x**.

Assume that $f_{x_i x_j}(x_i, x_j)$ is the joint density function of the random variables $\mathbf{x}_i$ and $\mathbf{x}_j$. The *correlation*, denoted by $R_{\mathbf{x}_i \mathbf{x}_j}$, between random variables $\mathbf{x}_i$ and $\mathbf{x}_j$ is defined by the joint moment as follows:

$$\begin{aligned}
R_{\mathbf{x}_i \mathbf{x}_j} &= E\{\mathbf{x}_i \mathbf{x}_j\} \\
&= \int_{-\infty}^{\infty} \int_{-\infty}^{\infty} x_i x_j f_{x_i x_j}(x_i, x_j) dx_i dx_j. \qquad (2.48)
\end{aligned}$$

The *covariance*, denoted by $C_{\mathbf{x}_i \mathbf{x}_j}$, of the random variables $\mathbf{x}_i$ and $\mathbf{x}_j$ is obtained by

$$C_{\mathbf{x}_i \mathbf{x}_j} = E\{(\mathbf{x}_i - m_i)(\mathbf{x}_j - m_j)\}$$

$$\begin{aligned}
&= \int_{-\infty}^{\infty} \int_{-\infty}^{\infty} (x_i - m_i)(x_j - m_j) f_{x_i x_j}(x_i, x_j) dx_i dx_j \\
&= \int_{-\infty}^{\infty} \int_{-\infty}^{\infty} x_i x_j f_{x_i x_j}(x_i, x_j) dx_i dx_j - m_i m_j \\
&= E\{\mathbf{x}_i \mathbf{x}_j\} - m_i m_j.
\end{aligned} \qquad (2.49)$$

It is also clear that the covariance function is equal to the correlation function, $C_{\mathbf{x}_i \mathbf{x}_j} = R_{\mathbf{x}_i \mathbf{x}_j}$, when $m_i = 0$ and/or $m_j = 0$.

There is a relation between the covariance and correlation known as the *correlation coefficient*. The correlation coefficient is defined to be the covariance divided by the standard deviations of random variables $\mathbf{x}_i$ and $\mathbf{x}_j$ as follows:

$$\rho_{x_i x_j} = \frac{C_{\mathbf{x}_i \mathbf{x}_j}}{\sigma_{x_i} \sigma_{x_j}}. \qquad (2.50)$$

If two random variables, $\mathbf{x}_i$ and $\mathbf{x}_j$, are uncorrelated, their covariance and their correlation coefficient are equal to zero.

Furthermore, if $E\{\mathbf{x}_i \mathbf{x}_j\} = E\{\mathbf{x}_i\} E\{\mathbf{x}_j\} = m_i m_j$, then the random variables $\mathbf{x}_i$ and $\mathbf{x}_j$ are said to be uncorrelated. This implies that the covariance $C_{\mathbf{x}_i \mathbf{x}_j} = 0$. If the random variables $\mathbf{x}_i$ and $\mathbf{x}_j$ are *statistically independent*, then they are uncorrelated, but if the random variables $\mathbf{x}_i$ and $\mathbf{x}_j$ are uncorrelated, they are not necessarily statistically independent. If the random variables $\mathbf{x}_i$ and $\mathbf{x}_j$ have $E\{\mathbf{x}_i \mathbf{x}_j\} = 0$, then they are called *orthogonal*. Sometimes, when the random variables $\mathbf{x}_i$ and $\mathbf{x}_j$ are uncorrelated, they are referred to as *linearly independent*.

2.4.5 Moment Generation Function

Another useful theorem is moments of a density function, which play an important role in theoretical and applied statistics. In some cases, if all the moments are known, the density function can be determined. Thus, it would be useful if we could find a function that would represent all the moments. We call such a function a *moment generation function*.

Let **x** be a random variable with density function $f_\mathbf{X}(\cdot)$. A moment generation function is defined to be

$$m_x(t) = E\{e^{t\mathbf{X}}\} = \int_{-\infty}^{\infty} e^{t\mathbf{X}} f_\mathbf{X}(x)dx, \qquad (2.51)$$

where the random variable **x** is continuous and every value of t is in some interval $-c < t < c$, and $c > 0$.

If the random variable **x** is a discrete-time random variable, a discrete-time moment generation function is defined as

$$m_x(t) = E\{e^{t\mathbf{X}}\} = \sum_\mathbf{X} e^{t\mathbf{X}} f_\mathbf{X}(x). \qquad (2.52)$$

In order to obtain the rth derivative of $m_x(t)$ in (2.51), we differentiate the moment generation function r times with respect to r as follows:

$$\frac{d^r m_x(t)}{dt^r} = \int_{-\infty}^{\infty} \mathbf{x}^r e^{t\mathbf{X}} f_\mathbf{X}(x)dx. \qquad (2.53)$$

Let $t \to 0$, (2.53) can be rewritten as follows:

$$\frac{d^r m_x(0)}{dt^r} = E\{\mathbf{x}^r\}, \qquad (2.54)$$

where the left side of (2.54) is the rth derivative of $m_x(t)$ when $t \to 0$. Therefore, the moments of a distribution function may be obtained by differentiation of the moment generation function.

Example 2.3

Assume that **x** is a random variable with a probability density function given by

$$f_\mathbf{X}(x) = \alpha e^{-\alpha x}, \quad 0 \le x < \infty. \qquad (2.55)$$

Using (2.51), the moment generation function can be written as

$$\begin{aligned} m_x(t) &= E\{e^{t\mathbf{X}}\} \\ &= \int_0^{\infty} e^{t\mathbf{X}} \alpha e^{-\alpha x} dx \\ &= \frac{\alpha}{\alpha - t}, \quad t < \alpha. \end{aligned} \qquad (2.56)$$

Differentiating (2.56) yields

$$m'_{\mathbf{x}}(t) = \frac{\alpha}{(\alpha - t)^2}, \qquad (2.57)$$

and

$$m''_{\mathbf{x}}(t) = \frac{2\alpha}{(\alpha - t)^3}. \qquad (2.58)$$

Hence, when $t \to 0$, we obtain the first moment

$$m'(0) = E\{\mathbf{x}\} = \frac{1}{\alpha}, \qquad (2.59)$$

and the second moment

$$m''(0) = E\{\mathbf{x}^2\} = \frac{2}{\alpha^2}. \qquad (2.60)$$

2.5 Functions of Random Variables

In this section, we further discuss functions of random variables, including sums and transformations of random variables, $\mathbf{x}_1, \mathbf{x}_2, ..., \mathbf{x}_n$.

2.5.1 Sums of Random Variables

Assume that we have n random variables, $\mathbf{x}_1, \mathbf{x}_2, ..., \mathbf{x}_n$. We can then discuss theory of the summations of n random variables and proof results as follows:

Theorem 1. Given n random variables, $\mathbf{x}_1, \mathbf{x}_2, ..., \mathbf{x}_n$, the mean of the sum is the sum of the means:

$$E\left\{\sum_{i=1}^{n} \mathbf{x}_i\right\} = \sum_{i=1}^{n} E\{\mathbf{x}_i\}, \qquad (2.61)$$

and also

$$\text{Var}\left\{\sum_{i=1}^{n} \mathbf{x}_i\right\} = \sum_{i=1}^{n} \text{Var}\{\mathbf{x}_i\} + 2 \sum_{i} \sum_{j(i<j)} \text{Cov}\{\mathbf{x}_i, \mathbf{x}_j\}, \qquad (2.62)$$

where the notations Var and Cov denote variance and covariance of the random variables, $\mathbf{x}_1, \mathbf{x}_2, ..., \mathbf{x}_n$.

Proof: Using (2.61), we can rewrite the right side of the equation as

$$E\left\{\sum_{i=1}^{n} \mathbf{x}_i\right\} = E\{\mathbf{x}_1 + \mathbf{x}_2 + \cdots + \mathbf{x}_n\}$$
$$= E\{\mathbf{x}_1\} + E\{\mathbf{x}_2\} + \cdots + E\{\mathbf{x}_n\}$$
$$= \sum_{i=1}^{n} E\{\mathbf{x}_i\}. \qquad (2.63)$$

Thus, we prove that the mean of the sum is the sum of the means given by (2.61). For proving (2.62), we can rewrite the left side of the equation as follows:

$$\operatorname{Var}\left\{\sum_{i=1}^{n} \mathbf{x}_i\right\} = E\left\{\left[\sum_{i=1}^{n} \mathbf{x}_i - E\left\{\sum_{i=1}^{n} \mathbf{x}_i\right\}\right]^2\right\}$$
$$= E\left\{\left[\sum_{i=1}^{n} (\mathbf{x}_i - E\{\mathbf{x}_i\})\right]^2\right\}$$
$$= E\left\{\sum_{i=1}^{n}\sum_{j=1}^{n} (\mathbf{x}_i - E\{\mathbf{x}_i\})(\mathbf{x}_j - E\{\mathbf{x}_j\})\right\}$$
$$= \sum_{i=1}^{n}\sum_{j=1}^{n} E\{(\mathbf{x}_i - E\{\mathbf{x}_i\})(\mathbf{x}_j - E\{\mathbf{x}_j\})\}$$
$$= \sum_{i=1}^{n} \operatorname{Var}\{\mathbf{x}_i\} + 2\sum_{i}\sum_{j(i<j)} \operatorname{Cov}\{\mathbf{x}_i, \mathbf{x}_j\}. \quad (2.64)$$

The result of the last line in (2.64) is due to the fact that the sum of the variances is the diagonal term while the sums of the covariances are the off-diagonal terms.

From theorem 1, for two random variables $\mathbf{x}_1$ and $\mathbf{x}_2$, it follows that the sum of means is obtained by

$$E\{\mathbf{x}_1 \pm \mathbf{x}_2\} = E\{\mathbf{x}_1\} \pm E\{\mathbf{x}_2\}, \qquad (2.65)$$

and the sum of the variance is obtained by

$$\operatorname{Var}\{\mathbf{x}_1 \pm \mathbf{x}_2\} = \operatorname{Var}\{\mathbf{x}_1\} + \operatorname{Var}\{\mathbf{x}_2\} \pm 2\operatorname{Cov}\{\mathbf{x}_1, \mathbf{x}_2\}. \qquad (2.66)$$

These provide the mean and variance of the sum or the difference of two random variables, $\mathbf{x}_1$ and $\mathbf{x}_2$.

Theorem 2. Consider n random variables, $\mathbf{x}_1, \mathbf{x}_2, ..., \mathbf{x}_n$. If $\mathbf{x}_1, \mathbf{x}_2, ..., \mathbf{x}_n$ are uncorrelated random variables, then the variance of the sum is the sum of the variance:

$$\text{Var}\left\{\sum_{i=1}^{n} \mathbf{x}_i\right\} = \sum_{i=1}^{n} \text{Var}\{\mathbf{x}_i\}. \tag{2.67}$$

Proof: This theory can be proved by using theorem 1. Note that the random variables $\mathbf{x}_1, \mathbf{x}_2, ..., \mathbf{x}_n$ are uncorrelated. Thus, then $cov\{\mathbf{x}_i, \mathbf{x}_j\} = 0$, for all $i, j, i \neq j$. As a result, the second term in (2.62) equals zero. Therefore, we prove the result of theorem 2,

$$\text{Var}\left\{\sum_{i=1}^{n} \mathbf{x}_i\right\} = \sum_{i=1}^{n} \text{Var}\{\mathbf{x}_i\}. \tag{2.68}$$

Theorem 3. Consider n random variables, $\mathbf{x}_1, \mathbf{x}_2, ..., \mathbf{x}_n$, and n constants, denoted by $c_1, c_2, ..., c_n$, then

$$\begin{aligned}\text{Var}\left\{\sum_{i=1}^{n} c_i \mathbf{x}_i\right\} &= \sum_{i=1}^{n}\sum_{j=1}^{n} c_i c_j \text{Cov}\{\mathbf{x}_i, \mathbf{x}_j\} \\ &= \sum_{i=1}^{n} c_i^2 \text{Var}\{\mathbf{x}_i\} \\ &\quad + \sum_{i}\sum_{j(j \neq i)} c_i c_j \text{Cov}\{\mathbf{x}_i, \mathbf{x}_j\}. \end{aligned} \tag{2.69}$$

This theorem states a result that is somewhat related to theorem 1. Hence, by using the proof method in theorem 1, this theorem can be proved as well.

Theorem 4. Given n random variables, $\mathbf{x}_1, \mathbf{x}_2, ..., \mathbf{x}_n$, if the random variables $\mathbf{x}_1, \mathbf{x}_2, ..., \mathbf{x}_n$ are independent and identically distributed with mean m_x and variance σ_x^2, and if

$$\bar{\mathbf{x}}_n = \frac{1}{n}\sum_{i=1}^{n}\mathbf{x}_i, \qquad (2.70)$$

then the mean of $\bar{\mathbf{x}}_n$ is given by

$$E\{\bar{\mathbf{x}}_n\} = m_x, \qquad (2.71)$$

and the variance of $\bar{\mathbf{x}}_n$ is given by

$$\mathrm{Var}\{\bar{\mathbf{x}}_n\} = \frac{\sigma_x^2}{n}. \qquad (2.72)$$

Proof: To prove (2.71), by using (2.70), we rewrite that

$$\begin{aligned}
E\{\bar{\mathbf{x}}_n\} &= E\left\{\frac{1}{n}\sum_{i=1}^{n}\mathbf{x}_i\right\} \\
&= \frac{1}{n}\sum_{i=1}^{n}E\{\mathbf{x}_i\} \\
&= \frac{1}{n}\sum_{i=1}^{n}m_x \\
&= m_x. \qquad (2.73)
\end{aligned}$$

To prove (2.72), by using (2.70), we obtain that

$$\begin{aligned}
\mathrm{Var}\{\bar{\mathbf{x}}_n\} &= \mathrm{Var}\left\{\frac{1}{n}\sum_{i=1}^{n}\mathbf{x}_i\right\} \\
&= \left(\frac{1}{n}\right)^2 \mathrm{Var}\left\{\sum_{i=1}^{n}\mathbf{x}_i\right\}. \qquad (2.74)
\end{aligned}$$

Since the random variables $\mathbf{x}_1, \mathbf{x}_2, ..., \mathbf{x}_n$ are independent and identically distributed, then the random variables $\mathbf{x}_1, \mathbf{x}_2, ..., \mathbf{x}_n$ are uncorrelated random variables. Therefore, by using theorem 2, we rewrite (2.74) as follows:

$$\begin{aligned} \text{Var}\{\bar{\mathbf{x}}_n\} &= \left(\frac{1}{n}\right)^2 \sum_{i=1}^{n} \text{Var}\{\mathbf{x}_i\} \\ &= \left(\frac{1}{n}\right)^2 \sum_{i=1}^{n} \sigma_x^2 \\ &= \frac{\sigma_x^2}{n}. \end{aligned} \qquad (2.75)$$

Hence, we complete the proof of this theorem.

2.5.2 Transformations of Random Variables

Assume that $\mathbf{x}_i$, $i = 1, 2, \cdots, n$, be jointly continuous random variables with density function $f_{\mathbf{X}}(x_1, x_2, \cdots, x_n)$, and $\mathbf{y}_i$, $i = 1, 2, \cdots, n$, be a set of random variables such that

$$\mathbf{y}_i = g_i(\mathbf{x}_1, \mathbf{x}_2, \cdots, \mathbf{x}_n), \quad i = 1, 2, \cdots, n. \qquad (2.76)$$

Equation (2.76) is a one-to-one transformation. Also assume that $\mathbf{x}_i$, $i = 1, 2, \cdots, n$, can be inverted and expressed in terms of functions of $\mathbf{y}_i$, $i = 1, 2, \cdots, n$, as follows:

$$\mathbf{x}_i = g_i^{-1}(\mathbf{y}_1, \mathbf{y}_2, \cdots, \mathbf{y}_n), \quad i = 1, 2, \cdots, n. \qquad (2.77)$$

The objective of the transformations of random variables is to determine the joint probability density function $f_{\mathbf{Y}}(y_1, y_2, \cdots, y_n)$, given the joint probability density function $f_{\mathbf{X}}(x_1, x_2, \cdots, x_n)$.

Let us define a transformation function as

$$J = \begin{vmatrix} \frac{\partial g_1^{-1}}{\partial y_1} & \frac{\partial g_2^{-1}}{\partial y_1} & \cdots & \frac{\partial g_n^{-1}}{\partial y_1} \\ \frac{\partial g_1^{-1}}{\partial y_2} & \frac{\partial g_2^{-1}}{\partial y_2} & \cdots & \frac{\partial g_n^{-1}}{\partial y_2} \\ \vdots & \vdots & \vdots & \vdots \\ \frac{\partial g_1^{-1}}{\partial y_n} & \frac{\partial g_2^{-1}}{\partial y_n} & \cdots & \frac{\partial g_n^{-1}}{\partial y_n} \end{vmatrix}, \qquad (2.78)$$

where J is referred to as the *Jacobian transformation*. Assume that all the partial derivatives in J are continuous and the determinant of J is nonzero. Then, we obtain the desired relation of the joint probability density function $f_\mathbf{y}(y_1, y_2, \cdots, y_n)$ as

$$f_\mathbf{y}(y_1, \cdots, y_n) = |J| f_\mathbf{x}(g_1^{-1}(\mathbf{y}_1, \cdots, \mathbf{y}_n), \cdots, g_n^{-1}(\mathbf{y}_1, \cdots, \mathbf{y}_n)). \tag{2.79}$$

Example 2.4

Let $\mathbf{x}_1$ and $\mathbf{x}_2$ be two independent random variables, and the linear transformation of random variables is $\mathbf{y}_1 = \mathbf{x}_1 + \mathbf{x}_2$ and $\mathbf{y}_2 = \frac{\mathbf{x}_1}{\mathbf{x}_2}$. Then we obtain

$$x_1 = g_1^{-1}(y_1, y_2) = \frac{y_1 y_2}{1 + y_2}, \tag{2.80}$$

$$x_2 = g_2^{-1}(y_1, y_2) = \frac{y_1}{1 + y_2}, \tag{2.81}$$

and the Jacobian transformation

$$J = \begin{vmatrix} \frac{y_2}{1+y_2} & \frac{1}{1+y_2} \\ \frac{y_1}{(1+y_2)^2} & \frac{-y_1}{(1+y_2)^2} \end{vmatrix}$$

$$= -\frac{y_1}{(1 + y_2)^2}. \tag{2.82}$$

Therefore, the joint probability density function $f_\mathbf{y}(y_1, y_2)$ is obtained by

$$f_\mathbf{y}(y_1, y_2) = |J| f_\mathbf{x}(g_1^{-1}(y_1, y_2) = \frac{y_1 y_2}{1 + y_2}, g_2^{-1}(y_1, y_2) = \frac{y_1}{1 + y_2}). \tag{2.83}$$

In order to find the marginal distribution of $\mathbf{y}_1$ and $\mathbf{y}_2$, we need to integrate out y_1

$$f_{\mathbf{y}_1}(y_1) = \int_{-\infty}^{\infty} f_\mathbf{y}(y_1, y_2) dy_1, \tag{2.84}$$

and to integrate out y_2

$$f_{\mathbf{y}_2}(y_2) = \int_{-\infty}^{\infty} f_\mathbf{y}(y_1, y_2) dy_2. \tag{2.85}$$

2.6 Discrete Distributions and Densities

In this section, we introduce several discrete densities and derive their means and variances.

2.6.1 Discrete Uniform Distribution

A random variable **x** is defined as a *discrete uniform distribution* if the discrete density function is given by

$$f_{\mathbf{x}}(x) = \begin{cases} \frac{1}{N} & \text{for } x = 1, 2, \cdots, N \\ 0 & \text{otherwise} \end{cases} \quad (2.86)$$

where the range of parameter N is over the positive integers. The random variable **x** is also called a discrete uniform random variable.

A theorem states that if a random variable **x** has a discrete uniform distribution, then the mean is given by

$$E\{\mathbf{x}\} = \frac{N+1}{2}, \quad (2.87)$$

the variance is given by

$$\text{Var}\{\mathbf{x}\} = \frac{N^2 - 1}{12}, \quad (2.88)$$

and the moment generation function is given by

$$m_x(t) = E\{e^{t\mathbf{X}}\} = \sum_{j=1}^{N} e^{jt} \left(\frac{1}{N}\right). \quad (2.89)$$

We can prove these theorem results as follows:

$$E\{\mathbf{x}\} = \sum_{j=1}^{N} \frac{j}{N} = \frac{N+1}{2}, \quad (2.90)$$

and

$$\text{Var}\{\mathbf{x}\} = E\{\mathbf{x}^2\} - (E\{\mathbf{x}\})^2 \quad (2.91)$$

$$= \sum_{j=1}^{N} \frac{j^2}{N} - \left(\frac{N+1}{2}\right)^2 \qquad (2.92)$$

$$= \frac{N(N+1)(2N+1)}{6N} - \left(\frac{N+1}{2}\right)^2 \qquad (2.93)$$

$$= \frac{(N+1)(N-1)}{12} \qquad (2.94)$$

$$= \frac{N^2 - 1}{12}. \qquad (2.95)$$

By using (2.52), we can obtain the moment generation function of the discrete uniform distribution as follows:

$$m_x(t) = E\{e^{t\mathbf{x}}\} = \sum_{j=1}^{N} e^{t\mathbf{x}} f_{\mathbf{x}}(x) = \sum_{j=1}^{N} e^{jt} \left(\frac{1}{N}\right). \qquad (2.96)$$

2.6.2 Binomial Distribution

A random variable $\mathbf{x}$ is defined as a *discrete binomial distribution* if the discrete density function of $\mathbf{x}$ is given by

$$f_{\mathbf{x}}(x) = \begin{cases} \binom{n}{x} p^x q^{n-x} & \text{for } x = 1, 2, \cdots, n \\ 0 & \text{otherwise} \end{cases} \qquad (2.97)$$

where the parameters p satisfy $0 \leq p \leq 1$, $q = 1 - p$, and n ranges over the positive integers. Figure 2.1 shows the binomial discrete density function $f_{\mathbf{x}}(x)$ of $\mathbf{x}$, with the parameters of $n = 30$, $p = 0.3$, and $q = 0.7$.

A theorem states that if a random variable $\mathbf{x}$ has a discrete binomial distribution, then the mean is

$$E\{\mathbf{x}\} = np, \qquad (2.98)$$

the variance is

$$\text{Var}\{\mathbf{x}\} = npq, \qquad (2.99)$$

and the moment generation function is

$$m_x(t) = (q + pe^t)^n. \qquad (2.100)$$

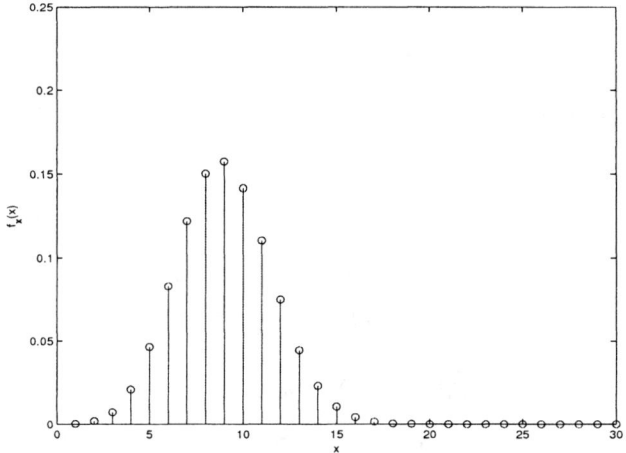

Figure 2.1 The binomial discrete density function $f_\mathbf{X}(x)$ of **x**, with $n = 30$, $p = 0.3$, and $q = 0.7$.

We now prove these theorem results starting first with the proof of the moment generation function as follows:

$$\begin{aligned}
m_x(t) &= E\{e^{t\mathbf{X}}\} \\
&= \sum_{x=1}^{N} e^{tx} \binom{n}{x} p^x q^{n-x} \\
&= \sum_{x=1}^{N} \binom{n}{x} (pe^t)^x q^{n-x} \\
&= (pe^t + q)^n.
\end{aligned} \qquad (2.101)$$

Taking the first and second derivatives of (2.101) obtains

$$m'_x(t) = npe^t(pe^t + q)^{n-1}, \qquad (2.102)$$

and

$$m''_x(t) = n(n-1)(pe^t)^2(pe^t + q)^{n-2} + npe^t(pe^t + q)^{n-1}. \qquad (2.103)$$

Thus, we obtain the mean

$$E\{\mathbf{x}\} = m'_x(0) = np, \qquad (2.104)$$

and the variance

$$\begin{aligned} \mathrm{Var}\{\mathbf{x}\} &= E\{\mathbf{x}^2\} - (E\{\mathbf{x}\})^2 \\ &= m''_x(0) - (np)^2 \\ &= n(n-1)p^2 + np - (np)^2 \\ &= np(1-p). \end{aligned} \qquad (2.105)$$

Note that if $n = 1$, then (2.97) becomes

$$f_\mathbf{x}(x) = \begin{cases} p^x q^{1-x} & \text{for } x = 0 \text{ or } 1 \\ 0 & \text{otherwise} \end{cases} \qquad (2.106)$$

where the parameters p satisfy $0 \le p \le 1$, and $q = 1 - p$. Equation (2.106) is referred to as a *discrete Bernoulli distribution*. In this case, the mean, variance, and moment are as follows: $E\{\mathbf{x}\} = p$, $\mathrm{Var}\{\mathbf{x}\} = pq$, and $m_x(t) = pe^t + q$.

2.6.3 Poisson Distribution

A random variable $\mathbf{x}$ is defined as a *discrete Poisson distribution* if the discrete density function of $\mathbf{x}$ is given by

$$f_\mathbf{x}(x) = \begin{cases} \frac{e^{-\lambda} \lambda^x}{x!} & \text{for } x = 1, 2, \cdots \\ 0 & \text{otherwise} \end{cases} \qquad (2.107)$$

where the parameter $\lambda > 0$. Figure 2.2 shows the Poisson discrete density function $f_\mathbf{x}(x)$ of $\mathbf{x}$ with the parameters of $n = 30$ and $\lambda = 4$.

A theorem states that if a random variable $\mathbf{x}$ has a discrete Poisson distribution, then the mean is

$$E\{\mathbf{x}\} = \lambda, \qquad (2.108)$$

the variance is

$$\mathrm{Var}\{\mathbf{x}\} = \lambda, \qquad (2.109)$$

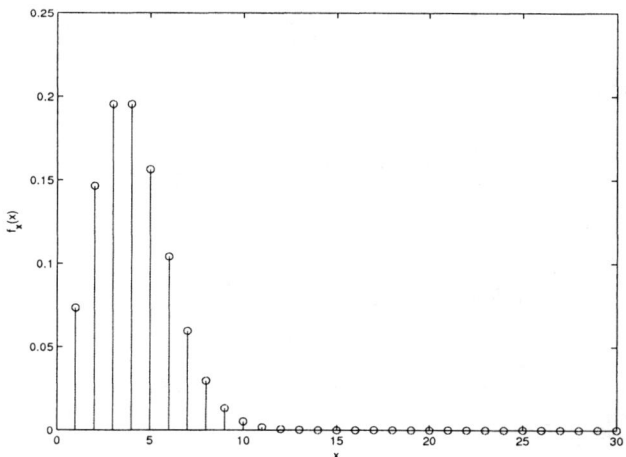

Figure 2.2 The Poisson discrete density function $f_\mathbf{X}(x)$ of $\mathbf{x}$ with $n = 30$ and $\lambda = 4$.

and the moment generation function is

$$m_x(t) = e^{\lambda(e^t - 1)}. \tag{2.110}$$

We prove these theorem results with the moment generation function as follows:

$$\begin{aligned} m_x(t) &= E\{e^{t\mathbf{X}}\} \\ &= \sum_{x=0}^{\infty} \frac{e^{tx} e^{-\lambda} \lambda^x}{x!} \\ &= e^{-\lambda} \sum_{x=0}^{\infty} \frac{(\lambda e^t)^x}{x!} \\ &= e^{-\lambda} e^{\lambda e^t}. \end{aligned} \tag{2.111}$$

Taking the first and second derivatives of (2.111) yields

$$m_x'(t) = \lambda e^{-\lambda} e^t e^{\lambda e^t}, \tag{2.112}$$

and
$$m_x''(t) = \lambda e^{-\lambda} e^t e^{\lambda e^t}[\lambda e^t + 1]. \tag{2.113}$$

Therefore, we obtain the mean
$$E\{\mathbf{x}\} = m_x'(0) = \lambda, \tag{2.114}$$

and the variance
$$\begin{aligned} \mathrm{Var}\{\mathbf{x}\} &= E\{\mathbf{x}^2\} - (E\{\mathbf{x}\})^2 \\ &= m_x''(0) - \lambda^2 \\ &= \lambda(\lambda + 1) - \lambda^2 \\ &= \lambda. \end{aligned} \tag{2.115}$$

The discrete Poisson distribution has many applications because it provides a realistic model for many random phenomena. Special application areas include modeling for fatal traffic accidents per day in a given state, the number of radioactive particle emissions per unit of time, the number of telephone calls per minute, the number of defects per unit of some material, the number of meteorites that collide with a test satellite during a single orbit, and so on. Mood et al. [5] pointed out that if certain assumptions regarding the phenomenon under observation are satisfied, the discrete Poisson model is the correct model.

Example 2.5

Assume that the average number of telephone calls arriving at the switchboard is 1,200 calls per hour. We want to determine as follows: (1) What is the probability that no calls will arrive in a period of 1 minute? (2) What is the probability that more than 100 calls will arrive in a period of 1 minute?

Note that 1,200 calls per hour equals to 20 calls per minute. Thus, the mean rate of occurrence is 20 per minute as well. The probability that no calls will arrive in a period of 1 minute is obtained by
$$P = e^{-vt} = e^{-(20)(1)} \approx 2.061 \times 10^{-9}, \tag{2.116}$$

and the probability that more than 100 calls will arrive in a period of 1 minute is given by

$$\begin{aligned} P &= \sum_{k=101}^{\infty} \frac{e^{-vt}(vt)^k}{k!} \\ &\approx ve^{-vt}\frac{(vt)^{n-1}}{(n-1)!} \\ &= 20e^{-(20)(1)}\frac{[(20)(1)]^{101-1}}{(101-1)!} \\ &= 5.6 \times 10^{-36}. \end{aligned} \qquad (2.117)$$

2.7 Continuous Distributions and Densities

In this section, we discuss a set of special continuous distributions and densities that belong to the parameter families of universal probability density functions.

2.7.1 Gaussian Density Function

A random variable **x** is known as *Gaussian* or *normal* if its density function is given by

$$f_\mathbf{x}(x) = \frac{1}{\sigma_x \sqrt{2\pi}} \exp\left\{-\frac{(x-m_x)^2}{2\sigma_x^2}\right\}, \qquad (2.118)$$

where m_x is its mean defined by (2.44) and σ_x^2 is its variance defined by (2.45). The symbol σ_x is also known as the *standard deviation*. The density function in (2.118) is symmetric around the mean m_x. Since $f_\mathbf{x}(x)$ is a density function, then we obtain

$$\int_{-\infty}^{\infty} f_x(x)dx = 1. \qquad (2.119)$$

The corresponding Gaussian distribution function can be obtained by taking the integral of the Gaussian density function in (2.118),

$$F_x(x) = \int_{-\infty}^{x} f_x(u)du$$

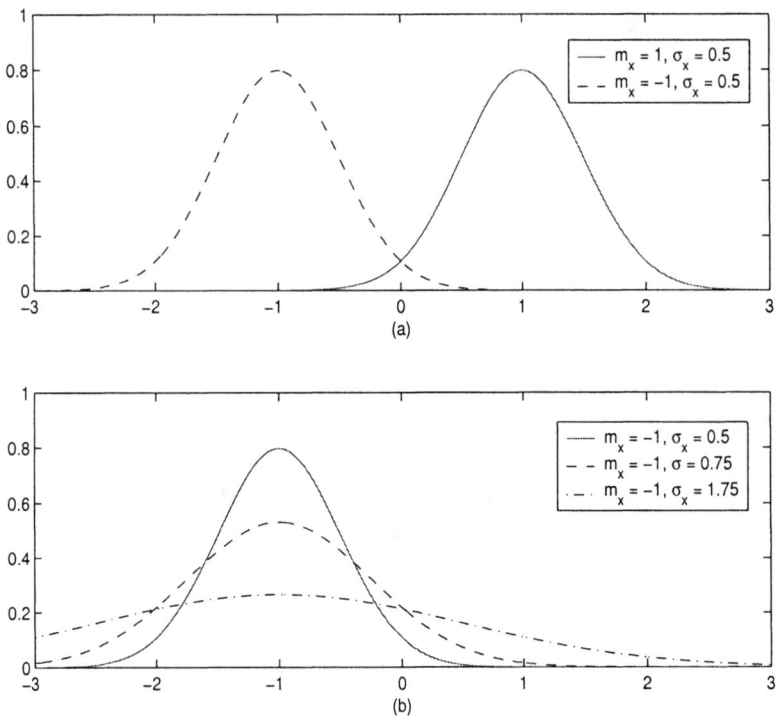

Figure 2.3 The Gaussian density functions $f_\mathbf{X}(x)$: (a) the different means and the same standard deviation and (b) the same standard deviation and the different means.

$$= \int_{-\infty}^{x} \frac{1}{\sigma_x \sqrt{2\pi}} \exp\left\{-\frac{(u - m_x)^2}{2\sigma_x^2}\right\} du. \quad (2.120)$$

The Gaussian density functions $f_\mathbf{X}(x)$, with the different means and the same standard deviation and the same standard deviation and the different means, are shown in Figure 2.3(a, b), respectively. Note that the mode of a Gaussian density occurs at $x = m_x$ and inflection points happen at $m_x - \sigma_x$ and $m_x + \sigma_x$. Also, with the same mean value, increasing the variance leads to a decrease in the peak value of the Gaussian distribution functions.

We often use a shorthand notation

$$\mathbf{x} \sim N(m_x, \sigma_x^2) \qquad (2.121)$$

to represent that a random variable **x** has a Gaussian density function. We can also use the notations $\phi_{m_x,\sigma_x^2}(x)$ and $\Phi_{m_x,\sigma_x^2}(x)$ for the Gaussian density function and the Gaussian distribution function, respectively.

If the Gaussian random variable **x** has zero mean and unit variance, it is called a standard or normalized Gaussian random variable. The normalized Gaussian density function is given by

$$\phi_{0,1}(x) = \frac{1}{\sqrt{2\pi}} \exp\left\{-\frac{x^2}{2}\right\}, \qquad (2.122)$$

and the corresponding normalized Gaussian distribution function is obtained by

$$\Phi_{0,1}(x) = \int_{-\infty}^{x} \phi_{0,1}(u)\,du. \qquad (2.123)$$

There are many important properties for the Gaussian random variables. We introduce the two most important properties as follows.

Property 1. If the random variable is the Gaussian random variable, denoted by $\mathbf{x} \sim N(m_x, \sigma_x^2)$, then the probability of the random variable **x** falls in the range interval $(a, b]$, given by [6]

$$p\{a < \mathbf{x} \le b\} = \Phi_{0,1}\left(\frac{b - m_x}{\sigma_x}\right) - \Phi_{0,1}\left(\frac{a - m_x}{\sigma_x}\right). \qquad (2.124)$$

This property can be proved as follows:

$$\begin{aligned}
p\{a < \mathbf{x} \le b\} &= \int_a^b \frac{1}{\sigma_x \sqrt{2\pi}} \exp\left\{-\frac{(u-m_x)^2}{2\sigma_x^2}\right\} du \\
&= \int_{(a-m_x)/\sigma_x}^{(b-m_x)/\sigma_x} \frac{1}{\sqrt{2\pi}} \exp\left\{-\frac{z^2}{2}\right\} dz \\
&= \Phi_{0,1}\left(\frac{b-m_x}{\sigma_x}\right) - \Phi_{0,1}\left(\frac{a-m_x}{\sigma_x}\right).
\end{aligned} \qquad (2.125)$$

Property 2. If a distribution function is the normalized Gaussian distribution function, then we have $\Phi_{0,1}(x) = 1 - \Phi_{0,1}(-x)$. This property can be approved by using the Gaussian symmetry.

The normalized Gaussian distribution appears to be a reasonable model of the behavior of certain random phenomena. A number of the processes in applications are Gaussian or approximately Gaussian as followed in central limit theorems, which will be discussed in Section 2.8.3.

2.7.2 Error Function

The *error function*, which is denoted by erf(x), is defined by

$$\text{erf}(x) = \frac{2}{\sqrt{\pi}} \int_0^x e^{-u^2} du. \tag{2.126}$$

Figure 2.4 shows a plot of the error function in (2.126), with the semilog and the parameter x from 0 to 5.

The corresponding *complementary error function*, which is denoted by erfc(x), is defined by

$$\begin{aligned}\text{erfc}(x) &= \frac{2}{\sqrt{\pi}} \int_x^\infty e^{-u^2} du. \\ &= 1 - \text{erf}(x).\end{aligned} \tag{2.127}$$

The error function has many properties as follows:

$$\begin{aligned}\text{erf}(-x) &= -\text{erf}(x) & (2.128) \\ \text{erfc}(-x) &= 2 - \text{erfc}(x) & (2.129) \\ \text{erf}(0) &= \text{erfc}(\infty) & (2.130) \\ \text{erf}(\infty) &= \text{erfc}(0) & (2.131) \\ \text{erfc}(\infty) &= 0 & (2.132) \\ \text{erfc}(0) &= 1. & (2.133)\end{aligned}$$

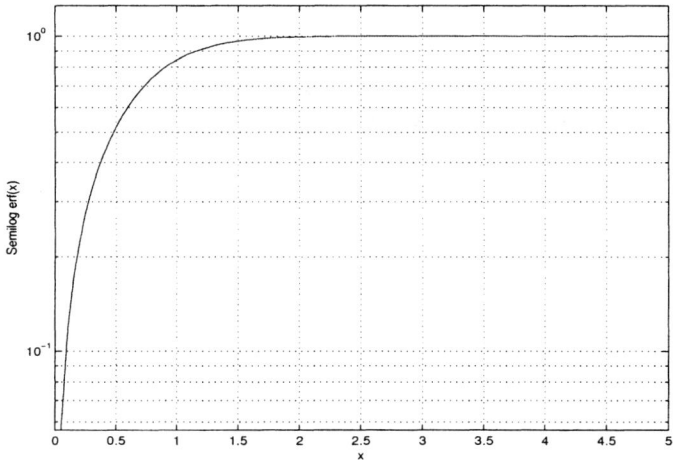

Figure 2.4 A plot of the error function erf(x), with the semilog and the parameter x from 0 to 5.

Note that the Gaussian distribution function given by (2.120) can be rewritten in terms of erf(x) as follows:

$$\begin{aligned}
F_x(x) &= \int_{-\infty}^{x} \frac{1}{\sigma_x\sqrt{2\pi}} \exp\left\{-\frac{(u-m_x)^2}{2\sigma_x^2}\right\} du \\
&= \frac{1}{\sqrt{\pi}} \int_{-\infty}^{x} \frac{1}{\sigma_x\sqrt{2\pi}} \exp\left\{-\frac{(u-m_x)^2}{2\sigma_x^2}\right\} du \\
&= \frac{1}{\sqrt{2\pi}} \int_{-\infty}^{0} e^{-\frac{t^2}{2}} dt + \frac{2}{\sqrt{\pi}} \int_{0}^{(x-m)/\sqrt{2}\sigma} e^{-t^2} dt \\
&= \frac{1}{2} + \frac{1}{2}\text{erf}\left(\frac{x-m}{\sqrt{2}\sigma}\right).
\end{aligned} \qquad (2.134)$$

Equation (2.134) can also be expressed in terms of erfc(x),

$$F_x(x) = 1 - \frac{1}{2}\text{erfc}\left(\frac{x-m}{\sqrt{2}\sigma}\right). \qquad (2.135)$$

If $x > 0$, the asymptotic expansion for calculating erfc(x) is given by [7]

$$\text{erfc}(x) \approx \frac{e^{-x^2}}{\sqrt{\pi}x}\left[1 - \frac{1}{2x^2} + \frac{1 \cdot 3}{2^2 x^4} - \cdots \pm \frac{1 \cdot 3 \cdots (2n-1)}{2^n x^{2n}}\right]. \tag{2.136}$$

If x is a large positive value, the successive terms of the right-hand side in (2.136) decrease very rapidly. Thus, the bounds on erfc(x) are obtained by

$$\frac{e^{-x^2}}{\sqrt{\pi}x}\left(1 - \frac{1}{2x^2}\right) < \text{erfc}(x) < \frac{e^{-x^2}}{\sqrt{\pi}x}. \tag{2.137}$$

Since erfc(x) = 1 − erf(x), we then obtain the bounds on erf(x)

$$1 - \frac{e^{-x^2}}{\sqrt{\pi}x} < \text{erf}(x) < 1 - \frac{e^{-x^2}}{\sqrt{\pi}x}\left(1 - \frac{1}{2x^2}\right). \tag{2.138}$$

2.7.3 Q-Function

Consider a standardized Gaussian random variable **x** with zero mean and unit variance given by (2.121). The *Q-function* is defined by

$$Q(x) = \frac{1}{\sqrt{2\pi}} \int_x^\infty e^{-\frac{x^2}{2}} dx. \tag{2.139}$$

Thus, the Q-function can be written in terms of erfc(x) as

$$Q(x) = \frac{1}{2}\text{erfc}\left(\frac{x}{\sqrt{2}}\right), \tag{2.140}$$

and in terms of $\Phi_{0,1}(x)$ as

$$Q(x) = 1 - \Phi_{0,1}(x). \tag{2.141}$$

Conversely, we can obtain erfc(x) in terms of the Q-function by using $u = x/\sqrt{2}$

$$\text{erfc}(u) = 2Q(\sqrt{2}u). \tag{2.142}$$

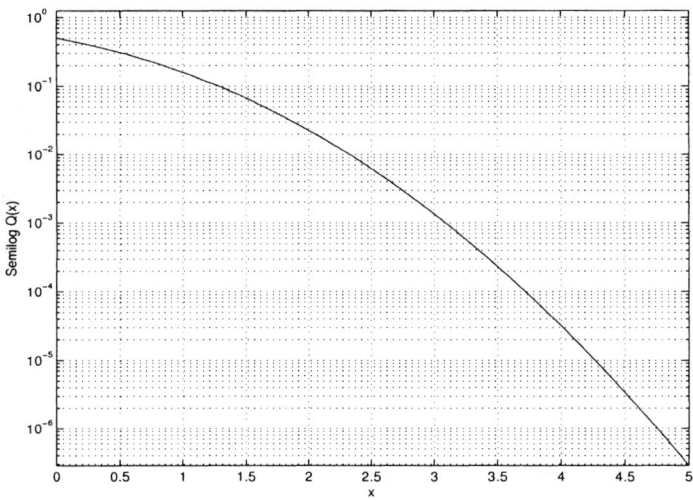

Figure 2.5 A plot of the Q-function $Q(x)$, with the semilog and the parameter x from 0 to 5.

Figure 2.5 shows a plot of the Q-function, with the semilog and the parameter x from 0 to 5.

Using (2.136), the asymptotic expansion of the Q-function $Q(x)$, with $x > 0$, is obtained by

$$Q(x) \approx \frac{e^{-(\frac{x}{\sqrt{2}})^2}}{\sqrt{2\pi}x} \left[1 - \frac{1}{2\left(\frac{x}{\sqrt{2}}\right)^2} + \frac{1 \cdot 3}{2^2 \left(\frac{x}{\sqrt{2}}\right)^4} - \cdots \right.$$

$$\left. \cdots \pm \frac{1 \cdot 3 \cdots (2n-1)}{2^n \left(\frac{x}{\sqrt{2}}\right)^{2n}} \right]. \quad (2.143)$$

If x is a large positive value, the successive terms of the right-hand side in (2.143) decrease very rapidly. Thus, the bounds on $Q(x)$ are given by

$$\frac{e^{-\frac{x^2}{2}}}{\sqrt{2\pi}x} \left(1 - \frac{1}{x^2}\right) < Q(x) < \frac{e^{-\frac{x^2}{2}}}{\sqrt{2\pi}x}. \quad (2.144)$$

2.7.4 Multivariate Gaussian Distribution

Consider the Gaussian random variables $\mathbf{x}_i$, $i = 1, 2, \cdots, n$, with means m_i, $i = 1, 2, \cdots, n$, variances σ_i^2, $i = 1, 2, \cdots, n$, covariances μ_{ij}, $i, j = 1, 2, \cdots, n$ and $\mu_{ii} = \sigma_i^2$, $i = 1, 2, \cdots, n$. Let $\mathbf{C}_x$ denote the $n \times n$ covariance matrix, $\mathbf{m}_x$ denote the $n \times 1$ column vector of mean values m_i, $i = 1, 2, \cdots, n$, and $\mathbf{X}$ denote the vector of the Gaussian random variables $\mathbf{x}_i$, $i = 1, 2, \cdots, n$. Then, the multivariate Gaussian density function is given by [1]

$$p(\mathbf{X}) = \frac{1}{(2\pi)^{n/2}|\mathbf{C}_x|^{1/2}} \exp\left[-\frac{1}{2}(\mathbf{X} - \mathbf{m}_x)^T \mathbf{C}_x^{-1}(\mathbf{X} - \mathbf{m}_x)\right], \quad (2.145)$$

where $\mathbf{C}_x^{-1}$ is the inverse of $\mathbf{C}_x$ and $\mathbf{m}_x^T$ is the transpose of $\mathbf{m}_x$.

In practice, due to finite sample sizes, only estimates of the mean $\hat{\mathbf{m}}_x$ and covariance $\hat{\mathbf{C}}_x$ can be obtained for (2.145) by using unbiased estimate methods as follows:

$$\hat{\mathbf{m}}_x = E[\mathbf{x}] = \frac{1}{N} \sum_{k=1}^{N} \mathbf{x}_k, \quad (2.146)$$

and

$$\hat{\mathbf{C}}_x = E[(\mathbf{x} - \mathbf{m}_x)(\mathbf{x} - \mathbf{m}_x)^T]$$
$$= \frac{1}{N-1} \sum_{k=1}^{N} (\mathbf{x}_k - \hat{\mathbf{m}}_x)(\mathbf{x}_k - \hat{\mathbf{m}}_x)^T, \quad (2.147)$$

where N is the total number of samples.

2.7.5 Uniform Density Function

A random variable $\mathbf{x}$ is known as a *uniform random variable* if its density function is a constant in the range $(a, b]$ and 0 elsewhere:

$$f_x(x) = \begin{cases} \frac{1}{b-a} & a < x \leq b, \text{ and } b > a \\ 0 & \text{otherwise.} \end{cases} \quad (2.148)$$

The corresponding distribution function of the random variable $\mathbf{x}$ is obtained by taking the integral of the density function in (2.148),

$$F_x(x) = \int_{-\infty}^{x} f_x(u) du$$

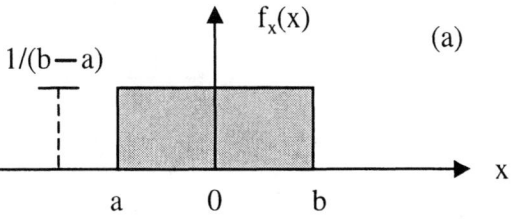

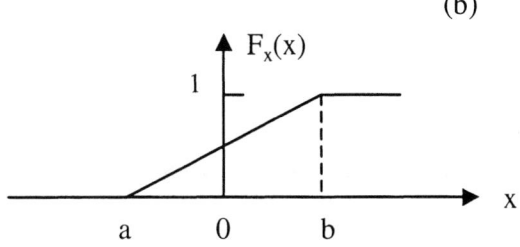

Figure 2.6 The uniform random variable **x**: (a) uniform density function $f_x(x)$ and (b) uniform distribution function $F_x(x)$, where $\|a\| = \|b\|$.

$$= \begin{cases} 0, & -\infty < x < a \\ \frac{x-a}{b-a}, & a < x \le b \\ 1, & \text{otherwise.} \end{cases} \quad (2.149)$$

The uniform density and corresponding distribution functions are shown in Figure 2.6(a, b), respectively.

The mean of the uniform random variable **x** is obtained by

$$\begin{aligned} m_x &= E\{\mathbf{x}\} \\ &= \int_a^b \frac{x}{b-a} dx \\ &= \frac{a+b}{2}, \end{aligned} \quad (2.150)$$

and the variance is obtained by

$$\sigma_x^2 = E\{\mathbf{x}^2\} - E^2\{\mathbf{x}\}$$

$$\begin{aligned}&= \int_a^b \frac{x^2}{b-a}dx - \frac{(a+b)^2}{4}\\&= \frac{b^3-a^3}{3(b-a)} - \frac{(a+b)^2}{4}\\&= \frac{(b-a)^2}{12}.\end{aligned} \qquad (2.151)$$

The uniform distribution function gets its name because its density function is constant over the interval range $(a, b]$. The uniform distribution function can also be referred to as the rectangular function due to the shape of its density function.

2.7.6 Chi-Square Distribution

A random variable **x** has *chi-square density* if its density function is obtained by

$$f_x(x) = \frac{1}{\Gamma(k/2)}\left(\frac{1}{2}\right)^{k/2} x^{k/2-1} e^{-\frac{1}{2}x}, \qquad (2.152)$$

where k is called the *degrees of freedom* and is a positive integer, and $\Gamma(\cdot)$ is the *gamma function*, defined by

$$\Gamma(t) = \int_0^\infty x^{t-1} e^{-x} dx, \quad t > 0. \qquad (2.153)$$

The gamma function has properties as follows:

$$\Gamma(t+1) = t\Gamma(t), \qquad (2.154)$$

and if $t = n$ and n is an integer, then the gamma function is

$$\Gamma(n+1) = n!, \qquad (2.155)$$

$$\Gamma\left(n+\frac{1}{2}\right) = \frac{1 \cdot 3 \cdot 5 \cdots (2n-1)}{2^n}\sqrt{\pi}, \qquad (2.156)$$

and, in particular,

$$\Gamma\left(\frac{1}{2}\right) = 2\Gamma\left(\frac{3}{2}\right) = \sqrt{\pi}. \qquad (2.157)$$

It can be shown that the mean of chi-square random variable **x** is
$$m_x = E\{\mathbf{x}\} = k, \tag{2.158}$$
and the variance is
$$\sigma_x^2 = 2k. \tag{2.159}$$

The corresponding chi-square distribution function of the random variable **x** is obtained by taking the integral of the density function,
$$F_x(x) = \int_0^x \frac{1}{\Gamma(k/2)} \left(\frac{1}{2}\right)^{k/2} u^{k/2-1} e^{-\frac{1}{2}u} du, \tag{2.160}$$
where $k = 1, 2, \cdots, 30$. For larger values of k ($k > 30$), an approximation of normal distribution with zero mean and unit variance can be used and is quite accurate. Thus, x_α (α is a fraction), the αth quantity point of the distribution, may be computed by
$$x_\alpha = \frac{1}{2}(z_\alpha + \sqrt{2k-1})^2, \tag{2.161}$$
where z_α is the αth quantity point of the standard normal distribution. Mood, Graybill, and Boes [5] have shown (2.161) in error by less than 1%.

2.7.7 $\mathcal{F}$ Distribution

Let **u** be a chi-square random variable with m degrees of freedom and **v** be a chi-square random variable with n degrees of freedom. Also assume that **u** and **v** are statistically independent. Then, the new random variable **x** in terms of **u**, **v**, m, and n is as follows:
$$\mathbf{x} = \frac{\mathbf{u}/m}{\mathbf{v}/n}. \tag{2.162}$$

Equation (2.162) is distributed as an $\mathcal{F}$ distribution with m and n degrees of freedom. Equation (2.162) is called the *variance ratio*. The density function of the random variable **x** is given by
$$f_\mathbf{x}(x) = \frac{\Gamma[(m+n)/2]}{\Gamma(m/2)\Gamma(n/2)} \left(\frac{m}{n}\right)^{m/2} \left\{\frac{x^{(m-2)/2}}{[1 + (m/n)x]^{(m+n)/2}}\right\}. \tag{2.163}$$

The mean of the $\mathcal{F}$ density function of the random variable $\mathbf{x}$ is obtained by

$$E\{\mathbf{x}\} = E\left\{\frac{\mathbf{u}/m}{\mathbf{v}/n}\right\}$$
$$= \left(\frac{n}{m}\right) E\{\mathbf{u}\} E\left\{\frac{1}{\mathbf{v}}\right\}. \qquad (2.164)$$

By using (2.158), we obtain $E\{\mathbf{u}\} = m$, and

$$E\left\{\frac{1}{\mathbf{v}}\right\} = \int_0^\infty \frac{1}{v} \frac{1}{\Gamma(n/2)} \left(\frac{1}{2}\right)^{n/2} v^{n/2-1} e^{-\frac{1}{2}v} dv$$
$$= \frac{1}{\Gamma(n/2)} \left(\frac{1}{2}\right)^{n/2} \int_0^\infty v^{(n-4)/2} e^{-\frac{1}{2}v} dv$$
$$= \frac{\Gamma[(n-2)/2]}{\Gamma(n/2)} \left(\frac{1}{2}\right)^{n/2} \left(\frac{1}{2}\right)^{-(n-2)/2}$$
$$= \frac{1}{n-2}. \qquad (2.165)$$

Therefore, the mean is obtained by

$$E\{\mathbf{x}\} = \left(\frac{n}{m}\right) E\{\mathbf{u}\} E\left\{\frac{1}{\mathbf{v}}\right\}$$
$$= \frac{n}{m} \frac{m}{n-2}$$
$$= \frac{n}{n-2}. \qquad (2.166)$$

In a similar way, the variance is derived to be

$$\sigma_x^2 = \frac{2n^2(m+n-2)}{m(n-2)^2(n-4)}, \quad n > 4. \qquad (2.167)$$

The corresponding $\mathcal{F}$ distribution function of the random variable $\mathbf{x}$ is obtained by taking the integral of the density function,

$$F_x(x) = \int_0^x \frac{\Gamma[(m+n)/2]}{\Gamma(m/2)\Gamma(n/2)} \left(\frac{m}{n}\right)^{m/2} \left\{\frac{u^{(m-2)/2}}{[1+(m/n)u]^{(m+n)/2}}\right\} du. \qquad (2.168)$$

Note that it is not easy to calculate (2.168) directly. However, to obtain the values of $\mathcal{F}$ distribution, we can use the table given in [5].

Further note that the mean of the $\mathcal{F}$ distribution depends only on the degrees of freedom of the denominator. This is because the $\mathcal{F}$ density function is not symmetrical in the degrees of freedom m and n. It is interesting to note that the $\mathcal{F}$ distribution is the ratio of two independent chi-square random variables **u** and **v** divided by their respective degrees of freedom m and n. We have found this $\mathcal{F}$ distribution form to be of practical interest in a beamformer and space-time signal processing.

2.7.8 Rayleigh Distribution

Assume that random variables $\mathbf{x}_1$ and $\mathbf{x}_2$ are independent standardized Gaussian with zero mean and unit variance. Also let a new random variable given by $\mathbf{y} = \sqrt{\mathbf{x}_1^2 + \mathbf{x}_2^2}$.

In order to find the probability density function and probability distribution function, we first define

$$\mathbf{v} = \tan^{-1}\left(\frac{\mathbf{x}_2}{\mathbf{x}_1}\right), \qquad (2.169)$$

and then obtain

$$\mathbf{x}_1 = \mathbf{y}\cos\mathbf{v}, \qquad (2.170)$$
$$\mathbf{x}_2 = \mathbf{y}\sin\mathbf{v}. \qquad (2.171)$$

Using the Jacobian transformation in (2.78) yields

$$\begin{aligned}
J &= \begin{vmatrix} \frac{\partial x_1}{\partial y} & \frac{\partial x_1}{\partial v} \\ \frac{\partial x_2}{\partial y} & \frac{\partial x_2}{\partial v} \end{vmatrix} \\
&= \begin{vmatrix} \cos v & y\sin v \\ \sin v & y\cos v \end{vmatrix} \\
&= y[\cos^2 v + \sin^2 v] \\
&= y. \qquad (2.172)
\end{aligned}$$

Therefore, the joint probability density function $f_{y,v}(y,v)$ is obtained by

$$f_{y,v}(y,v) = |J| f_{x_1,x_2}(y\cos v, y\sin v)$$

$$= y\left(\frac{1}{2\pi\sigma^2}e^{-\frac{x_1^2+x_2^2}{2\sigma^2}}\right)$$

$$= \frac{y}{2\pi\sigma^2}e^{-\frac{y^2}{2\sigma^2}}. \qquad (2.173)$$

It follows that the marginal probability density function of **y** is obtained by

$$f_y(y) = \int_0^{2\pi} f_{y,v}(y,v)dv$$

$$= \frac{y}{\sigma^2}e^{-\frac{y^2}{2\sigma^2}}, \ y \geq 0. \qquad (2.174)$$

The corresponding probability distribution function is obtained by

$$F_y(y) = 1 - e^{-\frac{y^2}{2\sigma^2}}, \ y \geq 0. \qquad (2.175)$$

Hence, a random variable having the probability density function given by (2.174) and (2.175) is said to be *Rayleigh distributed*.

Figure 2.7(a, b) shows the Rayleigh density function $f_y(y)$ and the corresponding distribution function $F_y(y)$ with $\sigma = 0.5$ and $\sigma = 1$, respectively. As can be seen, increasing the σ value in (2.174) and (2.175) decreases the peak value of the Rayleigh density function $f_y(y)$ and reduces the convergence speed of the Rayleigh distribution function $F_y(y)$ to the constant value of 1.

2.7.9 Rice Distribution

Assume that random variables $\mathbf{x}_1$ and $\mathbf{x}_2$ are independent Gaussian with nonzero means m_1, m_2, and nonunit variance σ^2. Also, let a new random variable given by $\mathbf{y} = \sqrt{\mathbf{x}_1^2 + \mathbf{x}_2^2}$.

In order to obtain the probability density function and probability distribution function for the new random variable **y**, we first define

$$\mathbf{v} = \tan^{-1}\left(\frac{\mathbf{x}_2}{\mathbf{x}_1}\right), \qquad (2.176)$$

and then obtain

$$\mathbf{x}_1 = \mathbf{y}\cos\mathbf{v}, \qquad (2.177)$$

$$\mathbf{x}_2 = \mathbf{y}\sin\mathbf{v}. \qquad (2.178)$$

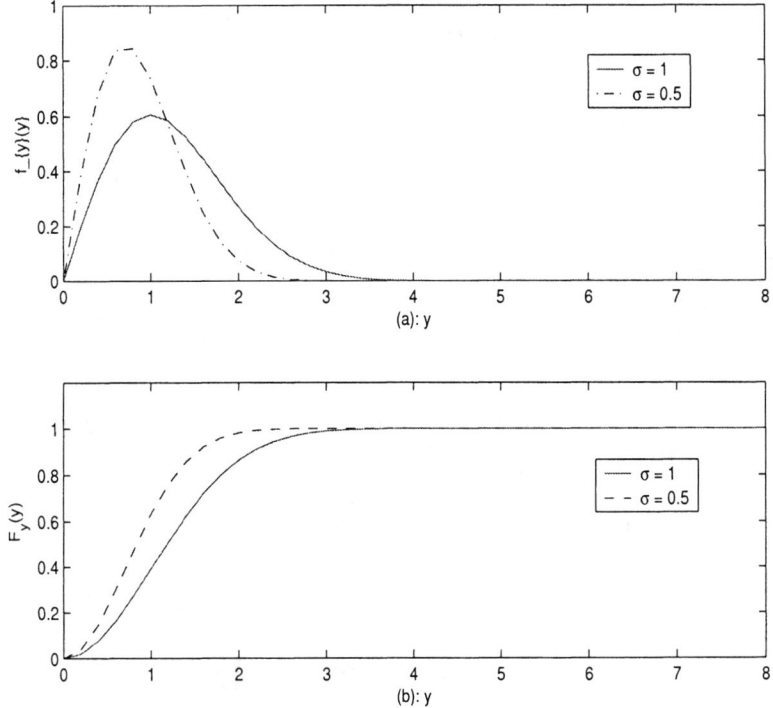

Figure 2.7 The Rayleigh random variable of **y**: (a) the density function and (b) the distribution function.

Using the Jacobian transformation result $J = y$ given by (2.172), we obtain the joint probability density function of the new random variable **y** as follows:

$$\begin{aligned} f_{y,v}(y,v) &= |J| f_{x_1,x_2}(x_1, x_2) \\ &= y f_{x_1,x_2}(y \cos v, y \sin v). \end{aligned} \quad (2.179)$$

The joint probability density function $f_{x_1,x_2}(y, v)$ in (2.179) is obtained by

$$f_{x_1,x_2}(y,v) = \frac{1}{2\pi\sigma^2} \exp\left[-\frac{(y\cos v - m_1)^2 + (y\sin v - m_2)^2}{2\sigma^2}\right]. \tag{2.180}$$

Now let $\mu = \sqrt{m_1^2 + m_2^2}$ and $\phi = \tan^{-1}\left(\frac{m_2}{m_1}\right)$, where $0 \leq \phi \leq 2\pi$, and then $m_1 = \mu\cos\phi$ and $m_2 = \mu\sin\phi$. Thus, (2.180) can be rewritten as

$$f_{x_1,x_2}(y,v) = \frac{1}{2\pi\sigma^2} \exp\left[-\frac{y^2 + \mu^2 - 2y\mu\cos(v-\phi)}{2\sigma^2}\right]. \tag{2.181}$$

The corresponding marginal probability density function of the random variable **y** is obtained by

$$\begin{aligned}
f_y(y) &= \int_0^{2\pi} f_{y,v}(y,v) dv \\
&= \left[\frac{y}{\sigma^2} e^{-\frac{y^2+\mu^2}{2\sigma^2}}\right]\left[\frac{1}{2\pi}\int_0^{2\pi} e^{-\frac{y\mu\cos(v-\phi)}{\sigma^2}} dv\right] \\
&= \frac{y}{\sigma^2} e^{-\frac{y^2+\mu^2}{2\sigma^2}} I_0\left(\frac{y\mu}{\sigma^2}\right), \quad y \geq 0,
\end{aligned} \tag{2.182}$$

where

$$I_0(z) = \frac{1}{2\pi}\int_0^{2\pi} e^{-z\cos\theta} d\theta. \tag{2.183}$$

Equation (2.183) is called the *zero-order modified Bessel function*. Hence, a random variable **y** having the probability density function given by (2.182) is said to be *Rice* or *Ricean distributed*.

Note that if $\mu = 0$, we have $I_0(0) = 1$. Thus, (2.182) is exactly equal to (2.174). Therefore, the Rayleigh distribution is a special case of the Rice distribution.

2.8 Upper Bounds on the Probability

In order to evaluate the performance of a digital communications system, it is necessary to measure the area under the tail of the probability density function. In this section, we introduce methods of upper bounds on the tail probability.

2.8.1 Chebyshev Inequality

Assume that **x** is a continuous random variable with probability density function $f_x(\cdot)$ and let $g(\cdot)$ be a nonnegative function with its domain as the real line. Then we have an upper bound on the tail of the probability as follows:

$$P\{g(\mathbf{x}) \geq k\} \leq \frac{E\{g(\mathbf{x})\}}{k}, \text{ for every } k > 0. \quad (2.184)$$

This upper bound on the tail of the probability is called the *Chebyshev inequality*.

Equation (2.184) can be proved as follows:

$$\begin{aligned}
E\{g(\mathbf{x})\} &= \int_{-\infty}^{\infty} g(x) f_x(x) dx \\
&= \int_{g(x) \geq k} g(x) f_x(x) dx + \int_{g(x) < k} g(x) f_x(x) dx \\
&\geq \int_{g(x) \geq k} g(x) f_x(x) dx \\
&\geq k \int_{g(x) \geq k} g(x) f_x(x) dx \\
&= k p\{g(\mathbf{x}) \geq k\}. \quad (2.185)
\end{aligned}$$

Now, dividing by k on both sides of (2.185) obtains the result of the Chebyshev inequality in (2.184).

If **x** is a random variable with finite variance, $g(x) = (x - m_x)^2$ and $k = \delta^2$, where δ is any positive number, then (2.184) can be rewritten as

$$\begin{aligned}
p\{|\mathbf{x} - m_x| \geq \delta\} &= P\{(\mathbf{x} - m_x)^2 \geq \delta^2\} \\
&\leq \frac{\sigma_x^2}{\delta^2}, \text{ for every } \delta > 0, \quad (2.186)
\end{aligned}$$

where m_x is the mean and σ_x is the variance. Equation (2.186) can be rewritten in another form as follows:

$$p\{|\mathbf{x} - m_x| < \delta\} \geq 1 - \frac{\sigma_x^2}{\delta^2}. \quad (2.187)$$

Note that (2.187) indicates:

$$p\{m_x - \delta < \mathbf{x} < m_x + \delta\} \geq 1 - \frac{\sigma_x^2}{\delta^2}. \qquad (2.188)$$

In other words, the probability that $\mathbf{x}$ falls within δ units of m_x is greater than or equal to $\left(1 - \frac{\sigma_x^2}{\delta^2}\right)$.

Further note that the Chebyshev inequality provides an upper bound, which does not require knowing the distribution of the random variable $\mathbf{x}$. For the probability of particular events, we only need to know the mean and variance of the random variable $\mathbf{x}$.

Example 2.6

Assume that $\delta = 2\sigma_x$ in (2.188). We then obtain the bound on the tail of the probability

$$p\{m_x - 2\sigma < \mathbf{x} < m_x + 2\sigma\} \geq 1 - \frac{\sigma_x^2}{4\sigma_x^2} = \frac{3}{4}. \qquad (2.189)$$

This says that for any random variable $\mathbf{x}$ having finite variance, the probability that $\mathbf{x}$ falls within two standard deviations of its mean is at least $\frac{3}{4}$.

2.8.2 Law of Large Numbers

Suppose that random variables $\mathbf{x}_i$, $i = 1, 2, \cdots, n$, are statistically independent and identically distributed, each having a finite mean m_x and a finite variance σ_x. Let $\mathbf{y}$ be defined as the sample mean as follows:

$$\mathbf{y} = \frac{1}{n} \sum_{i=1}^{n} \mathbf{x}_i, \qquad (2.190)$$

where the mean $m_y = m_x$ and $\sigma_y^2 = \frac{\sigma_x^2}{n}$.

The tail probability of the random variable $\mathbf{y}$ can be upper-bounded by using the Chebyshev inequality. Using (2.186), we obtain the bound for the random variable $\mathbf{y}$

$$p\{|\mathbf{y} - m_y| \geq \delta\} \leq \frac{\sigma_y^2}{\delta^2}. \qquad (2.191)$$

It follows that we obtain the Chebyshev inequality on the random variable $\mathbf{x}_i$, $i = 1, 2, \cdots, n$,

$$p\left\{\left|\frac{1}{n}\sum_{i=1}^{n}\mathbf{x}_i - m_x\right| \geq \delta\right\} \leq \frac{\sigma_x^2}{n\delta^2}. \tag{2.192}$$

Note that when $n \to \infty$, (2.192) becomes

$$\lim_{n\to\infty} p\left\{\left|\frac{1}{n}\sum_{i=1}^{n}\mathbf{x}_i - m_x\right| \geq \delta\right\} \leq \lim_{n\to\infty} \frac{\sigma_x^2}{n\delta^2} \approx 0. \tag{2.193}$$

Equation (2.193) is called the *law of large numbers* or the *weak law of large numbers*. This says that the probability that the estimate of the mean differs from the mean m_x by more than any positive value δ approaches zero when n approaches infinity.

2.8.3 Central Limit Theorem

The most widely applied central limit theorem states that if random variables $\mathbf{x}_1, \cdots, \mathbf{x}_n$ are independent and identically distributed with mean m_x and variance σ_x^2, then, for each random variable $\mathbf{z}_n$,

$$\begin{aligned}\mathbf{z}_n &= \frac{\bar{\mathbf{x}}_n - E\{\bar{\mathbf{x}}_n\}}{\sqrt{\text{Var}\{\bar{\mathbf{x}}_n\}}} \\ &= \frac{\bar{\mathbf{x}}_n - m_x}{\sigma_x/\sqrt{n}},\end{aligned} \tag{2.194}$$

where $\bar{\mathbf{x}}_n$ is defined and given by (2.70), the distribution function $F_{\mathbf{z}_n}(z)$ in (2.194) converges to the normalized Gaussian distribution function $\Phi_{0,1}(z)$, when n approaches ∞. Thus, the distribution function $F_{\mathbf{z}_n}(z)$ is referred to as the *central limit theorem*.

This central limit theorem states that the distribution function $F_{\mathbf{z}_n}(z)$ approximates a normalized Gaussian distribution function. In other words, this theorem tells us that $\bar{\mathbf{x}}_n$ is approximately, or asymptotically, distributed as a normalized distribution with mean m_x and variance σ_x^2/n.

Further note that this central limit theorem assumes nothing about the form of the original density function except that it has finite variance. If given large enough samples, whatever the distribution function with finite variance, the sample mean $\bar{\mathbf{x}}_n$ will have, approximately, the Gaussian distribution. In practice, most density functions encountered have finite variance.

The nature of this central limit theorem provides a specific error bound if the random variables $\mathbf{x}_1, \cdots, \mathbf{x}_n$ are independent and identically distributed with $n \geq 30$ for many applications [4]. However, in practice, we suggest using the value $n \geq 60$ for many applications. In some cases, we may need a large value of n for samples. Generally, this central limit theorem represents densities extremely well, with a few random variables of the mean.

In corollary, if random variables $\mathbf{x}_1, \cdots, \mathbf{x}_n$ are independent and identically distributed with common mean m_x and variance σ_x^2, we then have useful results as follows:

$$p\left[a < \frac{\bar{\mathbf{x}}_n - m_x}{\sigma_x/\sqrt{n}} < b\right] \approx \Phi_{0,1}(b) - \Phi_{0,1}(a), \quad (2.195)$$

$$p[c < \bar{\mathbf{x}}_n < d] \approx \Phi_{0,1}\left(\frac{d - m_x}{\sigma_x/\sqrt{n}}\right) - \Phi_{0,1}\left(\frac{c - m_x}{\sigma_x/\sqrt{n}}\right), \quad (2.196)$$

or

$$p\left[q < \sum_{i=1}^{n} \mathbf{x}_i < p\right] \approx \Phi_{0,1}\left(\frac{p - nm_x}{\sqrt{n}\sigma_x}\right) - \Phi_{0,1}\left(\frac{q - nm_x}{\sqrt{n}\sigma_x}\right). \quad (2.197)$$

Note that (2.195), (2.196), and (2.197) provide approximate values for the probabilities of certain events in terms of averages or sums. In fact, in practice, the central limit theorem is inherent in these approximations.

2.9 Stochastic Signal Processes

This section discusses stochastic processes for the characterization and analysis of *discrete-time random processes*. A discrete-time random process can be simply considered to be an indexed sequence of random variables. The extension of these random variable concepts as discussed in earlier sections to the discrete-time random processes is straightforward.

2.9.1 Definition of Discrete-Time Random Process

A discrete-time random process is a sequence of random variables. We denote the discrete-time random variable by $\mathbf{x}[n]$, where the brackets indicate that n is an integer. Since the discrete-time random process is an indexed sequence of random variables, the calculation of the mean of each of these random variables can be obtained by

$$m_x[n] = E\{\mathbf{x}[n]\}. \tag{2.198}$$

Similarly, the variance of each random variable $\mathbf{x}[n]$ is obtained by

$$\begin{aligned} \sigma_x^2[n] &= E\{(\mathbf{x}[n] - m_x[n])(\mathbf{x}[n] - m_x[n])^*\} \\ &= E\{\mathbf{x}^2[n]\} - m_x^2[n]. \end{aligned} \tag{2.199}$$

Two important functions in the study of discrete-time random processes are the autocorrelation

$$R_x[k, l] = E\{\mathbf{x}[k]\mathbf{x}^*[l]\} \tag{2.200}$$

and the autocovariance

$$C_x[k, l] = R[k, l] - m_x[k]m_x^*[l]. \tag{2.201}$$

If $m_x[k] = 0$ and/or $m_x[l] = 0$, the autocovariance and autocorrelation are then equal, $C_x[k, l] = R_x[k, l]$. Also if $k = l$, then the autocovariance function in (2.201) equals the variance in (2.199) given by

$$C_x[k, k] = \sigma_x^2[k]. \tag{2.202}$$

Therefore, the mean in (2.198) defines the average value of the discrete-time random process at index n, while the variance in (2.199) represents the average squared deviation of the discrete-time random process away from the mean at index n. Furthermore, the autocorrelation function in (2.200) and autocovariance function in (2.201) provide the degree of linear dependence between two discrete-time random variables, $\mathbf{x}[k]$ and $\mathbf{x}[l]$.

Example 2.7

Let us consider the complex harmonic discrete-time random process given by

$$\mathbf{x}[n] = Ce^{j(n\omega+\phi)}, \qquad (2.203)$$

where C and ω are fixed constants and ϕ is a phase random variable that is uniform over the interval $(-\pi, \pi]$.

The density function of ϕ is obtained by

$$f(\phi) = \begin{cases} \frac{1}{2\pi}, & -\pi \leq \phi \leq \pi \\ 0, & \text{otherwise.} \end{cases} \qquad (2.204)$$

The mean of the discrete-time random process is obtained by

$$\begin{aligned} m_x[n] &= E\{\mathbf{x}[n]\} \\ &= E\{Ce^{j(n\omega+\phi)}\} \\ &= \int_{-\infty}^{\infty} Ce^{j(n\omega+\phi)} f(\phi) d\phi \\ &= \int_{-\pi}^{\pi} \frac{C}{2\pi} e^{j(n\omega+\phi)} d\phi \\ &= 0. \end{aligned} \qquad (2.205)$$

Equation (2.205) indicates that the discrete-time random variable $\mathbf{x}[n]$ is a zero mean process for all n. The autocorrelation of the discrete-time random process is given by

$$\begin{aligned} R_x[k,l] &= E\{\mathbf{x}[k]\mathbf{x}^*[l]\} \\ &= E\{Ce^{j(k\omega+\phi)} C^* e^{-j(l\omega+\phi)}\} \\ &= |C|^2 E\{e^{j(k-l)\omega}\}. \end{aligned} \qquad (2.206)$$

As we can see, the mean of the discrete-time random variable is zero and the autocorrelation depends only on the difference between k and l. This says that the mean and autocorrelation do not change even if the discrete-time random process is shifted in time index. In fact, the complex harmonic discrete-time random process is a *wide-sense stationary* process, which will be discussed next.

2.9.2 Stationary Processes

A stochastic process of the discrete-time random process $\mathbf{x}[n]$ is called *stationary* if its statistical properties are invariant to a shift of the origin. This means that the discrete-time random processes $\mathbf{x}[n]$ and $\mathbf{x}[n+k]$ have the same statistics for any k.

Two discrete-time random processes $\mathbf{x}[n]$ and $\mathbf{y}[n]$ are called *jointly stationary process* if the joint statistics of $\mathbf{x}[n]$ and $\mathbf{y}[n]$ are the same as the statistics of $\mathbf{x}[n+k]$ and $\mathbf{y}[n+k]$ for any k.

In general, if the nth joint density of discrete-time random variables is given by

$$f_{\mathbf{x}_1[n],...,\mathbf{x}_n[n]}(l_1,...,l_n) = f_{\mathbf{x}_1[n+k],...,\mathbf{x}_n[n+k]}(l_1,...,l_n) \quad (2.207)$$

for all k and all n, then the stochastic process is said to be *stationary in the strict sense* (SSS). If (2.207) holds for values of n up to and including N, the stochastic process is Nth-order stationary.

A stochastic process $\mathbf{x}[n]$ is known as wide-sense stationary (WSS) if it is satisfied by the following conditions:

1. Its mean is a constant $E\{\mathbf{x}[n]\} = m_x$.
2. Its autocorrelation function depends only on the difference, $\tau = k - l$, so that

$$R_x[\tau] = R_x[k,l] = E\{\mathbf{x}[k]\mathbf{x}^*[l]\}, \quad (2.208)$$

where τ is known as the *lag*.

In general, if a stochastic process is stationary in the strict sense (SSS), then it is also WSS. The converse is not true. Note that WSS does not even imply first-order stationary. However, in the case of a Gaussian random process, WSS is equal to SSS. This is one of important properties of Gaussian random processes [3].

2.9.3 Estimated Functions

In practice, ensemble averages, including means, variances, autocorrelations, autocovariances and individual density functions, are not generally known. Thus, estimating these averages from a realization of a discrete-time random process becomes an important and necessary step. Therefore, this section presents estimation methods for the mean, variance, and autocorrelation of the discrete-time random process.

Consider the estimate problem for the ensemble averages of a discrete-time random process $\mathbf{x}[n]$. If a large number of samples N are available, then an estimated mean is obtained by

$$\hat{m}_x = \frac{1}{N} \sum_{n=0}^{N-1} \mathbf{x}[n], \tag{2.209}$$

and estimated variance is given by

$$\hat{\sigma}_x^2 = \frac{1}{N} \sum_{n=0}^{N-1} |\mathbf{x}[n] - \hat{m}_x|^2. \tag{2.210}$$

The autocorrelation can be estimated given by

$$\hat{R}_x[k] = \frac{1}{N} \sum_{n=0}^{N-1} \mathbf{x}[n]\mathbf{x}^*[n-k]. \tag{2.211}$$

Note that (2.209) and (2.210) are also called the *sample mean* and *sample variance*. To obtain these estimates to be accurate and unbiased, the processes must be stationary or, ideally, *ergodic*, which is a stronger condition than stationary. For a precise mathematical definition, a discussion of ergodicity is referred to by Papoulis [4].

2.9.4 Power Spectrum

This section presents a power spectrum or a power spectral density (PSD) for a discrete-time random process by using a discrete-time Fourier transform (DTFT). The DTFT plays an important role in the description and analysis for the discrete-time random processes.

Consider a WSS process $\mathbf{x}[n]$. We first determine the autocorrelation sequence $R_x[k]$ in the discrete-time domain. Since $R_x[k]$ is a deterministic sequence, we then compute its DTFT as follows:

$$P_x(e^{j\omega}) = \sum_{k=-\infty}^{\infty} R_x[k] e^{-jk\omega}. \tag{2.212}$$

Equation (2.212) is called the *power spectrum* or PSD for the discrete-time random process. In the reverse direction, given the power spectrum in (2.212), which is periodic with period 2π, the autocorrelation sequence $R_x[k]$ can be computed by taking the inverse discrete-time Fourier transform (IDTFT) of $P_x(e^{j\omega})$,

$$R_x[k] = \frac{1}{2\pi} \int_{-\pi}^{\pi} P_x(e^{j\omega}) e^{jk\omega} d\omega. \tag{2.213}$$

As can be seen, the power spectrum contributes a frequency domain description and analysis of the second-order moment of the discrete-time random process.

To represent the power spectrum, we can also use the z-transform instead of the DTFT as follows (see the z-transform in Appendix A):

$$P_x(z) = \sum_{k=-\infty}^{\infty} R_x[k] z^{-k}. \tag{2.214}$$

Equation (2.214) is also known as the power spectrum of the discrete-time random variable $\mathbf{x}[n]$.

If the discrete-time random variable $\mathbf{x}[n]$ is a real process, then we obtain the following:

$$R_x[-k] = R_x[k]. \tag{2.215}$$

Equation (2.215) indicates that $R_x[k]$ is real and even. Therefore, (2.212) yields the following:

$$P_x(e^{j\omega}) = R_x[0] + 2 \sum_{k=0}^{\infty} R_x[k] \cos(k\omega). \tag{2.216}$$

Equation (2.216) shows that the power spectrum of a real process is real and even. This is because the term $\cos(k\omega)$ is a real and even function.

Example 2.8

Assume that an autocorrelation function of the discrete-time random process $\mathbf{x}[n]$ is given by $R_x[k] = Aa^{|k|}$, where $a < 1$ is real, and A

is a constant. The power spectrum in (2.214) is obtained by

$$\begin{aligned}
P_x(z) &= \sum_{k=-\infty}^{\infty} R_x[k] z^{-k} \\
&= A \left(\sum_{k=-\infty}^{-1} a^{-k} z^{-k} + \sum_{k=0}^{\infty} a^k z^{-k} \right) \\
&= A \left(\frac{az}{1-az} + \frac{z}{z-a} \right) \\
&= \frac{A(a^{-1} - a)}{(a^{-1} + a) - (z^{-1} + z)}.
\end{aligned} \qquad (2.217)$$

Substituting $z = e^{j\omega}$ into (2.217), the power spectrum in (2.212) is then obtained as follows:

$$P_x(e^{j\omega}) = \frac{A(1-a^2)}{1 - 2a\cos\omega + a^2}. \qquad (2.218)$$

Thus, the power spectrum in (2.218) is real, even, and positive.

2.9.5 Stochastic Processes for Linear Systems

This section derives the relationship between the second-order statistics of a discrete-time random process, including mean, autocorrelation, and power spectrum, for a linear time-invariant (LTI) system.

Assume that a discrete-time random variable **x**[n] is a WSS random process and an input and $h[n]$ is the impulse response of a stable LTI system. The output **y**[n] is a discrete-time random process whose values are the discrete-time convolution of input **x**[n] with impulse response $h[n]$ given by

$$\begin{aligned}
\mathbf{y}[n] &= \mathbf{x}[n] * h[n] \\
&= \sum_{k=-\infty}^{\infty} h[k]\mathbf{x}[n-k].
\end{aligned} \qquad (2.219)$$

The mean of the output discrete-time random process can be expressed by taking the expected value of (2.219),

$$m_y[n] = E\{\mathbf{y}[n]\}$$

$$= \sum_{k=-\infty}^{\infty} h[k]E\{\mathbf{x}[n-k]\}$$

$$= m_x[n] \sum_{k=-\infty}^{\infty} h[k]. \qquad (2.220)$$

Equation (2.220) indicates that the mean $m_y[n]$ of the output discrete-time random process $\mathbf{y}[n]$ is a constant and is directly related to the mean $m_x[n]$ by a scale factor equal to the sum of the discrete-time impulse response values.

Alternatively, an equivalent expression in (2.220) in terms of frequency response is obtained by

$$m_y[n] = m_x[n]H(e^{j0}), \qquad (2.221)$$

where $H(e^{j0})$ is a frequency-domain transfer function at $\omega = 0$. This is to say that the mean $m_y[n]$ equals the mean $m_x[n]$ scaled by the value of the frequency response of the LTI system at $\omega = 0$.

The autocorrelation function of the output discrete-time random process is given by

$$\begin{aligned} R_y[n, n+l] &= E\{\mathbf{y}[n]\mathbf{y}^*[n+l]\} \\ &= \sum_{k=-\infty}^{\infty} h[k] \sum_{r=-\infty}^{\infty} h^*[r] \\ &\quad \cdot E\{\mathbf{x}[n-k]\mathbf{x}^*[n+l-r]\}. \end{aligned} \qquad (2.222)$$

Since the input discrete-time random process $\mathbf{x}[n]$ is WSS, the term of $E\{\mathbf{x}[n-k]\mathbf{x}^*[n+l-r]\}$ depends only on the index $l+k-r$. Thus, we can rewrite (2.222) as follows:

$$\begin{aligned} R_y[n, n+l] &= \sum_{k=-\infty}^{\infty} h[k] \sum_{r=-\infty}^{\infty} h^*[r] R_x[l+k-r] \\ &= R_y[l]. \end{aligned} \qquad (2.223)$$

Note that the autocorrelation function $R_y[l]$ in (2.223) depends only on the index difference of l. To obtain the variance of the output discrete-time random process, $\mathbf{y}[n]$, we can set $l = 0$ in (2.223) and get

$$\begin{aligned}
\sigma_y^2[n] &= R_y[0] \\
&= \sum_{k=-\infty}^{\infty} \sum_{r=-\infty}^{\infty} h[k]h^*[r]R_x[k-r]. \quad (2.224)
\end{aligned}$$

Hence, from (2.221), (2.223), and (2.224), if input discrete-time random variable $\mathbf{x}[n]$ is WSS, then the output discrete-time random process $\mathbf{y}[n]$ will also be WSS with the condition $\sigma_y^2[n] < \infty$ for the stable LTI system.

To obtain the power spectrum, we take the DTFT on both sides of (2.223),

$$\begin{aligned}
P_y(e^{j\omega}) &= H(e^{j\omega})H^*(e^{j\omega})P_x(e^{j\omega}) \\
&= |H(e^{j\omega})|^2 P_x(e^{j\omega}). \quad (2.225)
\end{aligned}$$

Note that if the discrete-time impulse response $h[n]$ is real, then we obtain

$$H^*(e^{j\omega}) = H(e^{-j\omega}). \quad (2.226)$$

In this case, the power spectrum of the output discrete-time random process is given in terms of the z-transform as follows:

$$P_y(z) = H(z)H(1/z)P_x(z). \quad (2.227)$$

Thus, (2.227) provides a convenient way for analysis of the power spectrum $P_y(z)$, specially in spectral factorization.

One interesting point is that the total average power in the output for the LTI system can be calculated by

$$\begin{aligned}
E\{\mathbf{y}^2[n]\} &= R_y[0] \\
&= \frac{1}{2\pi} \int_{-\pi}^{\pi} P_y(e^{j\omega}) d\omega \\
&= \frac{1}{2\pi} \int_{-\pi}^{\pi} |H(e^{j\omega})|^2 P_x(e^{j\omega}) d\omega. \quad (2.228)
\end{aligned}$$

This means that the area under the curve $|H(e^{j\omega})|^2 P_x(e^{j\omega})$ with a range interval $-\pi \leq \omega \leq \pi$ represents the mean-square value of the input discrete-time random process.

Another important subject is that the cross-correlation function between the input discrete-time random process and output discrete-time random process of an LTI system is used for applications of estimating the impulse response or frequency response of a linear system. The cross-correlation function is obtained by

$$\begin{aligned} R_{xy}[l] &= E\{\mathbf{x}[n]\mathbf{y}^*[n+l]\} \\ &= \sum_{k=-\infty}^{\infty} h[k] E\{\mathbf{x}[n]\mathbf{x}^*[n+l-k]\} \\ &= \sum_{k=-\infty}^{\infty} h[k] R_x[l-k]. \end{aligned} \qquad (2.229)$$

Equation (2.229) means that the cross-correlation function $R_{xy}[n]$ is the convolution of the discrete-time impulse response $h[n]$ with the autocorrelation function $R_x[n]$ of the input. If the input is white noise such that $R_x[l] = \sigma_x^2 \delta[l]$, then (2.229) becomes

$$R_{xy}[l] = \sigma_x^2 h[l]. \qquad (2.230)$$

This indicates that the cross-correlation function $R_{xy}[l]$ is proportional to the discrete-time impulse response of the LTI system when the input discrete-time random process is zero-mean white noise.

The power spectrum of the cross-correlation function $R_{xy}[l]$ can be obtained by taking the DTFT on both sides of (2.229),

$$P_{xy}(e^{j\omega}) = H(e^{j\omega}) P_x(e^{j\omega}). \qquad (2.231)$$

In a similar way, if the input discrete-time random process is a zero-mean white noise, then (2.231) can be rewritten as follows:

$$P_{xy}(e^{j\omega}) = \sigma_x^2 H(e^{j\omega}). \qquad (2.232)$$

Hence, (2.232) is proportional to the frequency response of the LTI system. Therefore, (2.230) and (2.231) can be used to estimate the discrete-time impulse response $h[n]$ and frequency response $H(e^{j\omega})$ of the LTI system when the input discrete-time random process is a zero-mean white noise.

Example 2.9

Consider a first-order discrete-time difference equation given by

$$\mathbf{y}[n] = a\mathbf{x}[n] - b\mathbf{x}[n-1], \tag{2.233}$$

where $\mathbf{x}[n]$ is an input discrete-time random process, $\mathbf{y}[n]$ is the output discrete-time random process of a linear system, and a, b are the constants. Also assume that the input discrete-time random process $\mathbf{x}[n]$ is white noise with $P_x(e^{j\omega}) = \sigma_x^2$.

The z-transform transfer function of the first-order discrete-time difference equation is obtained by

$$\begin{aligned} H(z) &= \frac{\mathbf{Y}(z)}{\mathbf{X}(z)} \\ &= a - bz^{-1}. \end{aligned} \tag{2.234}$$

Using (2.227), we obtain the output power spectrum for the LTI system as follows:

$$\begin{aligned} P_y(z) &= (a - bz^{-1})(a - bz)P_x(z) \\ &= (a^2 + b^2 - abz - abz^{-1})\sigma_x^2. \end{aligned} \tag{2.235}$$

The output power spectrum can also be expressed in the frequency domain by using $z = e^{j\omega}$ as follows:

$$\begin{aligned} P_y(e^{j\omega}) &= \sigma_x^2[a^2 + b^2 - ab(e^{j\omega} - e^{-j\omega})] \\ &= \sigma_x^2(a^2 + b^2 - 2ab\cos\omega). \end{aligned} \tag{2.236}$$

since the term of $\cos\omega$ is real and even, the output power spectrum in (2.236) is a real, even, and positive function. This can be shown as follows:

$$\begin{aligned} P_y(e^{j\omega}) &= \sigma_x^2(a^2 + b^2 - 2ab\cos\omega) \\ &\geq \sigma_x^2(a-b)^2 \\ &\geq 0. \end{aligned} \tag{2.237}$$

2.9.6 Mean Square Estimation

Consider $\mathbf{x}_1[n], ..., \mathbf{x}_n[n]$ as discrete-time random variables. Assume that $T = t(\mathbf{x}_1[n], ..., \mathbf{x}_n[n])$ is an estimator of $\tau(\eta)$. Then, the expression

$$\xi = E_\eta\{(T - \tau(\eta))^2\} \qquad (2.238)$$

is said to be the *mean square error* (MSE) of the estimator $T = t(\mathbf{x}_1[n], ..., \mathbf{x}_n[n])$.

An MSE estimator in (2.238) is a measure of goodness, with small values being better than large values. In other words, the ξ in (2.238) is a measure of the spread of function T value about $\tau(\eta)$ in such a way that the variance of a discrete-time random variable is a measure of its spread about its mean. Thus, we would prefer the estimator with the smallest MSE, known as the *minimum mean square error* (MMSE). In general, though, the MSE estimator depends on η.

An estimator $T = t(\mathbf{x}_1[n], ..., \mathbf{x}_n[n])$ is referred to as an *unbiased* estimator if and only if

$$\begin{aligned} E_\eta\{T\} &= E_\eta\{t(\mathbf{x}_1[n], ..., \mathbf{x}_n[n])\} \\ &= \tau(\eta). \end{aligned} \qquad (2.239)$$

This indicates that an estimator is unbiased if its mean is equal to the parameter being estimated. Therefore, the MSE in (2.238) can be rewritten as follows:

$$\begin{aligned} E_\eta\{(T - \tau(\eta))^2\} &= E_\eta\{[(T - E_\eta\{T\}) - (\tau(\eta) - E_\eta\{T\})]^2\} \\ &= E_\eta\{(T - E_\eta\{T\})^2\} \\ &\quad - 2(\tau(\eta) - E_\eta\{T\})E_\eta\{T - E_\eta\{T\}\} \\ &\quad + E_\eta\{(\tau(\eta) - E_\eta\{T\})^2\} \\ &= \text{Var}\{T\} + [\tau(\eta) - E_\eta\{T\}]^2. \end{aligned} \qquad (2.240)$$

Note that the second-term in (2.240) is known as the *bias* of the MSE estimator T and can be either positive, negative, or zero. Furthermore, the MSE in (2.240) is the sum of two positive quantities. Thus, (2.240) provides the relationship among the MSE, variance, and bias of an estimator.

If the estimator T is unbiased, then (2.240) yields

$$E_\eta\{(T - \tau(\eta))^2\} = \text{Var}\{T\}. \tag{2.241}$$

In this case, T is called the unbiased estimator.

Example 2.10

In this example, we consider the linear estimator $\hat{\mathbf{y}}[n]$ of $\mathbf{y}[n]$ in terms of the discrete-time random variables, $\mathbf{x}_1[n], ..., \mathbf{x}_n[n]$. The linear estimator $\hat{\mathbf{y}}[n]$ is given as follows:

$$\hat{\mathbf{y}}[n] = \sum_{i=1}^{n} A_i \mathbf{x}_i[n], \tag{2.242}$$

where A_i is constant. Thus, an MSE between the linear estimator $\hat{\mathbf{y}}[n]$ and $\mathbf{y}[n]$ is obtained by

$$\begin{aligned}\xi_{MSE} &= E\{(\hat{\mathbf{y}}[n] - \mathbf{y}[n])^2\} \\ &= E\left\{\left[\sum_{i=1}^{n} A_i \mathbf{x}_i[n] - \mathbf{y}[n]\right]^2\right\}\end{aligned} \tag{2.243}$$

Using the result in (2.240), (2.243) can be rewritten as follows:

$$\xi_{MSE} = \text{Var}\left\{\sum_{i=1}^{n} A_i \mathbf{x}_i[n]\right\} + \left[\mathbf{y}[n] - E_{\mathbf{y}[n]}\left\{\sum_{i=1}^{n} A_i \mathbf{x}_i[n]\right\}\right]^2. \tag{2.244}$$

Thus, the MMSE can be achieved when the linear estimator $\hat{\mathbf{y}}[n]$ is an unbiased estimator of $\mathbf{y}[n]$. In this case, (2.244) yields

$$\xi_{MMSE} = \text{Var}\left\{\sum_{i=1}^{n} A_i \mathbf{x}_i[n]\right\}. \tag{2.245}$$

Further note that if we use the result in (2.69), (2.245) yields

$$\xi_{MMSE} = \sum_{i=1}^{n} A_i^2 \text{Var}\{\mathbf{x}_i[n]\} + \sum_{i} \sum_{j(j \neq i)} A_i A_j \text{Cov}\{\mathbf{x}_i[n], \mathbf{x}_j[n]\}. \tag{2.246}$$

2.10 Detection Theory and Optimum Receivers

In this section, we introduce detection in digital communications in the presence of additive noise. This topic is of particular interest in determining an optimum design of a receiver.

Detection theory provides a set of rules for a decision-making method, which is used to observe the received signals and predict the transmitted signals. The results of detection are subject to transmission errors. However, the goal of the detection theory is to deal with transmission error such that an acceptable quality of performance can be obtained at the receiver, thereby leading to an optimum receiver.

2.10.1 Optimality Criterion

In order to describe an optimality criterion, we first elaborate a simple case of binary communications. In this case, a transmitter sends a specified signal $s_0(t)$ based on a bit value of "0" and the specified signal $s_1(t)$ based on a bit value of "1" over the bit interval $t \in [0, T]$. The received signal $r(t)$ corresponding to the first bit is then presented by using the hypotheses testing as follows:

$$H_0 : r(t) = s_0(t) + \eta(t) \qquad (2.247)$$
$$H_1 : r(t) = s_1(t) + \eta(t) \qquad (2.248)$$

where $0 < t < T$ and $\eta(t)$ corresponds to additive white Gaussian noise (AWGN) channel noise with variance σ^2. Thus, our objective is to determine the bit value at the most accurate hypothesis given by (2.247) and (2.248) from the received signal $r(t)$.

The optimality criterion of selecting bit value in digital communications is the total probability error, denoted by p_e, and given by

$$\begin{aligned} p_e = & \; p\{\text{bit value ``1''}|\text{``0'' transmitted}\}p\{\text{``0'' transmitted}\} + \\ & \; p\{\text{bit value ``0''}|\text{``1'' transmitted}\}p\{\text{``1'' transmitted}\}. \end{aligned}$$
$$(2.249)$$

The optimal representation of the specific waveforms in this application uses the Karhunen-Loève transform [8, 9].

The Karhunen-Loève transform can be used to represent a second-order random process in terms of a set of orthonormal basis functions scaled by a sequence of random variables. Assume that $r(t)$ is a zero-mean second-order random process over the bit interval $[0, T]$. We can then represent $r(t)$ as follows:

$$r(t) = \sum_{i=1}^{\infty} r_i \phi_i(t), \ 0 \leq t \leq T \quad (2.250)$$

where r_i is the mutually uncorrected random variable and given by

$$r_i = \int_0^T r(t)\phi_i(t)dt, \quad (2.251)$$

where ϕ_i is the basis function over the bit interval $[0, T]$. Thus, an equality, which is a limit MSE, is established such that

$$\lim_{N \to \infty} E\left\{\left[r(t) - \sum_{i=1}^{N} r_i \phi_i(t)\right]^2\right\} = 0, \ 0 \leq t \leq T. \quad (2.252)$$

In order to ensure that the basis function is orthonormal ϕ_i, we can use the Gramm-Schmidt approach [10], which is a deterministic algorithm that can convert an arbitrary set of basis functions into an equivalent set of orthonormal basis functions. Thus, we are able to obtain the full set of functions beginning with the specified signals $s_0(t)$ and $s_1(t)$, that is,

$$\int_0^T \phi_i(t)s_j(t)dt = 0, \ i > 2 \text{ and } j = 0, 1. \quad (2.253)$$

Now, (2.247) and (2.248) can be rewritten as an equivalent hypothesis as follows:

$$H_0 : \mathbf{r} = \mathbf{q}_0 + \eta \quad (2.254)$$
$$H_1 : \mathbf{r} = \mathbf{q}_1 + \eta \quad (2.255)$$

where

$$\mathbf{q}_0 = \begin{bmatrix} \int_0^T \phi_1(t)s_0(t)dt \\ \int_0^T \phi_2(t)s_0(t)dt \end{bmatrix}, \qquad (2.256)$$

$$\mathbf{q}_1 = \begin{bmatrix} \int_0^T \phi_1(t)s_1(t)dt \\ \int_0^T \phi_2(t)s_1(t)dt \end{bmatrix}, \qquad (2.257)$$

and

$$\eta = \begin{bmatrix} \eta_1 & \eta_2 \end{bmatrix}^T. \qquad (2.258)$$

Therefore, in this case, the design of the optimum receiver becomes a simple two-dimensional detection approach.

2.10.2 Maximum Likelihood Detector

An optimum detector can be developed based on Bayes' minimum risk classifers [1] or the maximum a posteriori rule to select

$$\text{largest}\left\{p_{H_i|\mathbf{r}}\{H_i|\mathbf{r}=\mathbf{v}\}\right\}, \quad i=0,1. \qquad (2.259)$$

The expression of (2.259) is used to determine the hypothesis that is most likely given the observation vector $\mathbf{v}$. Thus, the optimum detector can be established by using the likelihood ratio test

$$L(\mathbf{r}) = \frac{p_{\mathbf{r}|H_1}\{\mathbf{r}\}}{p_{\mathbf{r}|H_0}\{\mathbf{r}\}}. \qquad (2.260)$$

If $L(\mathbf{r}) > \frac{p_1}{p_2}$, $L(\mathbf{r})$ belongs to H_1, where p_1 and p_2 are the a priori probabilities of the hypotheses. In the same way, if $L(\mathbf{r}) < \frac{p_1}{p_2}$, $L(\mathbf{r})$ belongs to H_0. Now, assume that the noise is white and Gaussian with variance σ^2. Thus, (2.260) can be rewritten as follows:

$$L(\mathbf{r}) = \frac{\prod_{i=1}^{2} \frac{1}{\sqrt{2\pi}\sigma} \exp\left[-\frac{(r_i-s_{1,i})^2}{2\sigma^2}\right]}{\prod_{i=1}^{2} \frac{1}{\sqrt{2\pi}\sigma} \exp\left[-\frac{(r_i-s_{0,i})^2}{2\sigma^2}\right]}, \qquad (2.261)$$

where

$$s_{j,i} = \int_0^T \phi_i(t)s_j(t)dt. \qquad (2.262)$$

Let $L(\mathbf{r}) = \frac{p_1}{p_2}$. We take the logarithm on both sides of the equation, cancel the common terms, and then obtain the optimum receiver as follows:

$$D_i > \sigma^2 \ln\left(\frac{p_1}{p_2}\right), \quad \text{if } H_1 \text{ is true,} \quad (2.263)$$

$$D_i < \sigma^2 \ln\left(\frac{p_1}{p_2}\right), \quad \text{if } H_0 \text{ is true,} \quad (2.264)$$

where

$$D_i = \sum_{i=1}^{2} r_i(s_{1,i} - s_{0,i}) - \frac{1}{2}\sum_{i=1}^{2}(s_{1,i}^2 - s_{0,i}^2). \quad (2.265)$$

Equation (2.265) is the two-dimensional version of the optimum receiver, which can be converted back into a continuous time receiver by using

$$\sum_{i=1}^{2} r_i s_{k,i} = \int_0^T r(t) s_k(t) dt, \quad (2.266)$$

and

$$\sum_{i=1}^{2} s_{k,i}^2 = \int_0^T s_k^2(t) dt = E_k. \quad (2.267)$$

Substituting (2.266) and (2.267) into (2.265) obtains

$$D_i = \int_0^T r(t)[s_1(t) - s_0(t)] dt - \frac{1}{2}(E_1 - E_0). \quad (2.268)$$

where E_0 and E_1 are the energies of signals $s_1(t)$ and $s_2(t)$, respectively.

Thus, the optimum receiver in (2.268) first correlates the received signal $r(t)$ with the difference signal $[s_1(t) - s_2(t)]$ and then compares to a threshold of $\sigma^2 \ln\left(\frac{p_1}{p_2}\right)$. In other words, the optimum receiver identifies the signal $s_i(t)$ that best matches with the received signal $r(t)$. Therefore, the optimum receiver is referred to as the *correlation receiver* or the *matched filter receiver*.

By extending the results of the case, the optimum receiver for M-ary signaling schemes is given by [8] to select the largest

$$D_i = \int_0^T s_i(t)r(t)dt - \frac{E_i}{2} + \sigma^2 \ln(p_i), \quad i = 1, 2, \cdots, M, \quad (2.269)$$

$s_i(t)$ is the signal set and p_i is the corresponding a priori probabilities. The optimum receiver for M-ary signals is then to implement a bank of correlation or matched filters followed by selecting the largest decision.

2.10.3 Probability of Error

In Section 2.10.2, we derived the optimum receiver based on the assumption of AWGN. The statistics of the logarithm of likelihood ratio are Gaussian random variables. Hence, the probability of error can be calculated directly by using the Q-function given by (2.139)

$$\begin{aligned} P_e &= Q\left(\frac{||s_0 - s_1||}{\sqrt{N_0}}\right) \\ &= \frac{1}{2}\text{erfc}\left(\frac{||s_0 - s_1||}{\sqrt{2N_0}}\right). \end{aligned} \quad (2.270)$$

where N_0 is the PSD of noise and $||s_0 - s_1||$ denotes the Euclidean distance between the signal representation. Note that the Q-function or erfc function is monotonically decreasing with increasing $\frac{||s_0-s_1||}{\sqrt{N_0}}$. Therefore, the probability of error decreases when the Euclidean distance between the representation signals increases.

2.11 Summary

In this chapter, we presented an introduction to the fundamental theory of probability and random variables, distribution and density functions, stochastic signal processes, and detection theory and optimum receivers. Beginning with a short review of probability consisting of intuitive, axiomatic, and conditional methods, we then introduced total probability and Bayes' theorems, the independence concept, and the definition of a random variable. These discussions of basic concepts and theories provide a necessary step for understanding random processes in signal processing for digital communications.

Another important topic introduced in this chapter is distribution and density functions and the role they play in the field of probability, random variables, and stochastic processes. First, we discussed the definitions of probability distribution and density functions, as well as their relationships. Second, we presented joint distributions and densities based on two random variables. This led us to develop the generalization of joint distributions and densities for n random variables based on transformations of random variables. Next, we defined the mean, variance, correlation, covariance, and the correlation coefficient of random processes. Discrete and continuous distributions and densities were also introduced, with emphasis on density functions and properties. Furthermore, we developed results for a sum of n random variables that are useful in the analysis of random processes in terms of mean, variance, and covariance. Additionally, we introduced upper bounds on the tail probability, including Chebyshev inequality and the law of large number, which led to the central limit theorem.

Next we presented stochastic signal processes. We defined the concept of a discrete-time random process and introduced autocorrelation and autocovariance as well as ensemble averages. We outlined properties of stationary random processes including SSS and WSS. Of particular importance in this chapter is the analysis of the mean, autocorrelation, and power spectrum in the context of stable LTI systems, along with the mathematic results in both the discrete-time and frequency domains. We then established MSE, MMSE, and estimation concepts as well as the relationships among the MSE, variance, and bias of an estimator.

Finally, we derived detection theory that led to optimum receivers based on the concept of a maximum likelihood detector. The performance of the optimum receivers is evaluated by using the probability of error.

References

[1] Miao, G. J., and M. A. Clements, *Digital Signal Processing and Statistical Classification*, Artech House, Norwood, Massachusetts, 2002.

[2] Davenport, W. B., and W. L. Root, *Random Signals and Noise*, McGraw-Hill, New York, 1958.

[3] Davenport, W. B., *Probability and Random Processes*, McGraw-Hill, New York, 1970.

[4] Papoulis, A., *Probability, Random Variable, and Stochastic Processes*, McGraw-Hill, New York, 1991.

[5] Mood, A. M., F. A. Graybill, and D. C. Boes, *Introduction to the Theory of Statistics*, 3rd ed., McGraw-Hill Book Company, New York, 1974.

[6] Bickel, P. J., and K. A. Doksum, *Mathematical Statistics: Basic Ideas and Selected Topics*, Holden-Day, Oakland, California, 1977.

[7] Haykin, S., *Digital Communications*, John Wiley & Sons, New York, 1988.

[8] Gibson, J. D., (ed.), *The Mobile Communications Handbook*, 2nd ed., CRC Press, LLC, Boca Raton, Florida, 1999.

[9] Tou, J. T., and R. C. Gonzalez, *Pattern Recognition Principles*, Addison-Wesley Publishing Company, Reading, Massachusetts, 1974.

[10] Proakis, J. G., *Digital Communications*, 2nd ed., McGraw-Hill, New York, 1989.

3

Sampling Theory

3.1 Introduction

In digital communications, to transmit and receive continuous-time signals, the continuous-time analog signals have to be converted into digital signals. Digital transmission of the continuous-time analog signals involves a sampling process, which is the process of converting the continuous-time signal into a discrete-time sequence. The process is also known as analog-to-digital (A/D) converting or simply digitizing.

The sampling process has to obey a *sampling theorem*, sometimes referred to as the *Nyquist-Shannon sampling theorem* [1]. The sampling theorem states conditions so that the samples represent no loss of information and can therefore be used to reconstruct the original signal with arbitrarily good fidelity.

When the sampling process does not meet the sampling theorem, the signal frequencies will alias. This leads to the loss of information. However, aliasing can be used as an advantage with a technique called *undersampling*. In this case, a continuous-time signal is sampled at less than the Nyquist sampling rate. Undersampling has become a key technique often utilized in A/D converters for radio frequency (RF) communication transceivers.

The sampling process can be presented by either the frequency domain or time domain or both. The frequency-domain analysis is often used in digital communications. Thus, in this chapter, using Fourier transform techniques, we first introduce this remarkable sampling theorem and then discuss undersampling and the operation of sampling.

This chapter is organized as follows. A short introduction of sampling is presented in this section. In Section 3.2, we introduce the discrete-time sampled signal in which we discuss instantaneous sampling and an ideal sampled signal. In Section 3.3, we first focus on the Nyquist sampling theorem and then discuss time- and frequency-domain interpolation formulas as well as aliasing. Undersampling is presented in Section 3.4, along with discussions of obtaining minimum sampling rate and antialiasing bandpass filters. Then, in Section 3.5, we expand the sampling theorem into a stochastic sampling theorem. Finally, a brief summary is given in Section 3.6.

3.2 Discrete-Time Sampled Signals

Signals can usually be classified into three categories: (1) analog signals that are continuous both in time space and amplitude; (2) discrete-time sampled signals (or discrete-time signals), which consist of signals discrete in time space and continuous in amplitude; and (3) digital signals that are discrete in both time space and amplitude [2]. One common way to create digital or discrete-time sampled signals is by sampling continuous-time signals. Signal processing has to do with the representation, manipulation, implementation, and transformation of signals, and the information that they carry in the transformation media (or channel).

3.2.1 Instantaneous Sampling

Assume that the continuous-time signal $x_a(t)$ is sampled at a rate of $F_s = 1/T_s$ samples per second or hertz (Hz), where F_s is known as the *sampling frequency*, T_s is referred to as the *sampling interval* or *sampling period*, and its reciprocal $1/T_s$ is the *sampling rate*. Then, the discrete-time sampled signal $x[n]$ is equal to the value of the continuous-time signal $x_a(t)$ at time nT_s as follows:

$$x[n] = x_a(nT_s), \quad -\infty < n < \infty. \tag{3.1}$$

Note that the discrete-time sampled signal $x[n]$ in (3.1) is a mathematically indexed sequence of numbers, where n is an integer from $-\infty$ to ∞. This form of sampling process is called *instantaneous sampling*. Thus, the discrete-time sampled signals $x[n]$ can often be considered as a result of sampling a continuous-time signal $x_a(t)$ by using an A/D converter.

The A/D converter actually consists of the combination of two processes: (1) sampling, which converts the signal from continuous-time domain to discrete-time domain; and (2) quantization, which converts the signal amplitude from a continuous infinite value to a finite set of discrete values.

For example, a continuous-time sinusoid signal $s(t)$ is given by the mathematical formula

$$s(t) = A[\cos(2\pi f_0 t + \phi)], \tag{3.2}$$

where A is the amplitude, f_0 is the frequency in hertz, and ϕ is the phase offset. If the discrete-time sampled signal $s[n]$ is obtained by using the A/D converter, sampling the continuous-time sinusoid signal $s(t)$ at a sampling rate of $F_s = 1/T_s$, we then have the discrete-time sampled signal

$$\begin{aligned} s[n] &= s(t)\,|_{t=nT_s} \\ &= A[\cos(2\pi f_0 T_s n + \phi)] \\ &= A\left[\cos\left(2\pi \frac{f_0}{F_s} n + \phi\right)\right] \\ &= A[\cos(\omega_0 n + \phi)], \end{aligned} \tag{3.3}$$

where $\omega_0 = 2\pi f_0 T_s$ is the normalized frequency in radians.

3.2.2 Ideal Sampled Signal

The sampling process in (3.1) can be represented as

$$x[n] = x_a(nT_s) = \int_{-\infty}^{\infty} x_a(\tau)\delta(\tau - nT_s)d\tau, \qquad (3.4)$$

where $\delta(\tau)$ is the *Dirac delta function* or *impulse*. On the other hand, the discrete-time sampled signal $x[n]$ has a continuous-time *pulse amplitude modulation* (PAM) representation in terms of impulses. This means that a continuous-time signal can be reconstructed from a discrete-time sampled signal.

To illustrate this operation, we assume that $x_s(t)$ can be obtained by using multiplication of $x_a(t)$ with the unit impulse train $\delta_T(t)$ with period T_s as follows:

$$\begin{aligned}
x_s(t) &= x_a(t)\delta_T(t) \\
&= x_a(t) \sum_{n=-\infty}^{\infty} \delta(t - nT_s) \\
&= \sum_{n=-\infty}^{\infty} x_a(t)\delta(t - nT_s) \\
&= \sum_{n=-\infty}^{\infty} x_a(nT_s)\delta(t - nT_s), \qquad (3.5)
\end{aligned}$$

where $x_s(t)$ is referred to as the *ideal sampled signal* [3].

3.3 Nyquist Sampling Theorem

Sampling is the process of converting a continuous-time signal (or a continuous-time function) into a numeric sequence (or a discrete-time function). The condition of sampling, which needs to represent no loss of signal information and can therefore be used to reconstruct the original signal with arbitrarily good fidelity, states that the signal must be bandlimited and that the sampling frequency must be at least twice the signal bandwidth. This condition of sampling is known as the Nyquist-Shannon sampling theorem [1]. Sometimes, it is simply

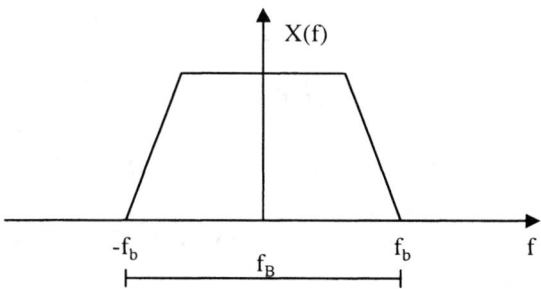

Figure 3.1 Frequency spectrum of a bandlimited signal.

referred to as the sampling theorem, which is a fundamental result in the field of information theory, in particular, telecommunications.

A bandlimited signal is constrained in terms of how fast it can change and therefore how much detail it can convey between discrete samples in time. The sampling theorem indicates that the discrete samples can completely represent the signal if the signal bandwidth is less than half the sampling rate F_s, which is referred to as the *Nyquist sampling frequency*.

In order to represent the concept of the sampling theorem, let $x(t)$ denote a real-value continuous-time signal and $X(f)$ be its unitary Fourier transform as follows:

$$X(f) = \int_{-\infty}^{\infty} x(t)e^{-j2\pi ft}dt, \qquad (3.6)$$

where $X(f)$ as shown in Figure 3.1 is a bandlimited signal with the highest frequency at f_b,

$$X(f) = 0, \quad |f| > f_b. \qquad (3.7)$$

The Nyquist sampling frequency F_s for alias-free components is given by

$$F_s \geq 2f_b, \qquad (3.8)$$

or equivalently
$$f_b \leq \frac{F_s}{2}. \tag{3.9}$$

The discrete-time sampled signal is obtained by
$$x[n] = x(nT_s), \quad -\infty < n < \infty, \tag{3.10}$$

where $T_s = \frac{1}{F_s}$ is the sampling interval, which is the constant time interval between successive samples.

3.3.1 Time-Domain Interpolation Formula

Equation (3.5) indicates multiplication in the time domain. However, the multiplication in the time domain corresponds to convolution in the frequency domain. By taking the Fourier transform of both sides of (3.5), we obtain the Fourier transform $X_s(\omega)$ of the ideal sampled signal $x_s(t)$ as follows:

$$\begin{aligned} X_s(\omega) &= \left[\frac{1}{2\pi} X_a(\omega)\right] * \left[\frac{2\pi}{T_s} \sum_{n=-\infty}^{\infty} \delta\left(\omega - \frac{2\pi n}{T_s}\right)\right] \\ &= \frac{1}{T_s} \sum_{n=-\infty}^{\infty} X_a\left(\omega - \frac{2\pi n}{T_s}\right), \end{aligned} \tag{3.11}$$

where "$*$" denotes convolution. Equation (3.11) indicates that sampling produces images of $X_a(\omega)$ in the frequency axis. Therefore, $X_s(\omega)$ will repeat periodically without overlapping if $F_s \geq 2f_b$.

For example, consider a case of a signal of frequency f_a sampled at a sampling frequency F_s using an ideal impulse sampler (or an ideal A/D converter) and assume $F_s > 2f_a$ as shown in Figure 3.2. The output of the ideal impulse sampler in the frequency domain produces *images* of the original signal at frequencies equal to $\pm kF_s \pm f_a$ for $k = 1, 2, 3, \cdots$.

The Nyquist frequency bandwidth is defined to be the frequency spectrum from DC to the half sampling frequency $\frac{F_s}{2}$. The frequency spectrum can be divided into a set of an infinite number of zones, known as *Nyquist zones*. Each of the Nyquist zones has a frequency

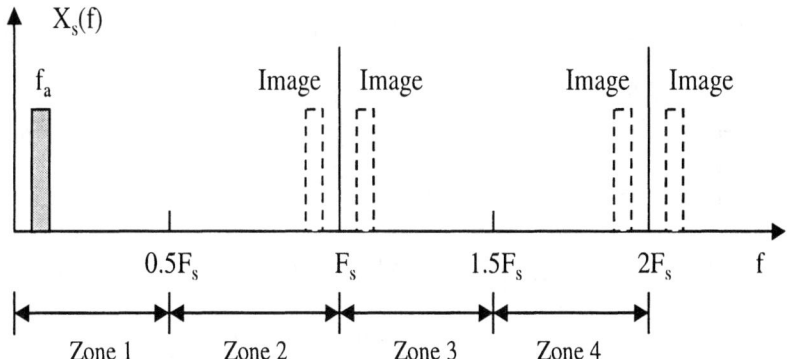

Figure 3.2 Frequency spectrum of an analog signal f_a sampled at F_s ($F_s > 2f_a$) using an ideal impulse sampler.

bandwidth equal to $\frac{F_s}{2}$. In this case, all of the signal of interest (or the bandwidth of sampled signals) is limited within the first Nyquist zone, and images of the original bandwidth of frequencies appear in each of the other Nyquist zones. Thus, in this case, the sampling process is referred to as *baseband sampling* in communication applications.

In the absence of aliasing distortion, the signal $x_a(t)$ can be reconstructed from its samples by using an ideal lowpass filter having a frequency response,

$$H(\omega) = \begin{cases} T_s, & |\omega| \leq \frac{\pi}{T_s} \\ 0, & \text{otherwise.} \end{cases} \quad (3.12)$$

Using the inverse Fourier transform of (3.12), we obtain the impulse response $h(t)$ of the ideal lowpass filter as follows:

$$h(t) = \frac{\sin(\pi t/T_s)}{\pi t/T_s}. \quad (3.13)$$

Using the property that convolution in time domain equals multiplication in frequency domain, the Fourier transform $X_a(\omega)$ of the signal $x_a(t)$ can be obtained by

$$X_a(\omega) = X_s(\omega) H(\omega). \tag{3.14}$$

For the signal $x_a(t)$ in the time domain, taking the inverse Fourier transform on both sides of (3.14) and using (3.5) and (3.13) yield

$$\begin{aligned} x_a(t) &= x_s(t) * h(t) \\ &= \sum_{n=-\infty}^{\infty} x_a(nT_s) \delta(t - nT_s) * \left[\frac{\sin(\pi t/T_s)}{\pi t/T_s} \right] \\ &= \sum_{n=-\infty}^{\infty} x_a(nT_s) \left\{ \frac{\sin[\pi(t - nT_s)/T_s]}{\pi(t - nT_s)/T_s} \right\}. \end{aligned} \tag{3.15}$$

Therefore, using (3.15), we are able to find the value of a bandlimited continuous-time function at any point in time given a set of samples of the function. Equation (3.15) is known as the *Nyquist-Shannon interpolation formula* or sometimes called the *time domain interpolation formula*.

The time-domain interpolation formula states as follows: Each sample is multiplied by a sinc-function. The width of each half-period of the sinc-function is scaled to match the sampling frequency, and the central point location of the sinc-function is shifted to the time of that sample. Then, all of these shifted and scaled functions are added together to reconstruct the original signal. Note that the result of this operation is indeed a continuous-time signal since the sinc-function is a continuous-time function.

In order to obtain the original continuous-time signal after the reconstructing process, we must also meet a critical condition on the sampling frequency, that is, the sampling frequency must be at least twice as large as the highest frequency component of the original continuous-time signal.

3.3.2 Frequency-Domain Interpolation Formula

A continuous-time signal $x(t)$ is called a *time-limited signal* if the condition is satisfied as follows:

$$x(t) = 0, \quad |t| > |T_0|. \tag{3.16}$$

The Fourier transform $X(\omega)$ of the time-limited signal $x(t)$ in (3.16) can be uniquely determined by using its values $X(n\omega_s)$ sampled at a uniform sampling frequency rate, where $\omega_s \leq \frac{\pi}{T_0}$. If $\omega_s = \frac{\pi}{T_0}$, then the Fourier transform $X(\omega)$ is given by [3]

$$X(\omega) = \sum_{n=-\infty}^{\infty} X(n\omega_s) \left[\frac{\sin[(\omega - n\omega_s)T_0]}{(\omega - n\omega_s)T_0} \right]. \qquad (3.17)$$

We refer to (3.17) as the *frequency-domain interpolation formula.*

Note that if a signal is frequency bandlimited, it cannot be time-limited and vice versa. In many communication applications, the signal to be sampled is usually time-limited and is not strictly frequency bandlimited. However, beyond certain defined frequency bandwidths, we see that the frequency components of physically occurring signals attenuate rapidly. Therefore, for practical applications, we treat these signals as being frequency bandlimited as well.

3.3.3 Aliasing

If frequency components are above the Nyquist sampling frequency, upon sampling the frequencies will overlap. The overlap is referred to as *aliasing*. Therefore, the discrete samples are subject to aliasing as well. The aliasing is undesirable in most signal processing and communication applications.

From the Nyquist sampling theorem, we must sample a signal at a sampling rate of at least twice the bandwidth of the signal. For baseband signals with frequency components starting from DC to a maximum frequency, this indicates that the sampling rate must be at least twice this maximum frequency. In such cases, the frequency bandwidth is the same as the maximum frequency.

In order to avoid aliasing, we can implement two methods as follows: (1) increase the sampling rate and/or (2) introduce an antialiasing filter or make the antialiasing filter more stringent. The antialiasing filter is used to restrict the signal bandwidth to satisfy the condition of sampling frequency. This holds in theory, but does not always work satisfactorily in practice. This is because a continuous-time signal will have some energy outside of the signal bandwidth. However, the energy may be small enough so that the aliasing effects are negligible.

3.4 Undersampling

Consider the case where the sampled signal bandwidth lies entirely within the other Nyquist zones rather than the first Nyquist zone. A real-value signal $x(t)$ is called a *bandpass signal* if the Fourier transform $X(f)$ satisfies the following condition:

$$X(f) \begin{cases} \neq 0, & f_1 < |f| < f_2 \\ = 0, & \text{otherwise.} \end{cases} \qquad (3.18)$$

Sampling for the bandpass signal is referred to as *bandpass sampling*. For example, a bandpass signal is the intermediate frequency (IF) output from a communication receiver with a center frequency at 71 MHz and a 4-MHz frequency bandwidth. In this case, if we use baseband sampling, we would need to use an A/D converter with a sampling rate about 146 MHz. But with bandpass sampling, we would only need to use an 8-MHz sampling rate. This method is a huge benefit compared to baseband sampling. The process of sampling a signal outside the first Nyquist zone is called *undersampling*. It is also known as *super-Nyquist sampling*, *harmonic sampling*, and *IF sampling* as well as direct IF to digital conversion.

Figure 3.3 shows the sampled signal restricted to different Nyquist zones by undersampling. Figure 3.3(a) is a case of baseband sampling, where the frequency bandwidth of sampled signals lies within the first Nyquist zone and images of the original bandwidth of frequencies display in each of the other Nyquist zones. Figure 3.3(b) shows a case where the sampled signal bandwidth is limited to be entirely within the second Nyquist zone. Note that when the signals are located in even Nyquist zones, the first Nyquist zone image contains all the information in the original signal, with the exception of its original location and frequency reversal. In

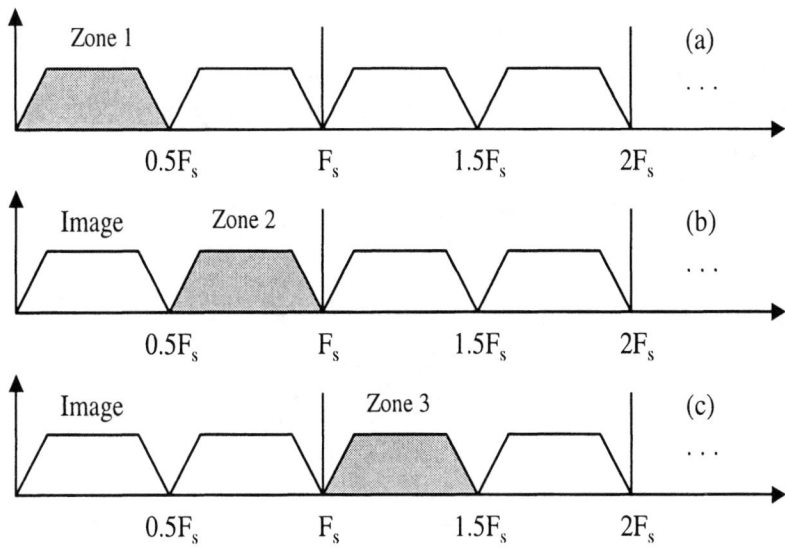

Figure 3.3 Undersampled signal restricted to the different Nyquist zones.

Figure 3.3(c), the sampled signal is restricted to the third Nyquist zone. In this case, the first Nyquist zone image has no frequency reversal and accurate representation of the original signal. Thus, all the signal information can be preserved as long as a signal is sampled at a sampling rate equal to or greater than twice its frequency bandwidth.

Undersampling, which samples signals with frequency components over the first Nyquist zone, has become popular in many communication receivers. It is becoming common practice to sample the bandpass signal directly, and then use digital signal processing techniques to deal with the signal, thereby eliminating the need for analog demodulation.

3.4.1 Minimum Sampling Rate

The minimum sampling rate for undersampling depends on f_1 and f_2 given by (3.18) and the frequency bandwidth $f_B = f_2 - f_1$. In order to avoid overlap, the sampling rate must satisfy the following constraints:

$$F_s \geq 2(f_2 - f_1), \qquad (3.19)$$
$$(n-1)F_s \leq 2f_1, \qquad (3.20)$$

and

$$nF_s \geq 2f_2, \qquad (3.21)$$

where n is an integer. Using $f_1 = f_2 - f_B$, these constraints can be rewritten as

$$F_s \geq 2f_B, \qquad (3.22)$$
$$F_s \leq \frac{2(f_2 - f_B)}{n-1}, \qquad (3.23)$$

and

$$F_s \geq \frac{2f_2}{n}. \qquad (3.24)$$

Thus, the minimum sampling rate is obtained by

$$\{F_s\}_{min} = \frac{2f_2}{k}, \qquad (3.25)$$

where k is the integer, but it does not exceed $\frac{f_2}{f_B}$. Further note that if the ratio of $\frac{f_2}{f_B}$ is an integer, then the minimum sampling rate is

$$\{F_s\}_{min} = 2f_B. \qquad (3.26)$$

Example 3.1

Consider a bandpass signal with the center frequency at 71 MHz, $f_1 = 69$ MHz, and $f_2 = 73$ MHz. Thus, the frequency bandwidth of the bandpass signal is equal to $f_B = 4$ MHz, $\frac{f_2}{f_B} = 18.25$, and $k = 18$ for the next lowest integer. Using (3.25), we obtain the minimum sampling rate $\{F_s\}_{min} = 8.1111$ MHz.

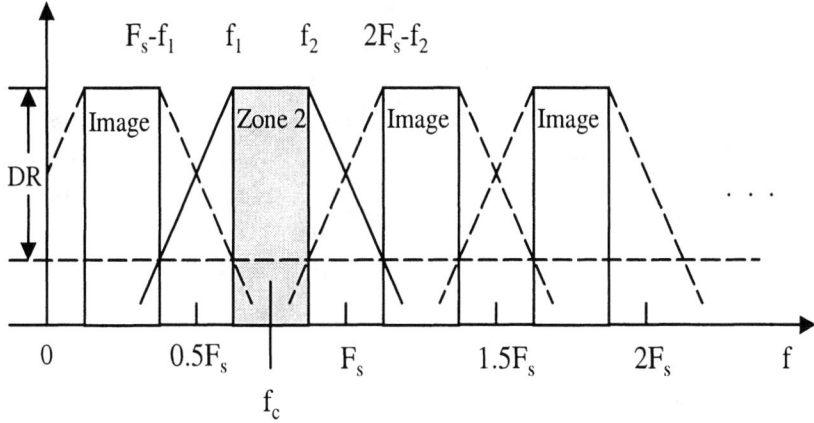

Figure 3.4 Antialiasing bandpass filter for undersampling a signal in the second Nyquist zone.

3.4.2 Antialiasing Bandpass Filter

In the undersampling process, the sampling rate is selected to be less than the bandpass signal's center frequency. This intentionally leads to aliasing the center frequency to a frequency less than the sample rate. Thus, to avoid interference we must assure that the continuous-time analog signal does not have frequency intervals aliasing onto any desired signal frequencies.

Figure 3.4 shows a continuous-time analog signal in the second Nyquist zone centered around a center (or carrier) frequency f_c, where lower and upper frequencies are f_1 and f_2, respectively. The frequency bandwidth of the continuous-time analog signal is equal to $f_B = f_2 - f_1$. An analog antialiasing filter is a bandpass filter, whose desired dynamic range is the filter stopband attenuation. The analog antialiasing filter is placed before an A/D converter. The upper transition band is the frequencies from f_2 to $2F_s - f_2$, and the lower transition band is the frequencies from f_1 to $F_s - f_1$. In the case of undersampling, increasing the sampling rate relaxes the antialiasing bandpass filter requirements.

In order to minimize the cost of the analog antialiasing bandpass filter, we can arrange the signal band of interest to alias to one-fourth of the selected sampling rate. The $\frac{k}{2} + \frac{1}{4}$ option aliases the signal to the positive one-quarter sampling rate while the $\frac{k}{2} - \frac{1}{4}$ option aliases the signal to the negative one-quarter sampling rate. To ensure that the carrier frequency f_c is placed in the center of a Nyquist zone, we use the two options to make a set of possible sampling rates as follows:

$$f_c = \left(\frac{k}{2} \pm \frac{1}{4}\right) F_s, \qquad (3.27)$$

where k is the integer and corresponds to the Nyquist zone. Equation (3.27) can be rewritten as

$$F_s = \frac{4f_c}{2k \pm 1}. \qquad (3.28)$$

In addition to (3.28), the Nyquist criteria must also be satisfied as follows:

$$F_s \geq 2f_B. \qquad (3.29)$$

Thus, we select k such that the Nyquist zone is chosen to be as large as possible while (3.29) is still satisfied. This method will help result in the minimum required sampling rate. Also note that if the Nyquist zone is selected to be an odd zone, then the carrier frequency f_c and the bandpass signal $x(t)$ will fall in the odd Nyquist zone. As a result, the image frequencies in the first Nyquist zone will not need to be reversed.

Example 3.2

Reconsider Example 3.1, given the signal bandwidth $f_B = 4$ MHz; the Nyquist sampling rate F_s is then equal to 8 MHz. Using $F_s = 8$ MHz and $f_c = 71$ MHz, we can solve (3.28) to obtain $k = 18.25$. Since k must be an integer, we round 18.25 to the lowest integer, 18. Then, we again substitute $k = 18$ and $f_c = 71$ MHz into (3.28) to yield $F_s = 8.1143$ MHz. Therefore, the final sampling frequency rate is $F_s = 8.1143$ MHz and $k = 18$ for the Nyquist zone.

Compared to the results obtained in Example 3.1, we see that the final sampling rate by using (3.28) in Example 3.2 is approximately the same value, about $F_s = 8.11$ MHz, by using (3.25) in Example 3.1. Therefore, both (3.25) and (3.28) will yield an approximate result of the sampling rate for undersampling processes.

3.5 Stochastic Sampling Theorem

In the earlier discussion of the sampling theorem, we showed that the bandlimited signal sampled at the Nyquist sampling rate can be reconstructed from its samples by using the Nyquist-Shannon interpolation formula given by (3.15). Equivalently, an analog continuous-time signal $x_a(t)$ can be reconstructed by passing the sampled signal $x[n]$ through an ideal lowpass filter with an impulse response given by (3.13).

In this section, we expand the sampling theorem to bandlimited stochastic processes. A stationary stochastic process $\mathbf{x}(t)$ is said to be a bandlimited stochastic process [4, 5] if its power spectrum

$$P_x(f) = 0, \text{ for } |f| > f_0 \qquad (3.30)$$

where f_0 is the highest frequency contained in the stationary stochastic process $\mathbf{x}(t)$. Note that $P_x(f)$ in (3.30) is the Fourier transform of the autocorrelation function $R(\tau)$. Thus, the corresponding autocorrelation function $R(\tau)$ can be obtained by

$$R(\tau) = \sum_{n=-\infty}^{\infty} R(nT_s) \left\{ \frac{\sin[\pi(t - nT_s)/T_s]}{\pi(t - nT_s)/T_s} \right\}, \qquad (3.31)$$

where T_s is the sampling interval equal to $T_s \geq \frac{1}{2f_0}$, and $R(nT_s)$ are the samples of $R(\tau)$ taken at $\tau = nT_s$, where $n = 0, \pm 1, \pm 2, \cdots$.

In a similar way, if $\mathbf{x}(t)$ is the bandlimited stationary stochastic process, then $\mathbf{x}(t)$ can be expressed as follows:

$$\mathbf{x}(t) = \sum_{n=-\infty}^{\infty} \mathbf{x}(nT_s) \left\{ \frac{\sin[\pi(t - nT_s)/T_s]}{\pi(t - nT_s)/T_s} \right\}, \qquad (3.32)$$

where T_s is the sampling interval equal to $T_s \geq \frac{1}{2f_0}$ and $\mathbf{x}(nT_s)$ is the sample of $\mathbf{x}(t)$ taken at $t = nT_s$, where $n = 0, \pm 1, \pm 2, \cdots$. Papoulis [4] proved the theoretical result in (3.32) by using the exponential $e^{j\omega t}$ as a function of ω, viewing t as a parameter, and expanding it into a Fourier series. Equations (3.31) and (3.32) are referred to as the *stochastic sampling theorem* of the interpolation formula for the stationary stochastic process or the *sampling expansion*. In this case, the samples are random variables, which can be described statistically by using appropriate joint probability density functions.

Proakis [5] showed that the equality relationship between the stochastic sampling representation given by (3.32) and the bandlimited stationary stochastic process $\mathbf{x}(t)$ holds in an optimal sense as follows:

$$E\left\{\left|\mathbf{x}(t) - \sum_{n=-\infty}^{\infty} \mathbf{x}(nT_s)\left\{\frac{\sin[\pi(t-nT_s)/T_s]}{\pi(t-nT_s)/T_s}\right\}\right|^2\right\} = 0. \quad (3.33)$$

Equation (3.33) is the mean square error (MSE), which is equal to 0.

3.6 Summary

In this chapter, we first introduced discrete-time sampled signals, including instantaneous sampling and the ideal sampled signal. Second, we focused on the fundamental theory of the sampling theorem, with an emphasis on the time- and frequency-domain interpolation formulas as well as the aliasing. Third, we addressed the undersampling techniques, the methods of determining the minimum sampling frequency rate, and the requirements of the analog antialiasing bandpass filter. We then expanded the sampling theorem into the stochastic signal processes. These discussions of the sampling theorem and its techniques provided a fundamental step to understanding sampling on signal processing for digital communications.

From a signal processing perspective, the sampling theorem describes two operations as follows: (1) a sampling process in which a continuous-time signal is converted to a discrete-time signal, and (2) a reconstruction process in which the continuous-time signal is recovered from the discrete-time signal by using the interpolation formula. However, in order to obtain the original continuous-time signal after the reconstruction process, a critical condition of the sampling rate must be met, that is, it must be at least twice as large as the highest frequency component of the original continuous-time signal. If the highest frequency component of the continuous-time signal that we sampled is known, the sampling theorem provides the lowest possible sampling rate so that perfect reconstruction can be ensured. In this case, the lowest possible sampling rate is referred to as the Nyquist sampling rate. On the other hand, if the sampling rate is given, the sampling theorem provides an upper bound for the frequencies of the continuous-time signal to assure perfect reconstruction. Both of these cases indicate that the continuous-time signal to be sampled should be bandlimited. In other words, for instance, any frequency component of the continuous-time signal that has a frequency component above a certain bound should be zero, or at least sufficiently close to zero.

In practice, the sampling theorem described in this chapter cannot be completely satisfied. This is because the ideal reconstruction process of using the sinc-functions is assumed. However, it cannot be implemented, since it indicates that each of samples contributes to the reconstructed continuous-time signal at almost all time points. Thus, we have to use some types of approximations of the sinc-functions, which are truncated to limited intervals. In this case, the approximations of the sinc-functions create an error that is sometimes referred to as the interpolation error. In addition, the sampled continuous-time signal can never be bandlimited exactly. This is to say that even if an ideal reconstruction process could be obtained, the reconstructed continuous-time signal would not be the sampled continuous-time signal exactly. In this case, the failure of band limitation produces an error for sampled signal.

Another important topic introduced in this chapter is undersampling and the role that it has played in the field of digital communication transceivers. Undersampling is used to sample a continuous-time signal outside the first Nyquist zone. In this case, the continuous-time signal must be sampled at a sampling rate equal to or greater than twice its bandwidth in order to preserve all the signal information. In addition, the frequency band of sampled signals should be restricted to a single Nyquist zone. In other words, the sampled continuous-time signals must not overlap at any multiple of half-sampling frequency rate $\frac{F_s}{2}$. Eventually, this can be achieved by using an analog antialiasing bandpass filter.

References

[1] Shannon, C. E., "Communication in the Presence of Noise," *Proceedings of the IEEE*, Vol. 86, No. 2, pp. 447–457, February 1998.

[2] Miao, G. J., and M. A. Clements, *Digital Signal Processing and Statistical Classification*, Artech House, Norwood, Massachusetts, 2002.

[3] Gibson, J. D., (ed.), *The Mobile Communications Handbook*, 2nd ed., CRC Press, LLC, Boca Raton, Florida, 1999.

[4] Papoulis, A., *Probability, Random Variable, and Stochastic Processes*, McGraw-Hill, New York, 1991.

[5] Proakis, J. G., *Digital Communications*, 2nd ed., McGraw-Hill Book Company, New York, 1989.

4

Channel Capacity

4.1 Introduction

A communication channel is a medium that is used to transmit signals from a transmitter to a receiver. It may be a pair of telephone wires, a coaxial cable, a band of radio frequencies, and so on. During a transmission, the signals at the receiver may be perturbed by noise along with channel distortions. However, the noise and channel distortions can be differentiated because the channel distortions are a fixed function applied to the signals while the noise has statistical and unpredictable perturbations. Therefore, the channel distortions can be corrected by using an inverse function of the channel distortions. On the other hand, the perturbations due to the noise cannot be eliminated because the signals do not usually undergo the same change during the transmission.

Assume that it is possible to reliably distinguish M different signal states in a period time of duration T over a communication channel. In other words, the channel can be used to transmit $\log_2 M$ bits in the period time of duration T. The rate of transmission R is then expressed as

$$R = \frac{\log_2 M}{T}. \tag{4.1}$$

However, more precisely, *channel capacity* is defined as

$$C = \lim_{T \to \infty} \left(\frac{\log_2 M}{T} \right). \tag{4.2}$$

Equation (4.2) was derived by Shannon [1–3] in 1949. Therefore, in this chapter, we first discuss Shannon's theorem of the channel capacity and then expand it into different channels, including Gaussian, bandlimited, single-input multiple-output (SIMO), multiple-input single-output (MISO), and multiple-input multiple-output (MIMO) channels.

This chapter is organized as follows. A short introduction of channel capacity is presented in this section. In Section 4.2, we introduce a Gaussian channel capacity in which we discuss a fundamental Shannon's theorem for a continuous channel in the presence of additive Gaussian noise. In Section 3.3, we focus on approaching a problem of communication transmission over a bandlimited channel given the Nyquist-Shannon sampling theorem and lead to the bandlimited channel capacity. MIMO channel capacity is presented in Section 4.4, along with discussions of obtaining capacity in terms of the sum of each of the nonzero eigenvalues. Then, we expand the MIMO channel capacity into SIMO and MISO channel capacities in Sections 4.5 and 4.6, respectively. Finally, a brief summary is given in Section 4.6.

4.2 Gaussian Channel Capacity

The most important communication channel is the Gaussian channel, which has been used to model many practical channels, including wireline, radio, and satellite links. Figure 4.1 shows a discrete-time Gaussian channel model with the output $y[n]$ at time index of n, where $y[n]$ is the sum of the input $x[n]$ and the noise $v[n]$. Thus, we obtain the output of the discrete-time Gaussian channel as follows:

$$y[n] = x[n] + v[n], \tag{4.3}$$

where the noise in (4.3) is assumed to be independent of the input signal $x[n]$, and satisfies an independent and identically distributed

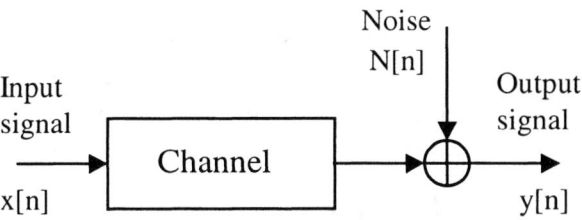

Figure 4.1 A general discrete-time Gaussian channel model.

normal (Gaussian) distribution with zero mean and variance of N.

Shannon [1–3] showed that if $T \to \infty$ in (4.2) and if the rate of transmission approaches the channel capacity, the information (or channel) capacity of a Gaussian channel is then obtained by

$$\begin{aligned} R \leq C &= \lim_{T \to \infty} \left(\frac{\log_2 M}{T} \right) \\ &= \frac{1}{2} \log_2 \left(1 + \frac{P}{N} \right), \end{aligned} \quad (4.4)$$

where R is the rate of transmission, C is the channel capacity in terms of the number of bits per transmission, P is the signal power, and N is the noise variance. In this case, the probability of error, denoted by P_e, approaches a *standard normal distribution* with zero mean and unit variance [4]:

$$P_e = \Phi_{0,1} \left[\sqrt{T} \sqrt{\frac{2P(P+N)}{N(P+2N)}} (R - C) \right], \quad (4.5)$$

where the standard normal distribution is given by

$$\Phi_{0,1}(x) = \int_{-\infty}^{x} \frac{1}{\sqrt{2\pi}} e^{-\frac{u^2}{2}} du. \quad (4.6)$$

Equation (4.5) expresses the fundamental Shannon's theorem for a continuous channel in the presence of additive Gaussian noise.

4.3 Bandlimited Channel Capacity

A common channel model for transmission of information messages over a band of radio frequencies or a telephone line is a bandlimited channel with white and Gaussian noise. In the time domain, an output of the bandlimited channel can be expressed as

$$y(t) = [x(t) + v(t)] * h(t), \qquad (4.7)$$

where "$*$" denotes the convolution operation, $x(t)$ is the input, $v(t)$ is the noise, and $h(t)$ is the time-domain impulse response of an ideal bandpass filter, which cuts out all frequencies greater than W. By using the sample with $t = nT$, where T is the sample interval, the output of the bandlimited channel in (4.7) can be rewritten as

$$y[n] = (x[n] + v[n]) * h[n], \qquad (4.8)$$

where $x[n]$ is the discrete-time input signal, $v[n]$ is the discrete-time noise, $h[n]$ is the discrete-time ideal bandpass filter, and $y[n]$ is the discrete-time output signal.

If a function $f(t)$ has a bandlimit with no frequencies higher than W, then the function is completely reconstructed by a set of samples of the function spaced at $\frac{1}{2W}$ seconds apart. In other words, if the function $f(t)$ contains no frequencies greater than W, it cannot change by a substantial amount in a time less than one-half cycle of the highest frequency $\frac{1}{2W}$. Mathematically, this statement about the bandlimited function $f(t)$ can be proved as follows. Let $F(\omega)$ be the frequency spectrum of the function $f(t)$. Then we obtain

$$\begin{aligned} f(t) &= \frac{1}{2\pi} \int_{-\infty}^{\infty} F(\omega) e^{j\omega t} d\omega \\ &= \frac{1}{2\pi} \int_{-2\pi W}^{2\pi W} F(\omega) e^{j\omega t} d\omega, \end{aligned} \qquad (4.9)$$

where $F(\omega)$ is assumed to be zero outside the frequency band of W, $-2\pi W \leq \omega \leq 2\pi W$. If we assume that

$$t = \frac{n}{2W}, \qquad (4.10)$$

where n is the integer, substituting (4.10) into (4.9) obtains

$$f\left(\frac{n}{2W}\right) = \frac{1}{2\pi} \int_{-2\pi W}^{2\pi W} F(\omega) e^{j\frac{\omega n}{2W}} d\omega. \qquad (4.11)$$

Note that on the left side of (4.11) is the sample value $f\left(\frac{n}{2W}\right)$ of the function $f(t)$. The integral on the right side of (4.11) is the definition of the nth coefficient in a Fourier series expansion of the periodic extension of the function $F(\omega)$, taking the interval from $-2\pi W$ to $2\pi W$ as the fundamental period. Therefore, this indicates that the sample values $f\left(\frac{n}{2W}\right)$ determine the Fourier coefficients in terms of the expansion function $F(\omega)$. Since $F(\omega)$ is zero for frequencies greater than W, the function $F(\omega)$ can be uniquely determined by using the samples. On the other hand, the function $F(\omega)$ determines the function $f(t)$ completely if its spectrum is given. Thereby, using the samples determines the function $f(t)$ completely.

Consider a pulse of the type function

$$h(t) = \frac{\sin(2\pi W t)}{2\pi W t}. \qquad (4.12)$$

This function $h(t)$, which is called the *sinc* function, is equal to 1 at $t = 0$ and is 0 for $t = \frac{n}{2W}$, $n \neq 0$. The frequency spectrum of the sinc function is constant in the frequency band from $-W$ to W, and is zero outside. Then the function $f(t)$ can be expressed by

$$f(t) = \sum_{n=-\infty}^{\infty} f\left(\frac{n}{2W}\right) \frac{\sin[\pi(2Wt - n)]}{\pi(2Wt - n)}. \qquad (4.13)$$

Equation (4.13) is referred to as the *Nyquist-Shannon sampling theorem* or *sampling theorem* [3]. It shows that a bandlimited function has only $2W$ degrees of freedom per second. The values of the function at the sampling points can be selected independently to specify the entire function.

Given the Nyquist-Shannon sampling theorem in (4.13), we can now approach the problem of communication transmission over a bandlimited channel. Assume that the bandlimited channel has a bandwidth of W. Both of the input and output signals are measured with samples taken $\frac{1}{2W}$ second apart. Note that each of the input samples is corrupted by noise to produce the corresponding output samples. Each of the noise samples is an independent and identically distributed Gaussian random variable because the noise is white and has a Gaussian distribution [5]. Further assume that the bandlimited channel is used over the time period of interval $[0, T]$. Thus, in this case, the power per sample is equal to

$$\frac{PT}{2WT} = \frac{P}{2W}. \tag{4.14}$$

If the noise has the power spectral density $\frac{N_0}{2}$ and bandwidth of W, the noise then has the power

$$\left(\frac{N_0}{2}\right) 2W = N_0 W. \tag{4.15}$$

Thus, with each of the $2WT$ noise samples in the time period T and (4.15), the noise variance per sample is equal to

$$N_0 W \frac{T}{2WT} = \frac{N_0}{2}. \tag{4.16}$$

Therefore, using (4.14) and (4.16) and substituting them into (4.4), we obtain the channel capacity for the bandlimited channel as follows:

$$\begin{aligned} C &= \frac{1}{2} \log_2 \left(1 + \frac{P/2W}{N_0/2}\right) \\ &= \frac{1}{2} \log_2 \left(1 + \frac{P}{N_0 W}\right), \end{aligned} \tag{4.17}$$

where C is the number of bits per sample. Since there are $2W$ samples per second, then the channel capacity of the bandlimited channel can be rewritten as

$$C = W \log_2 \left(1 + \frac{P}{N_0 W}\right), \tag{4.18}$$

where C is the number of bits per second (bps). Equation (4.18) is the channel capacity of the bandlimited channel with the power P in watts and the noise spectral density $\frac{N_0}{2}$ in watts per hertz (Hz).

Equation (4.18) further indicates that the channel capacity of the bandlimited channel increases monotonically by increasing in the SNR,

$$\text{SNR} = 10 \log \left(\frac{P}{N_0 W} \right). \quad (4.19)$$

In other words, with a fixed bandwidth, the channel capacity of the bandlimited channel increases as the transmitted signal power increases. On the other hand, if the transmitted signal power is fixed, the channel capacity is also increased by increasing the bandwidth of W.

Further note that if W approaches infinity in (4.18), the channel capacity of the bandlimited channel approaches the asymptotic value

$$\begin{aligned} C &= \frac{P}{N_0} \log_2 e \\ &= \frac{P}{N \ln 2} \\ &\approx \frac{P}{0.693 N} \text{ (bps)}. \end{aligned} \quad (4.20)$$

Therefore, for an infinite bandwidth channel, the channel capacity grows linearly as the transmitted signal power increases. Furthermore, the channel capacity increases rapidly as we increase the frequency band until the total noise power is approximately equal to the signal power. Then, the channel capacity increases slowly, and it approaches an asymptotic value at 1.443.

In any electrical conductor above the absolute zero of a temperature, the electrons are in a state of random motion. This leads to *thermal noise*, which produces an open circuit noise voltage as follows [6]:

$$\overline{V^2} = 4kTR \int_{f_1}^{f_2} \left(\frac{hf}{kT} \right) \left(\frac{1}{e^{\frac{hf}{kT}} - 1} \right) df, \quad (4.21)$$

where R is the resistance of the resistor measured in ohms, k is the Boltzmann's constant equal to $k = 1.38 \times 10^{-23}$ (joules/degree), T is the temperature of the resistor in degrees of Kelvin, when absolute zero is $-273°$ C or $-459°$ F, f is the frequency in hertz, and $\overline{V^2}$ is the mean square noise voltage, which is the average value of the square of the noise voltage across the resistor.

The expression of thermal noise or *Johnson noise* in (4.21) can be simplified to the system bandwidth of W given by [7]

$$\overline{V^2} = 4kTRW, \qquad (4.22)$$

where W is the bandwidth of the noise in cycles per second. Thus, the most noise power is given by

$$N = kTW. \qquad (4.23)$$

If we substitute (4.23) into (4.4), the channel capacity of the bandlimited channel becomes

$$C = W \log_2 \left(1 + \frac{P}{kTW}\right). \qquad (4.24)$$

If the W in (4.24) approaches infinity, the channel capacity of the bandlimited channel approaches the asymptotic value

$$C \approx \frac{P}{0.693kT} \text{ (bps)}. \qquad (4.25)$$

Thus, we can rewrite (4.25) as follows:

$$P \cong 0.693kTC. \qquad (4.26)$$

Equation (4.26) says that we need at least a power of $0.693kT$ joules per second to transmit one bit per second, no matter how wide the bandwidth used. In other words, on the average, we must at least use an energy of $0.693kT$ joules to transmit each bit of an information message. In practice, most communication systems require much more energy per bit for transmission over a bandlimited channel.

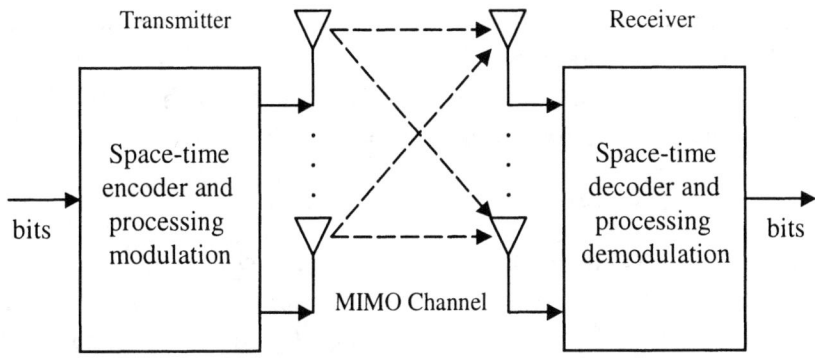

Figure 4.2 A general block diagram of a wireless MIMO communication system with multiple antenna elements at both the transmitter and the receiver.

4.4 MIMO Channel Capacity

Digital communication systems based on a MIMO channel have recently emerged as one of the most important technical breakthroughs for wireless communications. For an arbitrary wireless communication system, a communication link for which a transmitter and a receiver are equipped with multiple antenna elements is considered a MIMO system. Figure 4.2 shows a general block diagram of a wireless MIMO communication system, with multiple antenna elements at both the transmitter and the receiver.

The idea behind MIMO systems is to use space-time signal processing in which the natural dimension of digital communication data is complemented with the spatial dimension by using the multiple spatially distributed antennas. The MIMO systems are capable of turning multipath propagation into a benefit for the user. This is because the MIMO systems are able to provide *spatial diversity*, *time diversity*, and *frequency diversity* by coherently combining the use of the transmitter antennas at one end and the receiver antennas at the other end. Thereby, enhancing wireless transmission over the MIMO channel improves the channel capacity and the quality of bit error rate (BER).

Consider a continuous-time MIMO channel with n_T transmitter antennas and n_R receiver antennas. We let

$$\mathbf{x}(t) = [x_1(t), x_2(t), \cdots, x_{n_T}(t)]^T \qquad (4.27)$$

be the $n_T \times 1$ vector of transmitted signals, where $[\cdot]^T$ denotes the vector transpose, and $\mathbf{H}(\tau, t)$ be the channel impulse response. The the received signal $\mathbf{y}(t)$ is then obtained by the convolution of the channel impulse response $\mathbf{H}(\tau, t)$ and the transmitted signals $\mathbf{x}(t)$ as follows:

$$\mathbf{y}(t) = \int_{-\infty}^{\infty} \mathbf{H}(\tau, t)\mathbf{x}(\tau - t)d\tau + \mathbf{n}(t), \qquad (4.28)$$

where $\mathbf{n}(t)$ is assumed to be the $n_R \times 1$ Gaussian noise vector, and the received signal $\mathbf{y}(t)$ is given by

$$\mathbf{y}(t) = [y_1(t), y_2(t), \cdots, y_{n_R}(t)]^T. \qquad (4.29)$$

Equation (4.28) can be rewritten in the discrete-time representation by sampling the received signal $\mathbf{y}(t)$ at $t = nT$, where T is the sampling interval. Now let $\mathbf{y}[n] = \mathbf{y}[nT]$. Then the discrete-time MIMO channel can be expressed as

$$\mathbf{y}[n] = \sum_{k=-\infty}^{\infty} \mathbf{H}[k, n]\mathbf{x}[n - k] + \mathbf{n}[n]. \qquad (4.30)$$

Equation (4.30) can be simplified in terms of matrix form

$$\mathbf{y} = \mathbf{Hx} + \mathbf{n}, \qquad (4.31)$$

where the channel matrix is the $n_R \times n_T$ matrix given by

$$\mathbf{H} = \begin{bmatrix} h_{11} & h_{12} & \cdots & h_{1n_T} \\ h_{21} & h_{22} & \cdots & h_{2n_T} \\ \vdots & \vdots & \ddots & \vdots \\ h_{n_R 1} & h_{n_R 2} & \cdots & h_{n_R n_T} \end{bmatrix}, \qquad (4.32)$$

where elements h_{ij}, $i = 1, 2, \cdots, n_R$ and $j = 1, 2, \cdots, n_T$, represent the constant gain of the channel between the jth transmitter antenna and the ith receiver antenna over a symbol period.

Assume that Ω is a covariance matrix of the transmitter vector **x**, with n_T transmitter antennas and n_R receiver antennas, the channel capacity of the MIMO system is then given by [8, 9]

$$C = \log_2 \left[\det(\mathbf{I}_{n_R} + \mathbf{H}\Omega\mathbf{H}^H) \right], \quad (4.33)$$

where the "det" means determinant, $\mathbf{I}_{n_R}$ is the $n_R \times n_T$ identity matrix, $(\cdot)^H$ means the *Hermitian transpose* (or *transpose conjugate*), the channel capacity C is bps per Hz, and $tr(\Omega) \leq \rho$ holds to provide a power constraint, where ρ is the signal-to-noise ratio (SNR) at any receiver antennas.

If equal power transmission is used, the covariance matrix Ω is equal to

$$\Omega = \frac{\rho}{N}\mathbf{I}_{n_R}, \quad (4.34)$$

where $N = n_T$ is used to normalize a fixed total transmitter power. Substituting (4.34) into (4.33) now obtains the famous channel capacity equation for the MIMO system with n_R transmitter antennas and n_T receiver antennas as follows:

$$C = \log_2 \left[\det \left(\mathbf{I}_{n_R} + \frac{\rho}{N}\mathbf{H}\mathbf{H}^H \right) \right]. \quad (4.35)$$

Equation (4.35) was contributed by Foschini [10] in 1996, Foschini and Gans [11] in 1998, and Telatar [8] in 1999. Foschini [10] also demonstrated that the channel capacity for the MIMO system in (4.35) grows linearly for a case of $M = n_R = n_T$ rather than logarithmically. This is because the determinant operator in (4.35) produces a product of n nonzero eigenvalues of the channel matrix **H**. Each of the eigenvalues corresponds to the SNR over a channel eigenfunction, which is based on the transmission using a pair of right and left singular vectors of the channel matrix as transmitter antenna and receiver antenna weights, respectively. Also, because of the properties of the $\log_2$ function, the overall channel capacity is then the sum of capacities of each of these eigenfunctions, thereby resulting in the effect of channel capacity multiplication.

To illustrate this concept mathematically, we can decompose the MIMO channel into M equivalent parallel *single-input single-output* (SISO) channels by using a singular value decomposition (SVD) method for the channel matrix $\mathbf{H}$ (see the proof of SVD theory in Appendix B). Thus, the channel matrix $\mathbf{H}$ can be decomposed as follows:

$$\mathbf{H} = \mathbf{UDV}^*, \tag{4.36}$$

where $\mathbf{U}$ and $\mathbf{V}$ are unitary, and $\mathbf{D}$ is a diagonal matrix given by $\mathbf{D} = \text{diag}[\sqrt{\lambda_1}, \sqrt{\lambda_2}, \cdots, \sqrt{\lambda_M}, 0, \cdots, 0]$. The eigenvalues λ_i, $i = 1, 2, \cdots, M$, are the nonzero eigenvalues of the matrix $\mathbf{W}$, where $\mathbf{W} = \mathbf{HH}^*$. Thus, the channel model in (4.31) can be rewritten as

$$\mathbf{r} = \mathbf{Ds} + \mathbf{q}, \tag{4.37}$$

where $\mathbf{r} = \mathbf{U}^*\mathbf{y}$, $\mathbf{s} = \mathbf{V}^*\mathbf{x}$, and $\mathbf{q} = \mathbf{U}^*\mathbf{n}$. It is interesting to see that (4.37) is the MIMO system, but with M equivalent to parallel SISO channels in terms of signal powers given by the nonzero eigenvalues, λ_i, $i = 1, 2, \cdots, M$. Therefore, the channel capacity of the MIMO system in (4.35) can be rewritten in terms of the nonzero eigenvalues as follows:

$$C = \sum_{i=1}^{M} \log_2\left(1 + \frac{\rho}{N}\lambda_i\right) \text{ (bps/Hz)}, \tag{4.38}$$

where $\sum_{i=1}^{M} \lambda_i = M$. With this constraint, the channel capacity is maximum if all the singular values λ_i have the same value and it is minimum if there is only a single value. Thus, practical channels have capacities with values in between these two extremes.

Equation (4.38) is one of the most important formulas for channel capacities since it indicates that the MIMO channel capacity grows linearly and the overall capacity is the sum of capacities of each of the eigenvalues.

There are several tutorial papers in this area. We refer the interested readers to Foschini [10], Gesbert et al. [9], and Foschini et al. [12].

4.5 SIMO Channel Capacity

The discrete-time MIMO channel model that we discussed in Section 4.4 can be changed to a discrete-time SIMO channel model if we use a single transmitter antenna at one end and n_T receiver antennas at the other end. In this case, the MIMO channel model in (4.31) can be rewritten as

$$\mathbf{y} = \mathbf{h}x + \mathbf{n}, \qquad (4.39)$$

where the channel matrix $\mathbf{h}$ is the $n_R \times 1$ matrix given by

$$\mathbf{h} = [h_1, h_2, \cdots, h_{n_R}]^T, \qquad (4.40)$$

and elements h_i, $i = 1, 2, \cdots, n_R$, represent the constant gain of the channel between the single transmitter antenna and the ith receiver antenna over a symbol period. Equation (4.39) is known as the discrete-time SIMO channel.

A channel capacity of the discrete-time SIMO channel model is given by [9]

$$C = \log_2\left(1 + \rho \sum_{i=1}^{n_R} |h_i|\right) \text{ (bps/Hz)} \qquad (4.41)$$

where h_i is the constant gain for ith receiver antenna, $i = 1, 2, \cdots, n_R$. Note that the SIMO channel does not have transmitter diversity since the SIMO system only uses the single antenna at the transmitter end. Furthermore, increasing the number of n_R antennas at the receiver end only results in a logarithmic increase in average capacity.

4.6 MISO Channel Capacity

The discrete-time MIMO channel model can also be changed to a discrete-time MISO channel model if we use n_T transmitter antennas at one end and a single receiver antenna at the other end. In this case, we can rewrite the MIMO channel model in (4.31) as a MISO channel as follows:

$$y = \mathbf{hx} + v, \qquad (4.42)$$

where v is Gaussian noise, and the channel matrix **h** is the $1 \times n_T$ matrix given by

$$\mathbf{h} = [h_1, h_2, \cdots, h_{n_T}], \qquad (4.43)$$

and

$$\mathbf{x} = [x_1, x_2, \cdots, x_{n_T}]^T, \qquad (4.44)$$

where elements h_i, $i = 1, 2, \cdots, n_T$, represent the constant gain of the channel between the ith transmitter antenna and the single receiver antenna over a symbol period. Equation (4.42) is referred to as the discrete-time MISO channel.

The channel capacity of the discrete-time MISO channel model is obtained by [9, 13]

$$C = \log_2 \left(1 + \frac{\rho}{n_T} \sum_{i=1}^{n_T} |h_i| \right) \text{ (bps/Hz)}, \qquad (4.45)$$

where h_i is the constant gain for ith transmitter antenna, $i = 1, 2, \cdots, n_T$, and the total transmitter power is normalized by using the number of n_T transmitter antennas. Note that the MISO channel does not have receiver diversity because the MISO system only uses the single antenna at the receiver end. In addition, we see that increasing the number of n_T antennas at the transmitter end also results in a logarithmic increase relationship of n_T in average capacity.

4.7 Summary

In this chapter, we have first introduced a channel capacity definition that was contributed by Shannon in 1949. Shannon's theoretical work of the channel capacity has ignited subsequent developments in the field of communications.

Beginning with a short review of channel capacity, we have then focused on the fundamental theory of channel capacity for the Gaussian channels, which are the most important communication channels, with respect to signal power and noise variance as well as corresponding probability of error. We have also addressed common channel models and capacities for transmission of information message over a bandlimited wireless channel of radio frequencies or a bandlimited wireline channel with white and Gaussian noise. We have then obtained a famous channel capacity for bandlimited channels in terms of signal power, noise, and frequency bandwidth. In an extreme case, the channel capacity of the bandlimited channels approaches an asymptotic value if the frequency bandwidth moves toward infinity. Sequentially, we have presented a MIMO channel capacity based on a MIMO antenna system. This leads to a famous formula of the MIMO channel capacity with respect to a channel matrix and SNR at any receiver antennas. We have then demonstrated that the MIMO channel capacity grows linearly and overall capacity is the sum of capacities of each of the eigenvalues. In addition, with some modifications, we have expanded the theory of the MIMO channel capacity into SIMO and MISO channel capacities for SIMO and MISO antenna systems, respectively.

Unlike the MIMO channel capacity, the SIMO and MISO channel capacities can only result in a logarithmic increase in average capacity even if we increase the number of antennas at the receiver or the number of antennas at the transmitter, respectively. This is because the SIMO channel does not have transmitter diversity and the MISO channel does not have receiver diversity. Therefore, a MIMO channel system is superior to a SIMO or a MISO channel system in terms of channel capacity.

References

[1] Shannon, C. E., "A Mathematical Theory of Communications," *Bell Systems Technology Journal*, Vol. 27, pp. 379–423, July 1948a.

[2] Shannon, C. E., "A Mathematical Theory of Communications," *Bell Systems Technology Journal*, Vol. 27, pp. 623–656, October 1948b.

[3] Shannon, C. E., "Communication in the Presence of Noise," *Proceedings of the IEEE*, Vol. 86, No. 2, pp. 447–457, February 1998 (this paper is reprinted from the *proceedings of the IRE*, Vol. 37, No. 1, pp. 10–21, January 1949).

[4] Reza, F. M., *An Introduction to Information Theory*, Dover Publications, New York, 1994.

[5] Cover, T. M., and J. A. Thomas, *Elements of Information Theory*, John Wiley & Sons, New York, 1991.

[6] Waggener, W. N., *Pulse Code Modulation Systems Design*, Artech House, Norwood, Massachusetts, 1999.

[7] Pierce, J. R., *An Introduction to Information Theory: Symbols, Signals, and Noise*, 2nd ed., Dover Publications, New York, 1980.

[8] Telatar, I. E., "Capacity of Multiantenna Gaussian Channels," *European Transactions on Communications*, Vol. 10, No. 6, pp. 585–595, 1999.

[9] Gesbert, D., et al., "From Theory to Practice: An Overview of MIMO Space-Time Coded Wireless Systems," *IEEE Journal on Selected Areas in Communications*, Vol. 21, No. 3, pp. 281–302, April 2003.

[10] Foschini, G. J., "Layered Space-Time Architecture for Wireless Communication in a Fading Environment When Using Multi-Element Antennas," *Bell Labs Technical Journal*, pp. 41–59, Autumn 1996.

[11] Foschini, G. J., and M. J. Gans, "On Limits of Wireless Communications in a Fading Environment When Using Multiple Antennas," *Wireless Personal Communications*, Vol. 6, pp. 311–335, March 1998.

[12] Foschini, G. J., et al., "Analysis and Performance of Some Basic Space-Time Architectures," *IEEE Journal on Selected Areas in Communications*, Vol. 21, No. 3, pp. 303–320, April 2003.

[13] Papadias, C. B., and G. J. Foschini, "Capacity-Approaching Space-Time Codes for Systems Employing Four Transmitter Antennas," *IEEE Trans. on Information Theory*, Vol. 49, No. 3, pp. 726–733, March 2003.

5
Smart Antenna Systems

5.1 Introduction

In Chapter 4, we discussed multiple-input multiple-output (MIMO) systems with a core concept of using space-time signal processing, where the natural dimension of digital communication data is matched with the spatial dimension inherent in the use of multiple spatially distributed antennas. Thus, the MIMO systems can be viewed as an extension of the smart antennas that are a popular technology using an antenna array for improving wireless communications.

The antenna array contains many distributed antenna elements whose outputs are usually combined or selected to enhance system performance. In fact, the smart antenna is an antenna array system, but it has advanced signal processing algorithms to adapt to different signal environments. In other words, the smart antenna combines multiple antenna elements with an adaptive signal-processing capability to optimize its radiation and/or reception pattern automatically in response to the signal environment [1]. Thus, the smart antenna is able to mitigate fading through diversity reception and adaptive beamforming in addition to minimizing interference through spatial filtering, thereby enhancing both analog and digital systems.

This chapter is organized as follows. This section presents a short introduction about smart antenna systems. In Section 5.2, we begin with a review of smart antennas and beamforming structures, with an emphasis on switched, delay-and-sum, space-time, and interpolation beamforming. In Section 5.3, we introduce beamforming algorithms of MMSE, maximum signal-to-noise ratio (SNR), and minimum variance beamformer that are derived by using optimization constraint methods. The beamforming algorithms are designed to focus on specific signals while suppressing unwanted others at the same time. A brief summary is finally given in Section 5.4.

5.2 Smart Antennas and Beamforming Structures

A smart antenna system can be customarily classified as one of five beamformers, including switched beamformer, delay-and-sum beamformer, space-time beamformer, interpolation beamformer, or adaptive array beamformer. The first four beamformers employ a finite number of fixed, predefined patterns, multirate sampling, or combining strategies, while the adaptive array beamforming uses an infinite number of patterns that are adjusted in a real-time operation.

5.2.1 Switched Beamforming

A block diagram of a *switched beamforming* for the smart antenna system is shown in Figure 5.1, which includes multiple antennas, a fixed beamforming pattern with a set of the predetermined weight vectors, N receivers, and a switch controller.

The switched beamforming system tries to form a multiple fixed beamforming pattern with heightened sensitivity in particular directions. Control logic of the switch beamforming system detects signal strength, chooses from one of several predetermined fixed beamformings, and switches from one beamforming to another beamforming by using a switch controller to connect with one of the parallel outputs from the fixed beamforming pattern. The switched beamforming system combines the outputs of multiple

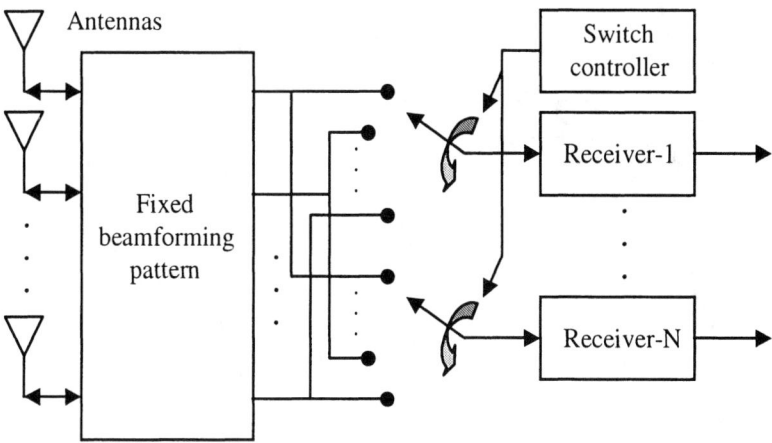

Figure 5.1 A block diagram of a switch beamforming for the smart antenna system.

antennas such that finely directional beams can be formed with more spatial selectivity than in a conventional and single-element antenna approach. In other words, the switched beamforming system selects one of several predetermined fixed-beamforming patterns, which are based on weighted combinations of antenna outputs, with the greatest output power in the remote user's channel. Furthermore, the switched beamforming system can switch its beamforming in different directions throughout space by changing the phase differences of the signals. Thus, the N receivers mitigate multipath components, which arrive at the receivers at different times, by exploiting the low correlation between temporal signal fading at different antennas. This structure of the switched beamforming system is able to enhance coverage and range extension. However, at close angle of arrival [2], the switched beamforming system cannot discriminate between multipath components. This leads to the inability to combine multipath components coherently.

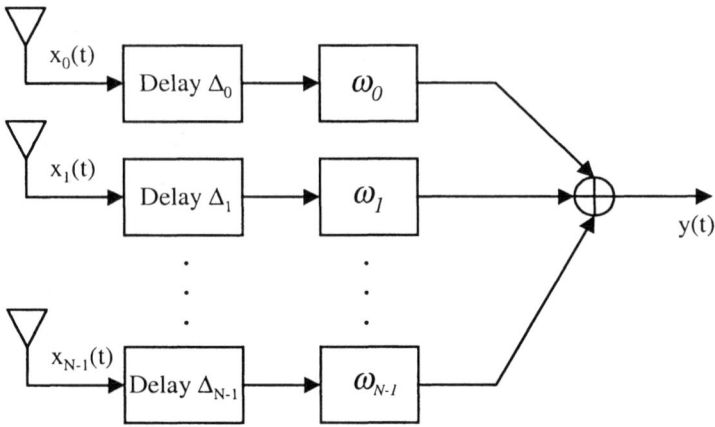

Figure 5.2 A block diagram of a delay-and-sum beamforming for the smart antenna system.

5.2.2 Delay-and-Sum Beamforming

One of the oldest and simplest beamforming is called the *delay-and-sum beamforming*, which still remains a powerful approach today. Figure 5.2 shows a block diagram of the delay-and-sum beamforming for the smart antenna system, which includes multiple antennas, a delay Δ_i, and a weighting value w_i, where $i = 0, 1, 2, \cdots, N-1$, and a sum of the resulting signals. The delay Δ_i is used to reduce mismatched delay because the mismatched delay decreases the SNR at the beamforming output and thereby reduces the array gain. The idea behind it is that if a propagating signal is present in multiple antennas, the outputs of the multiple antennas, delayed by appropriate amounts and added together, reinforce the signal with respect to noise in different directions. Thus, the output signal of the delay-and-sum beamforming is obtained by

$$y(t) = \sum_{i=0}^{N-1} w_i x_i(t - \Delta_i), \quad (5.1)$$

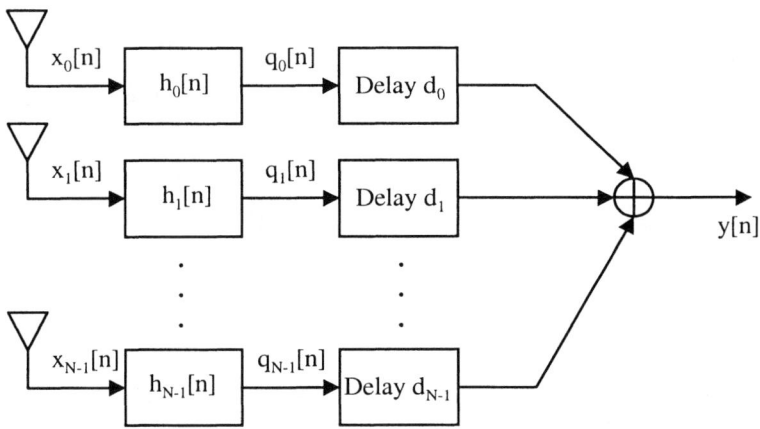

Figure 5.3 A block diagram of a space-time beamforming for the smart antenna system.

where Δ_i is the ith delay, and the weight value w_i is sometimes referred to as the beamforming shading, which is used to enhance the beamforming shape and reduce sidelobe levels. The weight value w_i can also be calculated by using one of many adaptive algorithms [3]. In this case, the delay-and-sum beamforming can be referred to as adaptive delay-and-sum beamforming.

5.2.3 Space-Time Beamforming

In the previous discussion of delay-and-sum beamforming, we assumed that the multiple antennas receive signals without distortion or spatial filtering. In practice, we should realize that more than one signal may be present in the propagation field received by the antennas and that noise can disturb the received signal. In order to remove unwanted disturbances, we need to insert additional linear filtering for the signal of interest in only a narrow wanted frequency band, as shown in Figure 5.3. Filtering the output signal of the multiple antennas considers both temporally and spatially in space-time signal processing. Thus, combining these filter and delay outputs by using a sum operation to form a beamforming is known as *space-time beamforming* or *filter-and-sum beamforming*.

The output signal $y[n]$ of the discrete-time space-time beamforming is obtained by

$$q_i[n] = \sum_{k=0}^{M-1} h_i[k]x_i[n-k], \quad \text{for } i = 0, 1, 2, \cdots, N-1, \quad (5.2)$$

where $q_i[n]$ is the ith output signal of the ith spatial filter, $h_i[n]$ is the ith linear-phase finite impulse response (FIR) filter, and $x_i[n]$ is the ith output signal of the ith antenna, and then

$$\begin{aligned} y[n] &= \sum_{i=0}^{N-1} q_i[n-d_i] \\ &= \sum_{i=0}^{N-1} \sum_{k=0}^{M-1} h_i[k]x_i[n-k-d_i], \end{aligned} \quad (5.3)$$

where d_i is the ith delay for ith antenna. Note that the linear FIR filters are placed on the antenna outputs to concentrate the later delay-and-sum beamforming operations into a range of temporal frequencies and to include spatial filtering of antenna signal. Each passband of the linear FIR filters corresponds to the desired signal spectrum. Furthermore, these linear FIR filters can be optimally related to signal spectrum and to noise characteristics. In addition, all the taps of the linear FIR filters can be updated in real time by using the adaptive signal processing algorithm. If this is the case, the space-time beamforming is referred to as *adaptive space-time beamforming*.

5.2.4 Interpolation Beamforming

In order to reduce the aberrations due to delay quantization, an interpolation method in a beamforming can be introduced between the samples of the antenna signals. Figure 5.4 shows a block diagram of an interpolation beamforming. Each antenna's output is passed through an upsampler, which adds $M-1$ zeros between the samples,

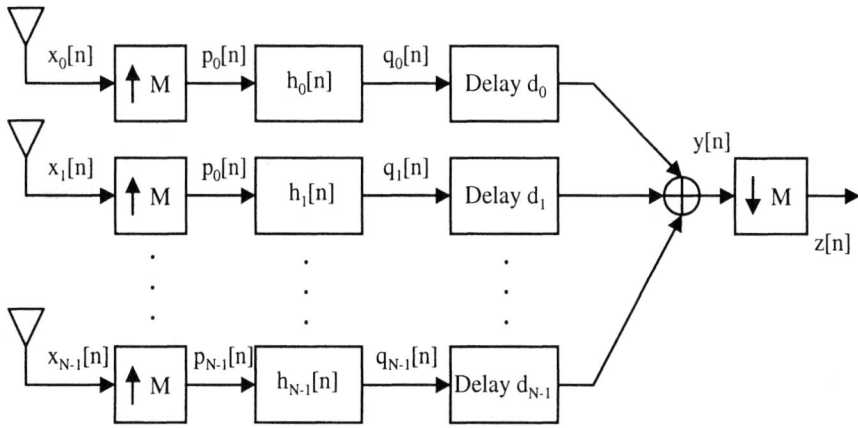

Figure 5.4 A block diagram of an interpolation beamforming for the smart antenna system.

and a lowpass FIR interpolation filter with a cutoff frequency at $\frac{\pi}{M}$. The output signal of the lowpass FIR interpolation filter with a higher sampling rate is then subjected to the delay and shading to form a beamforming. The output signal of the beamforming can be downsampled by a factor of M to obtain the signal with original sampling rate.

Upsampling by an integer factor of M involves an operation relationship between input and output given by

$$p_i[n] = \begin{cases} x_i[\frac{n}{M}], & n = 0, \pm M, \pm 2M, \cdots \\ 0 & \text{otherwise,} \end{cases} \quad (5.4)$$

where $i = 0, 1, 2, \cdots, N-1$ for the ith antenna. This is equivalent to

$$p_i[n] = \sum_{k=-\infty}^{\infty} x_i[k]\delta[n - kM]. \quad (5.5)$$

Equations (5.4) and (5.5) indicate that the discrete-time output sequence $p_i[n]$ is obtained by inserting $M - 1$ zero-valued samples between adjacent samples of the input sequence $x_i[n]$.

Using (5.3) and (5.5), we obtain the output signal of the interpolation beamforming as follows:

$$y[n] = \sum_{i=0}^{N-1}\sum_{k=0}^{M-1} h_i[k]p_i[n-k-d_i]$$

$$= \sum_{l=-\infty}^{\infty}\sum_{i=0}^{N-1}\sum_{k=0}^{M-1} h_i[k]x_i[l]\delta[n-k-d_i-lM],$$

(5.6)

where M is the integer factor determined by the downsampling processing.

Note that the signal before engaging the space-time beamforming has an effective sampling rate that is M times higher than the original sampling rate. However, it does not affect the beamforming calculation operations including filtering, time delays, shading, and summing. Inserting $M-1$ zeros between samples in the time domain causes periodic images of the signal spectrum in the frequency domain. However, lowpass FIR filtering can ideally eliminate all the images except the original one at baseband from $-\pi/M$ to π/M. Thus, the output signal of the interpolation beamforming can be downsampled by a factor of M because the original sampled signal did not induce temporal aliasing. Downsampling takes a discrete-time input signal sequence $y[n]$ and generates a new discrete-time output signal sequence $z[n]$ given by

$$z[n] = y[Mn], \qquad (5.7)$$

where M is an integer factor. Substituting (5.6) into (5.7) obtains the discrete-time output signal sequence $z[n]$ as follows:

$$z[n] = \sum_{l=-\infty}^{\infty}\sum_{i=0}^{N-1}\sum_{k=0}^{M-1} h_i[k]x_i[l]\delta[(n-l)M-k-d_i]. \qquad (5.8)$$

After downsampling, the interpolation beamforming output has the same sampling rate as the original signals. This complete process of interspersing zeros, lowpass FIR filtering, delaying, summing, and downsampling is also referred to as *time domain interpolation beamforming* [4]. In addition, the taps of all the lowpass FIR filters can be updated in real-time operation if we use an adaptive algorithm. In this case, we refer to it as *adaptive interpolation beamforming*. Furthermore, each of the combining upsamplers and lowpass FIR filters can be efficiently implemented by using polyphase interpolation decomposition or multistage interpolation filters [3, 5].

Another type of beamformer that we would like to mention here was developed for array sensors based on the frequency domain by using short-time Fourier analysis or the discrete Fourier transform (DFT), known as the *frequency-domain beamformer*. With some modifications, it is also possible to use the frequency-domain beamformer for the smart antenna system. The interested reader may refer to [4].

5.3 Beamforming Algorithms

There are many beamformer algorithms derived by using optimization constraint methods [2, 4, 6, 7]. The optimization constraint methods for the beamformer are used to focus on specific signals while suppressing unwanted others at the same time. In an optimal sense, we want to determine a set of optimal weight coefficients to reinforce any signal propagating across the beamformer with a set of delays. On the other hand, adaptive beamformer algorithms can vary the shading and increase SNR through the temporal, frequency, and spatial filtering based on the signal and noise characteristics at the time of observation.

5.3.1 MMSE Beamformer

This section describes a class of optimum linear FIR filters known as *Wiener filters* [3, 8] for a beamformer when a training signal is available. We focus on developing the derivation of the Wiener-Hopf equation, which provides the optimum linear FIR filter coefficients in the optimal sense of mimimum mean square error (MMSE) for the beamformer.

Assume that the training signal (an ideal signal) is denoted by $d[n]$, the input signal of the beamformer is denoted by $x[n]$, and the output signal of the beamformer is denoted by $y[n]$. The optimum linear FIR filter is used to estimate the MMSE for the beamformer, $y[n]$ for $d[n]$, given two wide-sense stationary processes jointly, $x[n]$ and $d[n]$, which are statistically related to each other. We also define the autocorrelation functions, $r_x(k)$ and $r_d(k)$, and the cross-correlation function, $r_{dx}(k)$.

In the discrete-time domain, the error signal $e[n]$ is written as

$$\begin{aligned} e[n] &= d[n] - y[n] \\ &= d[n] - \sum_{k=0}^{M-1} w[k]x[n-k], \end{aligned} \quad (5.9)$$

where $w[k]$ is the filter coefficients, or in vector form,

$$e[n] = d[n] - \mathbf{w}^T[n]\mathbf{x}[n]. \quad (5.10)$$

The mean square error (MSE) is a function of the chosen filter coefficient vector $\mathbf{w}[n]$, and can be written as

$$\begin{aligned} \xi &= E\{e[n]e^*[n]\} \\ &= E\{(d[n] - \mathbf{w}^T[n]\mathbf{x}[n])^2\} \\ &= E\{d[n]d^*[n] - d[n]\mathbf{x}^T[n]\mathbf{w}[n] \\ &\quad - \mathbf{w}^T[n]\mathbf{x}[n]d^*[n] + \mathbf{w}^T[n]\mathbf{x}[n]\mathbf{x}^T[n]\mathbf{w}[n]\} \\ &= \text{Var}\{d[n]\} - \mathbf{r}_{dx}[n]\mathbf{w}[n] - \mathbf{w}^T[n]\mathbf{r}_{dx}[n] \\ &\quad + \mathbf{w}^T[n]\mathbf{R}_x[n]\mathbf{w}[n], \end{aligned} \quad (5.11)$$

where $\mathbf{r}_{dx}[n] = E\{\mathbf{x}[n]d^*[n]\}$ is the product of the cross-correlation function and $\mathbf{R}_x[n] = E\{\mathbf{x}[n]\mathbf{x}^T[n]\}$ is the autocorrelation function.

To minimize ξ in (5.11), we make the derivative of ξ with respect to $\mathbf{w}[n]$ equal to zero and obtain the result as follows:

$$\mathbf{R}_x[n]\mathbf{w}[n] = \mathbf{r}_{dx}[n]. \quad (5.12)$$

where $\mathbf{R}_x[n]$ is an $M \times M$ *Hermitian Toeplitz* matrix of autocorrelation, $\mathbf{w}[n]$ is the filter coefficient vector, and $\mathbf{r}_{dx}[n]$ is the vector of cross-correlation between the ideal signed $d[n]$ and the input signal $x[n]$. Equation (5.12) is referred to as the Wiener-Hopf equation.

If we further assume that the matrix $\mathbf{R}_x[n]$ is invertible, then $\mathbf{w}[n]$ in (5.12) can be solved by

$$\mathbf{w}[n] = \mathbf{R}_x^{-1}[n]\mathbf{r}_{dx}[n]. \qquad (5.13)$$

Equation (5.13) is known as the *normal equation*. This is because the error signal $e[n]$ is orthogonal to each element of the input signal $x[n]$ given by

$$E\{\mathbf{x}[n]e^*[n]\} = 0. \qquad (5.14)$$

The corresponding MMSE for the beamformer is given by [3]

$$\xi_{MMSE} = R_d(0) - \mathbf{R}_{dx}^T[n]\mathbf{w}[n], \qquad (5.15)$$

or in another form

$$\xi_{MMSE} = R_d(0) - \mathbf{r}_{dx}^T[n]\mathbf{R}_x^{-1}[n]\mathbf{r}_{dx}[n]. \qquad (5.16)$$

Note that (5.15) or (5.16) is a single-step algorithm but involves heavy computation because of computing the inverse autocorrelation matrix $\mathbf{R}_x^{-1}[n]$ and multiplying the cross-correlation matrices $\mathbf{r}_{dx}[n]$ and $\mathbf{r}_{dx}^T[n]$.

5.3.2 Maximum SNR of the Beamformer

In this section, we discuss another optimization criterion based on eigenvalue analysis for the maximum beamformer gain. Assume that an input signal $\mathbf{x}[n]$ for a beamformer consists of a desired signal $\mathbf{s}[n]$ and interference and noise $\mathbf{n}[n]$ given by

$$\mathbf{x}[n] = \mathbf{s}[n] + \mathbf{n}[n]. \qquad (5.17)$$

Then the beamformer output can be written as follows:

$$\begin{aligned} y[n] &= \mathbf{w}^H[n]\mathbf{x}[n] \\ &= \mathbf{w}^H[n]\mathbf{s}[n] + \mathbf{w}^H[n]\mathbf{n}[n]. \end{aligned} \qquad (5.18)$$

The first and second terms on the right side of (5.18) represent a signal component and an interference and/or noise component, respectively. Then the SNR of the beamformer output can be formed by using an F_{SNR} ratio (see Miao [6]),

$$\text{SNR} = \frac{\mathbf{w}^H[n]\mathbf{R}_s[n]\mathbf{w}[n]}{\mathbf{w}^H[n]\mathbf{R}_n[n]\mathbf{w}[n]}, \quad (5.19)$$

where $\mathbf{R}_s[n]$ is the autocorrelation function of the desired signal $\mathbf{s}[n]$ and $\mathbf{R}_n[n]$ is the autocorrelation function of the interference and noise. This F_{SNR} criterion is intuitively attractive because it is easier to maximize SNR for the beamformer output.

In order to determine the beamformer output $y[n]$, we need to determine the values of the elements of the filter coefficient vector $\mathbf{w}[n]$ that maximizes the ratio of the F_{SNR} criterion in (5.19). This is also called the *canonical correlation function*, the *canonical discriminant function* (see Miao [6]), the *optimization array gain* (see Johnson and Dudgeon [4]), the *Rayleigh-Ritz ratio*, or the *signal-to-interference-and-noise ratio* (see Reed [2]).

The determination of the filter coefficient vector of the beamformer that maximizes the F_{SNR} criterion involves solving the eigenvalue and eigenvector Equation (5.19). To maximize the F_{SNR} criterion, we need to take the partial derivative of the F_{SNR} with respect to the filter coefficient vector $\mathbf{w}[n]$ and set it equal to zero

$$\begin{aligned}\frac{\partial F_{SNR}}{\partial \mathbf{w}[n]} &= \frac{2\mathbf{R}_s[n]\mathbf{w}[n]}{\mathbf{w}^H[n]\mathbf{R}_n[n]\mathbf{w}[n]} \\ &\quad - \left(\frac{2\mathbf{R}_n[n]\mathbf{w}[n]}{\mathbf{w}^H[n]\mathbf{R}_n[n]\mathbf{w}[n]}\right)\left(\frac{\mathbf{w}^H[n]\mathbf{R}_s[n]\mathbf{w}[n]}{\mathbf{w}^H[n]\mathbf{R}_n[n]\mathbf{w}[n]}\right) \\ &= 0. \quad (5.20)\end{aligned}$$

Simplying (5.20), we obtain

$$\frac{\mathbf{w}^H[n]\mathbf{R}_s[n]\mathbf{w}[n]}{\mathbf{w}^H[n]\mathbf{R}_n[n]\mathbf{w}[n]} = \frac{\mathbf{R}_s[n]\mathbf{w}[n]}{\mathbf{R}_n[n]\mathbf{w}[n]} = \lambda. \quad (5.21)$$

Therefore, it can be seen that the maximum value of λ is the largest eigenvalue of the matrix of $\mathbf{R}_n^{-1}[n]\mathbf{R}_s[n]$ and $\mathbf{w}[n]$ is the corresponding eigenvector as follows:

$$\left(\mathbf{R}_n^{-1}[n]\mathbf{R}_s[n] - \lambda\mathbf{I}\right)\mathbf{w}[n] = 0. \qquad (5.22)$$

The corresponding SNR for the beamformer is then obtained by

$$\{F_{SNR}\}_{max} = \lambda_{max}, \qquad (5.23)$$

where λ_{max} is the largest eigenvalue of the matrix $\mathbf{R}_n^{-1}[n]\mathbf{R}_s[n]$. Thus, (5.23) indicates that the maximum of SNR for the beamformer is equal to the maximum eigenvalue of the matrix $\mathbf{R}_n^{-1}[n]\mathbf{R}_s[n]$.

5.3.3 Minimum Variance Beamformer

Another famous optimization criterion for the beamformer is known as the *minimum variance beamformer* [4, 6]. In this method, we choose the normalization of the filter coefficient vector $\mathbf{w}[n]$ such that

$$\mathbf{w}^H[n]\mathbf{R}_n[n]\mathbf{w}[n] = 1. \qquad (5.24)$$

Equation (5.24) ensures that the ideal signal passes to the beamformer output having unity gain. This further implies that the pool covariance matrix of the output signal $y[n]$ has a unity variance.

To maximize the F_{SNR} ratio in (5.19), we form the expression (see Miao [6]) as follows:

$$F_{SNR} = \max_{\mathbf{w}[n]}\{\mathbf{w}^T[n]\mathbf{R}_s[n]\mathbf{w}[n]\}, \qquad (5.25)$$

subject to

$$\mathbf{w}^T[n]\mathbf{R}_n[n]\mathbf{w}[n] = 1. \qquad (5.26)$$

Now, we let $\mathbf{w}[n] = \mathbf{R}_n^{-1/2}[n]\mathbf{a}[n]$, where $\mathbf{a}[n]$ is a new vector, and then write a new expression of F_{SNR} ratio in terms of the vector $\mathbf{a}[n]$ as follows:

$$\begin{aligned} F_{SNR} &= \mathbf{w}^T[n]\mathbf{R}_s[n]\mathbf{w}[n] - \lambda(\mathbf{w}^T[n]\mathbf{R}_n[n]\mathbf{w}[n] - 1) \\ &= \mathbf{a}^T[n]\mathbf{R}_n^{-1/2}[n]\mathbf{R}_s[n]\mathbf{R}_n^{-1/2}[n]\mathbf{a}[n] \\ &\quad -\lambda(\mathbf{a}^T[n]\mathbf{a}[n] - 1), \end{aligned} \qquad (5.27)$$

where λ is a nonzero *Lagrange* multiplier. We take the partial derivative with respect to the vector $\mathbf{a}[n]$ and set it equal to zero

$$\frac{\partial F_{SNR}}{\partial \mathbf{a}[n]} = 2\mathbf{R}_n^{-1/2}[n]\mathbf{R}_s[n]\mathbf{R}_n^{-1/2}[n]\mathbf{a}[n] - 2\lambda\mathbf{a}[n] = 0. \quad (5.28)$$

Therefore, the eigenvalue and eigenvector problem of maximizing SNR for the beamformer output is obtained by

$$\left(\mathbf{R}_n^{-1/2}[n]\mathbf{R}_s[n]\mathbf{R}_n^{-1/2}[n] - \lambda\mathbf{I}\right)\mathbf{a}[n] = 0. \quad (5.29)$$

Solving (5.29) for the eigenvalue λ gives

$$\lambda = \frac{1}{\mathbf{R}_n^{1/2}[n]\mathbf{R}_s^{-1}[n]\mathbf{R}_n^{1/2}[n]}. \quad (5.30)$$

Equation (5.30) is referred to as the maximum variance beamformer, which is used to minimize the output power of the beamformer subject to the constraint with the constant gain at the angle of arrival of the desired signal.

Solving (5.29), we can also obtain the vector $\mathbf{a}[n]$. Thus, the corresponding filter coefficient vector $\mathbf{w}[n]$ for the beamformer is obtained by

$$\mathbf{w}[n] = \mathbf{R}_n^{-1/2}[n]\mathbf{a}[n]. \quad (5.31)$$

It can be shown [6] that the term of $\mathbf{R}_n^{-1}[n]\mathbf{R}_s[n]$ in (5.22) and the term of $\mathbf{R}_n^{-1/2}[n]\mathbf{R}_s[n]\mathbf{R}_n^{-1/2}[n]$ in (5.29) have the same eigenvalue by using the maximization and minimization theorem (see Appendix B).

It is also possible to further optimize the filter coefficients for the beamformer by using other approaches, which are developed based on these optimization concepts but with the additional condition of orthogonality and scalar spread effects, which allow selecting the greatest output power in the remote user's channel. The interested reader may refer to Miao [6] to explore further the optimization methods based on the generalized optimal declustering analysis for the beamformer.

We have discussed selecting the filter coefficients for the beamformer without regard to the nature of the field into which the beamformer is placed. The filter coefficients of the beamformer yielding the optimal gain depend on the characteristics of the interference and noise. This means that a good set of filter coefficients for the beamformer should adapt to the beamformer environment and should not be fixed before placing the beamformer. This leads to the idea of adapting signal processing algorithms for the beamformer to the signal environment. For adaptive beamformer algorithms with an application of a sidelobe canceller based on a linear predictive approach to array signal processing, we refer the interested reader to Johnson and Dudgeon [4].

5.4 Summary

In this chapter, we first introduced a smart antenna system in which an antenna array system with signal processing algorithms adapts to different signal conditions. This led to mitigating fading through diversity reception and adaptive beamforming. We then focused on fundamental beamforming structures and discussed their benefits with respect to different signal environments. These discussions have led to the necessary step of understanding how a beamforming structure affects performances of the smart antenna system, thereby providing a key guideline of designing the smart antenna system for digital communication systems.

Of particular importance in the evaluation of smart antenna system performance are beamforming algorithms that laid mathematical foundations based on optimization constraint methods, which have rapidly developed over the last two decades along with practical applications to the digital communications systems. The systems that exist today and those currently under development certainly reflect these recent advances in the smart antenna system.

With applications of using beamformer structures and algorithms, the smart antenna system can further enhance performance of digital radio frequency (RF) systems and mitigate fading through diversity reception and space-time processing while minimizing interference through beamformer spatial filtering. Advanced development in the smart antenna system will be a key to future digital communication systems with evolving higher data rates and spectral efficiencies. Therefore, by discussing smart antenna system technologies, we have placed technical foundations to integrate those technologies into more advanced development of signal processing technologies for digital communication systems.

References

[1] International Engineering Consortium, "Smart Antenna Systems," http://www.iec.org/online/tutorials/smart-ant/.

[2] Reed, J. H., *Software Radio: A Modern Approach to Radio Engineering*, Prentice Hall, Upper Saddle River, New Jersey, 2002.

[3] Miao, G. J., and M. A. Clements, *Digital Signal Processing and Statistical Classification*, Artech House, Norwood, Massachusetts, 2002.

[4] Johnson, D. H., and D. E. Dudgeon, *Array Signal Processing*, Prentice Hall, Englewood Cliffs, New Jersey, 1993.

[5] Vaidyanathan, P. P., "Multirate Digital Filters, Filter Banks, Polyphase Networks, and Applications: A Tutorial," *Proc. of IEEE*, Vol. 78, No. 1, pp. 56–93, 1990.

[6] Miao, J., *Component Feature-Based Digital Waveform Analysis and Classification*, UMI, A Bell & Howell Information Company, Ann Arbor, Michigan, 1996.

[7] Haykin, S., *Adaptive Filter Theory*, 3rd ed., Prentice Hall, Upper Saddle River, New Jersey, 1996.

[8] Kailath, T., *Lectures on Wiener and Kalman Filtering*, 2nd ed., Springer-Verlag, New York, 1981.

6

Channel Characterization and Distortion

6.1 Introduction

Communication channels can be wireless or wired physical channels, or a combination of both. In any case, those physical channels usually introduce linear and nonlinear distortions, random noise, and deterministic interference. Therefore, efficient and successful communication of messages via imperfect channels is still one of the major triumphs of information technology.

In wireless communication, a physical channel affects propagation of radio signals on both forward and reverse links. A signal propagating through the physical channel usually arrives at the destination along with a number of different paths that are referred to as *multipath*. These paths come from scattering, reflection, refraction, or diffraction of radiated energy off of objects that are located in the environment. The received signal is much weaker than the transmitted signal because of phenomena such as mean propagation loss, slow fading, and fast fading. In addition, digital communication through a bandlimited wireless channel with multipath phenomena is subject to *intersymbol interference* (ISI). This problem can be so severe that the correct reception of transmitted sequence is not feasible anymore without including a specific device in a receiver.

In wired communication, for example, a twisted-pair copper telephone line that was originally intended to carry voice signals at about a 4-kHz frequency bandwidth is now used to transmit from several megabits (Mb) to 52 Mb of data per second by using a *digital subscriber loop* (DSL), such as HDSL, ADSL, VDSL, and so forth. This has been possible because of the efficient use of the high-frequency propagation of copper wires, which suffer from a great deal of line attenuation and noise. The twisted-pair copper wire channel usually has different segments with different gauges ranging from 19 American wire gauges (AWG) to 26 AWG, multiple bridge taps, a wire drop, and lumped elements. In addition, far- and near-end crosstalk between pairs in a multipair cable is the dominant impairment in any DSL system. Therefore, with the limitation of frequency bandwidth, twisted-pair copper channel has severe distortion and is subject to ISI.

Communication systems usually operate with limitations of frequency bandwidth. This causes ISI and out-of-band radiation in adjacent channels. Pulse-shaping techniques can be designed to reduce the ISI effects and the spectral width of a modulated signal simultaneously. This leads to matched filtering at the receiver that compensates for the ISI caused by multipath within time-dispersive wireless channels or bandlimited wired channels.

This chapter is organized as follows. In this section, a short overview and the background of communication channels are briefly presented. Section 6.2 introduces characteristics of wireless channels including propagation, multipath, and fading. Section 6.3 describes wired channels that focus on the transmission loop, its crosstalk, and the simulation loop-model. Section 6.4 presents channel distortion with emphases on ISI, eye diagrams, and Nyquist criterion. Subsequently, pulse-shaping techniques that are designed to reduce and suppress out-of-band radiation while eliminating ISI effects are given in Section 6.5. Section 6.6 addresses matched filtering and its method on how to increase signal-to-noise ratio

(SNR) and reduce ISI and noise interference. A brief summary of this chapter is given in Section 6.7.

6.2 Wireless Channels

Wireless channels place fundamental limitations on the performance of radio signals on both forward and reverse links for communication systems in several ways. The transmission path between a transmitter and receiver can vary from a simple *line-of-sight* (LOS) path to a *nonline-of-sight* (NLOS) path, which is severely obstructed by buildings, mountains, and foliage. Additionally, when a mobile terminal moves in space, the speed of motion impacts how rapidly the signal level fades.

6.2.1 Free Space Propagation

In ideal free-space propagation received signal strength can be predicted when the transmitter and receiver have a clear and unobstructed LOS path. The free-space power received by a receiver antenna is given by the Friis free-space equation,

$$P_r(d) = \frac{P_t G_t G_r}{L} \left(\frac{\lambda}{4\pi d}\right)^2, \quad (6.1)$$

where P_t and $P_r(d)$ are the transmitted and received power, respectively, G_t and G_r are the transmitter and receiver antenna gain, d is the distance separation between the transmitter and receiver, λ is the wavelength in meters, and L is the system loss factor. When $L = 1$, this indicates no loss in the communication system implementation. When $L \geq 1$, there are extra losses due to effects such as transmission line attenuation, filter, and antenna losses. The wavelength λ is related to the carrier frequency by

$$\lambda = \frac{c}{f} = \frac{2\pi c}{\omega_c}, \quad (6.2)$$

where c is the speed of light in meters per second, f is the carrier frequency in hertz, and ω_c is the carrier frequency in radians per second.

The *path loss* for the free-space propagation is defined as the difference (in decibels) between the effective transmitted power and the received power given by

$$PL = 10 \log_{10} \left(\frac{P_t}{P_r}\right)$$
$$= -10 \log_{10} \left[\frac{G_t G_r}{L} \left(\frac{\lambda}{4\pi d}\right)^2\right] \text{ (dB)}. \quad (6.3)$$

Note that (6.1) and (6.3) do not hold for the distance $d = 0$. For this reason, the received power $P_r(d)$ uses a close-in distance d_0 as a known received power reference point to relate to $P_r(d)$ at the close-in distance d_0. Thus, the received power in the free-space propagation at a distance greater than d_0 is obtained by

$$P_r(d) = P_r(d_0) \left(\frac{d_0}{d}\right)^2, \quad d \geq d_0 \geq d_f, \quad (6.4)$$

where d_f is known as the *Fraunhofer* distance (or far-field distance) given by

$$d_f = \frac{2D^2}{\lambda}, \quad (6.5)$$

where D is the largest physical linear dimension of the antenna and $d_f \gg \lambda$.

Equation (6.4) can be rewritten in units of dBm if we take the logarithm of both sides and multiply by 10,

$$P_r(d) = 10 \log_{10} \left[\frac{P_r(d_0)}{0.001}\right] + 20 \log_{10} \left(\frac{d_0}{d}\right) \text{ (dBm)}, \quad (6.6)$$

where $d \geq d_0 \geq d_f$, $P_r(d)$ and $P_r(d_0)$ are in units of dBm and Watts, respectively.

Example 6.1

Assume that an antenna with a unity gain has a maximum dimension of 2 meters, and a transmitter generates 40 Watts of power at a

500-MHz carrier frequency. If a receiver antenna has a unity gain, determine (a) a far-field distance d_f, (b) the transmitter power in dBm and dBW, and (c) the receiver power in dBm at the far-field distance d_f and the receiver power in dBm at the distance of 5 km.

Given the maximum dimension of the transmitter antenna $D = 2$ meters, the carrier frequency $f = 500$ MHz, and the transmitter power $P_t = 40$ Watts, we now answer the questions as follows:

(a) Using (6.2) yields the wavelength λ,

$$\begin{aligned} \lambda &= \frac{c}{f} \\ &= \frac{3 \times 10^8 \text{ (m/s)}}{500 \times 10^6 \text{ (Hz)}} \\ &= 0.6 \text{ (m)}. \end{aligned} \qquad (6.7)$$

Then using (6.5), we obtain the far-field distance as

$$d_f = \frac{2D^2}{\lambda} = \frac{2(2)^2}{0.6} = 13.33 \text{ (m)}. \qquad (6.8)$$

(b) The transmitter power in dBm is

$$\begin{aligned} P_t &= 10 \log_{10} \left[\frac{P_t \text{ (mW)}}{1 \text{ (mW)}} \right] \\ &= 10 \log_{10} \left(40 \times 10^3 \right) \\ &= 46.02 \text{ (dBm)} \end{aligned} \qquad (6.9)$$

and the transmitter power in dBW is

$$\begin{aligned} P_t &= 10 \log_{10} \left[\frac{P_t \text{ (W)}}{1 \text{ (W)}} \right] \\ &= 10 \log_{10} (40) \\ &= 16.02 \text{ (dBW)}. \end{aligned} \qquad (6.10)$$

(c) Using (6.1), assume that $L = 1$ with no loss in the communication system for implementation. We obtain the receiver power at the far-field distance d_f,

$$\begin{aligned} P_r(d_f) &= \frac{P_t G_t G_r}{L}\left(\frac{\lambda}{4\pi d_f}\right)^2 \\ &= \frac{40 \times 1 \times 1}{1}\left(\frac{0.6}{4\pi \times 13.33}\right)^2 \\ &= 5.132 \times 10^{-4} \text{ (W)}. \end{aligned} \qquad (6.11)$$

Thus, the receiver power in dBm is obtained by

$$\begin{aligned} P_r(d_f) &= 10\log_{10} P_r(d_f) \text{ (mW)} \\ &= 10\log_{10}\left[5.132 \times 10^{-1} \text{ (mW)}\right] \\ &= -2.897 \text{ (dBm)}. \end{aligned} \qquad (6.12)$$

Using (6.6) with $d_0 = d_f = 13.33$ (m), the received power at $d = 5$ (km) is obtained by

$$\begin{aligned} P_r(d) &= 10\log_{10}\left[\frac{P_r(d_0)}{0.001}\right] + 20\log_{10}\left(\frac{d_0}{d}\right) \\ &= 10\log_{10}\left[\frac{5.132 \times 10^{-4}}{0.001}\right] + 20\log_{10}\left(\frac{13.33}{5 \times 10^3}\right) \\ &= -2.897 \text{ (dBm)} - 51.483 \text{ (dB)} \\ &= -54.380 \text{ (dBm)}. \end{aligned} \qquad (6.13)$$

6.2.2 Flat Surface Propagation

The path loss model in (6.3) is developed under a condition of the free-space propagation. However, transmitted signals in the wireless communications, such as land mobile radio applications, usually do not experience the free-space propagation. In environments of the land mobile radio applications, a main path is often accompanied by a flat surface-reflected path, which may destructively interfere with the primary path. In this case, the received power of a more appropriate theoretical model over the flat surface-reflected path [1, 2] is given by

$$P_r(d) = \frac{4P_t G_t G_r}{L}\left(\frac{\lambda}{4\pi d}\right)^2 \left[\sin\left(\frac{2\pi h_t h_r}{\lambda d}\right)\right]^2, \qquad (6.14)$$

where h_t and h_r are the heights of the transmitter and receiver antennas, respectively. If $d \gg h_t h_r$, using the approximation $\sin(x) \approx x$ for small x, (6.14) can be rewritten as

$$P_r(d) = \frac{P_t G_t G_r}{L}\left(\frac{h_t h_r}{d^2}\right)^2. \qquad (6.15)$$

Note that the propagation model of (6.15) over a flat surface-reflected path differs from the free-space propagation of (6.1) in three ways: (1) there is a direct influence of heights at both ends of the link, (2) the path loss is not frequency dependent, and (3) the received power decays with the fourth power rather than the square of the distance. Thus, at large distance $d \gg h_t h_r$, the received power of (6.15) falls off with distance at a rate of 40 dB/decade.

Using (6.14), the path loss for the propagation over a flat reflecting surface is defined as the difference (in decibels) between the effective transmitted power and the received power given by

$$\begin{aligned}PL &= 10\log_{10}\left(\frac{P_t}{P_r}\right) \\ &= -10\log_{10}\left\{\frac{4G_t G_r}{L}\left(\frac{\lambda}{4\pi d}\right)^2\left[\sin\left(\frac{2\pi h_t h_r}{\lambda d}\right)\right]^2\right\} \text{ (dB)}.\end{aligned}$$
$$(6.16)$$

Figure 6.1 shows a propagation path loss over the flat reflecting surface against the distance of d, with a carrier frequency at 1,800 MHz, $h_t = 1$ m, $h_r = 7$ m, unity gain of the transmitted antenna, $G_r = 2$, and no loss for the communication system implementation, $L = 1$. In that case, we note that the propagation path loss (or the received power) has alternate minima and maxima when the path distance d is relatively small. The last local maximum in the

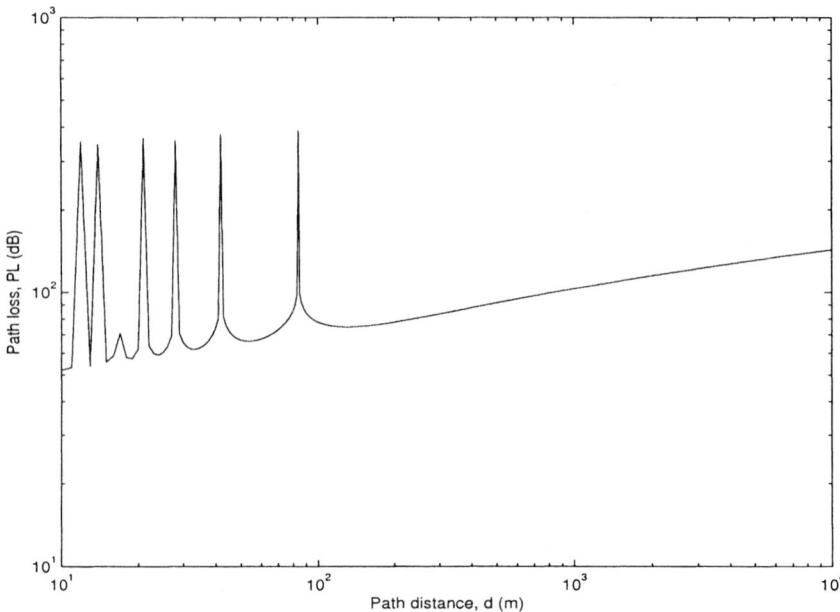

Figure 6.1 Propagation path loss over a flat reflecting surface against distance: $f_c = 1{,}800$ MHz, $h_t = 1$ m, $h_r = 7$ m, $G_t = 1$, $G_r = 2$, and $L = 1$.

propagation path loss can be obtained near the far field boundary via the following equation:

$$\frac{2\pi h_t h_r}{\lambda d} = \frac{\pi}{2}. \tag{6.17}$$

Since

$$\lambda = \frac{c}{f_c} = \frac{3 \times 10^8}{1{,}800 \times 10^6} = 0.167 \text{ (m)}, \tag{6.18}$$

substituting λ, h_t, and h_r into (6.17) and solving yields $d = 168$ m.

6.2.3 Multipath Propagation

In a wireless mobile communication system, a transmitted signal can travel to a receiver over multiple paths. This phenomenon is called

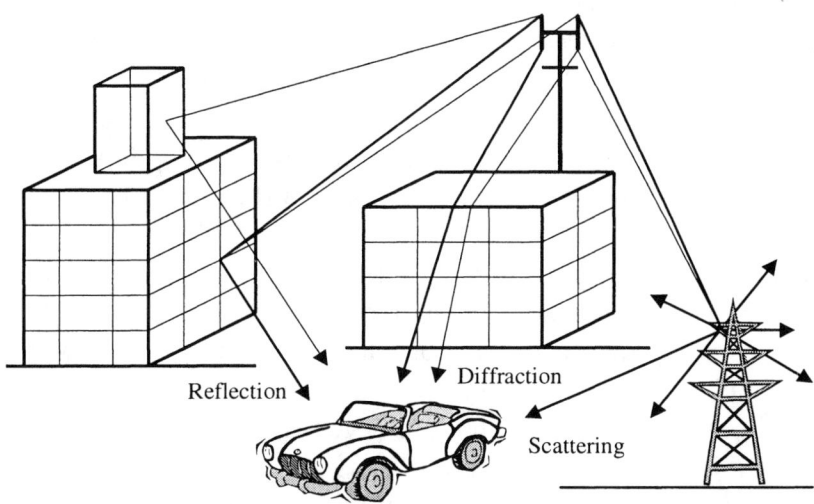

Figure 6.2 Multipaths based on the three propagation phenomena of reflection, diffraction, and scattering.

multipath propagation. There are three basic multipath propagation phenomena that impact propagation in a mobile communication system, including reflection, diffraction, and scattering. The multipaths formed by the reflectors, diffractors, and scatterers add up at a receiver antenna to produce the received signal. Figure 6.2 shows multipaths based on the three basic propagation phenomena of reflection, diffraction, and scattering.

Reflection takes place when a propagating electromagnetic wave (or radio wave) impinges upon an object including a building, a wall, and even the surface of the Earth, and so forth, which has very large dimensions when compared to the wavelength of the propagating wave. In other words, when an electromagnetic wave propagating in one medium impinges up another medium having different electrical properties, the electromagnetic wave is partially reflected and partially transmitted.

Diffraction takes place when an object having sharp irregularities (or edges) blocks the radio wave. This is to say that the diffraction allows radio waves to propagate around the curved surface of the Earth, beyond the horizon, and to propagate behind obstructions. When a receiver moves deeper into the obstructed (or shadowed) region, the received field strength decreases rapidly. However, the diffraction field still exists and usually has sufficient strength to produce a useful signal for the receiver.

Scattering takes place when the radio wave strokes objects with dimensions that are small compared to the wavelength. The reflected energy is spread out in all directions due to scattering when the radio wave impinges on a rough surface. Objects such as foliage, street signs, lampposts, and trees tend to scatter energy in all directions, thereby providing additional radio energy at the receiver.

The received power (or its reciprocal, path loss) is the most important parameter predicted by large-scale and small-scale fading and propagation models based on the physics of the reflection, diffraction, and scattering. The mathematical treatments on the three basic propagation schemes refer to [1, 3, 4].

6.2.4 Parameters of Multipath Channels

In a multipath propagation environment, several time-shifted, Doppler spread, and scaled versions of the transmitted signal arrive at the receiver. This phenomenon creates a so-called *multipath-intensity profile*, which describes relationships among parameters of multipath channels.

A relationship between a delayed power $S(\tau)$ in time domain and a Doppler power spectrum $S(v)$ in frequency domain for the multipath channels is shown in Figure 6.3(a, b), where τ is the delay and v is the Doppler shift. Figure 6.4(a, b) also shows a relationship between a space-frequency correlation function $R(\Delta f)$ in frequency domain and a space-time correlation function $R(\Delta t)$ in time domain for the multipath channels, where Δf and Δt are the frequency and the time, respectively. In addition, the delayed power $S(\tau)$ and the Doppler power spectrum $S(v)$ can be converted

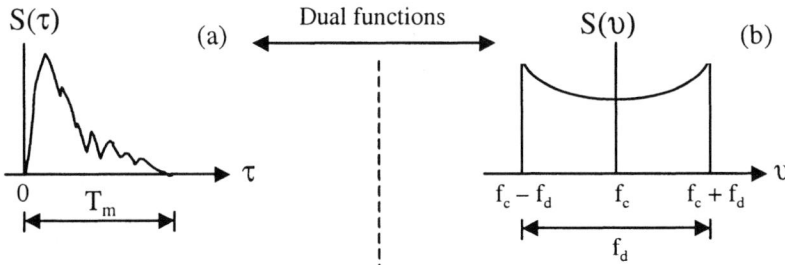

Figure 6.3 A relationship between the delayed power $S(\tau)$ and the Doppler power spectrum $S(v)$ for the multipath channels.

to the space-frequency correlation function $R(\Delta f)$ and the space-time correlation function $R(\Delta t)$, respectively, by using the Fourier transform.

Time Delay

The term of *time delay* is used to refer to the *excess delay*, which represents the delay of the signal propagation that exceeds the delay of the first signal arrival at the receiver. Figure 6.3(a) shows a relationship of the delayed power $S(\tau)$ versus the time delay τ. For a transmitted signal impulse, the time T_m between the first and last received component represents the *maximum excess delay*. The maximum excess delay is defined to be the time delay during which multipath energy falls some threshold level below the strongest component. Generally, the threshold level can be selected at 10 dB or 20 dB below the level of the strongest component.

In addition to the maximum excess delay, there are other terms, *mean excess delay* and *rms delay spread*, that are also used for the parameters of the multipath channels. The mean excess delay is

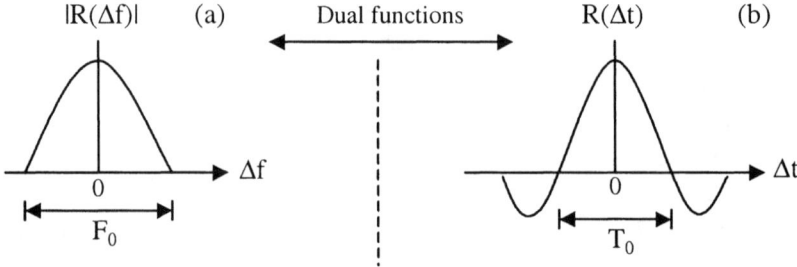

Figure 6.4 A relationship between the space-frequency correlation function $R(\Delta f)$ and the space-time correlation function $R(\Delta t)$ for the multipath channels.

defined by

$$\bar{\tau} = \frac{\sum_{k=1}^{N} S(\tau_k)\tau_k}{\sum_{k=1}^{N} S(\tau_k)}. \tag{6.19}$$

The rms delay spread is defined to be the square root of the second central moment of the power delay $S(\tau)$, given by

$$\sigma_\tau = \sqrt{\overline{\tau^2} - (\bar{\tau})^2}, \tag{6.20}$$

where $(\bar{\tau})^2$ is the mean squared and $\overline{\tau^2}$ is the second moment given by

$$\overline{\tau^2} = \frac{\sum_{k=1}^{N} S(\tau_k)\tau_k^2}{\sum_{k=1}^{N} S(\tau_k)}. \tag{6.21}$$

Generally, values of (6.20) are on the order of microseconds in outdoor mobile radio channels and on the order of nanoseconds in indoor wireless channels.

Coherence Bandwidth

A relation between the power delay $S(\tau)$ in Figure 6.3(a) and the magnitude frequency response (or the space-frequency correlation function) $R(\Delta f)$ in Figure 6.4(a) can be established by using the

Fourier transform. Both parameters of $S(\tau)$ and $R(\Delta f)$ are related as follows:

$$R(\Delta f) = \int_{-\infty}^{\infty} S(\tau) \exp(-j2\pi \Delta f \tau) d\tau. \qquad (6.22)$$

Therefore, it is possible to obtain an equivalent description of the multipath channel in the frequency domain using its frequency response characteristics.

Coherence bandwidth F_0 is inversely proportional to the maximum excess delay T_m, given by [4–6],

$$F_0 \approx \frac{1}{T_m}. \qquad (6.23)$$

The coherence bandwidth in (6.23), which is used to characterize the multipath channel in the frequency domain, is a statistical measure of the range of frequencies over which the multipath channel passes all spectral components with approximately equal gain and linear phase. In other words, the coherence bandwidth represents a frequency range over which frequency components have a strong potential for amplitude correlation and is a measure of the channel's frequency selectivity.

The coherence bandwidth F_0 can also be defined as a relation derived from the rms delay spread. If the coherence bandwidth uses the bandwidth over which the frequency correlation function is above 0.9, the coherence bandwidth is then approximately [1],

$$F_0 \approx \frac{1}{50\sigma_\tau}, \qquad (6.24)$$

where σ_τ is the rms delay spread given by (6.20). If the frequency correlation function is above 0.5, the coherence bandwidth can be approximately

$$F_0 \approx \frac{1}{5\sigma_\tau}. \qquad (6.25)$$

It should be noted that an exact relationship between the coherence bandwidth and the maximum excess delay or rms delay spread does not exist. The definition of the coherence bandwidth often differs from one reference to another [7], and tends to depend on the extent of the correlation, determined subjectively, over which the two signals are correlated.

Doppler Spread

Doppler spread D_s is a measure of spectral broadening f_d, as shown in Figure 6.3(b), caused by the time rate of change of the mobile radio channel. The Doppler spread is defined as the range of frequencies over the received Doppler power spectrum $S(v)$, with components in the range from $f_c - f_d$ to $f_c + f_d$, where f_c is the transmitted pure sinusoidal tone of frequency and f_d is the *Doppler shift*.

The Doppler power spectrum $S(v)$ was developed by Gans [8] based on Clarke's model [9]. Assume that scatterers have a uniform distribution of signals arriving at all arrival angles throughout the range $[0, 2\pi)$. The Doppler power spectrum $S(v)$ at the antenna terminals is then given by

$$S(v) = \begin{cases} \frac{3}{2\pi f_m} \left[1 - \left(\frac{v-f_c}{f_m}\right)^2\right]^{-1/2}, & |f - f_c| < f_m \\ 0, & \text{otherwise,} \end{cases} \quad (6.26)$$

where v is the Doppler frequency shift and f_m is the maximum Doppler shift given by

$$f_m = \max(|f_d|) = \max\left(\frac{V}{\lambda}|\cos\theta|\right) = \frac{V}{\lambda}, \quad (6.27)$$

where f_d can be either positive or negative depending on the arrival angle θ, V is the relative mobile velocity, and λ is the signal wavelength. Thus, in this case, the Doppler spread is obtained by

$$D_s = f_m. \quad (6.28)$$

Note that the Doppler power spectrum $S(v)$ in (6.26) is infinite when $f = f_c \pm f_m$. This indicates that the Doppler components arriving at exactly 0 and 180 degrees have an infinite Doppler power spectrum. However, this isn't a problem since the angle of arrival θ is continuously distributed and the probability of Doppler components arriving at exactly these angles is zero [1, 8, 9]. Thus, the knowledge of the Doppler power spectrum $S(v)$ in (6.26) allows us to determine how much spectral broadening is imposed on the signal as a function of the rate of change in the multipath channel state.

Coherence Time

A space-time correlation function $R(\Delta t)$ versus the time separation Δt is plotted in Figure 6.4(b). It is the correlation function between the multipath channel's response to a sinusoid signal sent at time t_1 and the response to a similar sinusoid sent at time t_2, where $\Delta t = t_2 - t_1$. *Coherence time* T_0 is a statistical measure of the time duration over which the multipath channel impulse response is invariant and quantifies the similarity of the multipath channel response at different times. This is to say that the coherence time is the time duration over which two received signals have a strong potential for amplitude correlation and the time domain dual of Doppler spread. Coherence time T_0 and Doppler spread f_m are inversely proportional to one another,

$$T_0 \approx \frac{1}{f_m}, \quad (6.29)$$

where f_m is the maximum Doppler shift defined in (6.27).

The coherence time T_0 can also be defined as the geometric mean if the coherence time is over which the time correlation function is above 0.5 [1],

$$T_0 = \sqrt{\frac{9}{16\pi f_m^2}} = \frac{0.423}{f_m}. \quad (6.30)$$

The definition in (6.30) implies that two signals arriving with a time separation Δt greater than the coherence time T_0 are affected differently by the multipath channel.

The relationship of the Doppler power spectrum $S(v)$ in Figure 6.3(b) and the space-time correlation function $R(\Delta t)$ in

Figure 6.4(b) is related by using the Fourier transform,

$$S(v) = \int_{-\infty}^{\infty} R(\Delta t) \exp(-j2\pi v \Delta t) d\Delta t. \qquad (6.31)$$

Thus, (6.31) indicates that we can obtain an equivalent description of the multipath channel in the frequency domain using its Doppler power spectrum characteristics.

Duality Concept

In both Figure 6.3 and Figure 6.4, a duality function is introduced. Two operations are called a duality function [5, 6] if the behavior of one with reference to a time-related domain (such as time or time delay) is identical to the behavior of the other with reference to the corresponding frequency-related domain (such as frequency or Doppler frequency shift).

Delay spread and coherence bandwidth, as developed earlier, are parameters of multipath channels that describe the time-dispersive nature of the multipath channels in a local area. But both of these parameters do not provide information about the time-varying nature of the multipath channels caused either by relative motion between the transmitter and receiver, or by the movement of objects in the multipath channels. On the other hand, Doppler spread and coherence time are also parameters of the multipath channels that describe the time varying nature of the multipath channel in a small-scale region. Depending on the parameters of wireless mobile multipath channels, different transmitted signals will experience different types of fading over a travel distance from the transmitter to the receiver.

6.2.5 Fading Characteristics

In a wireless mobile communication channel, in addition to the path loss as discussed previously, the received signal exhibits rapid fluctuations of the amplitude over a travel distance from the transmitter. Thus, a mobile user will usually experience signal variation in time. This phenomenon is referred to as *fading*. Fading is caused by interference between two or more versions of the transmitted signal (or multipath waves) that arrive at the receiver at slightly different times. The receiver antenna combines the multipath waves to provide a resulting signal that can vary widely in amplitude and phase, depending on the distribution of the transmitted energy, the propagation times of the radio waves, and the bandwidth of the transmitted signal.

At any time t, an instantaneously received fading signal can be expressed $s(t)$ [7]

$$s(t) = r(t)e^{j\phi(t)}, \tag{6.32}$$

where $\phi(t)$ and $r(t)$ are the phase and the envelope of the received fading signal $s(t)$. In addition, $r(t)$ can be rewritten as

$$r(t) = r_s(t)r_f(t), \tag{6.33}$$

where $r_s(t)$ is the long-term (or large-scale) fading and $r_f(t)$ is the short-term (or small-scale) fading. The *large-scale fading* represents the average signal power attenuation or the propagation path loss because of motion over large areas. The *small-scale fading* represents the dramatic changes of the amplitude and phase that can be experienced as a result of small changes in the spatial separation between a receiver and transmitter due to the multipath propagation. If no multipath fading is present, then the propagation path loss is the only major factor that must be considered in the wireless mobile communication environment. However, a mobile radio roaming over a large area must process signals that experience both types of fading with the small-scale fading superimposed on the large-scale fading.

6.2.6 Large-Scale Fading

For the large-scale fading in the wireless mobile communication environment, the mean path loss $\overline{PL}(d)$, which is a function of distance d between the transmitter and receiver, is proportionally expressed to an nth power of the distance d relative to a reference distance d_0 given by Rappaport [1],

$$\overline{PL}(d) \propto \left(\frac{d}{d_0}\right)^n. \tag{6.34}$$

The mean path loss $\overline{PL}(d)$ is often expressed in decibels, as shown here,

$$\overline{PL}(d) = \overline{PL}(d_0) + 10n \log_{10}\left(\frac{d}{d_0}\right) \quad \text{(dB)}, \qquad (6.35)$$

where n is called the path loss exponent, which means the rate at which the path loss increases with distance. The value of the path loss exponent n depends on the carrier frequency, antenna heights, and propagation environment. In free space, n is equal to 2, as seen in (6.3). When obstructions are present, the value of the path loss exponent n will be increased accordingly.

Choosing an appropriate free-space reference distance d_0 is also important for the propagation environment. Generally, the value of the reference distance d_0 is taken to be 1,000 meters for large cells, 100 meters for microcells, and 1 meter for indoor channels. The reference path loss $\overline{PL}(d_0)$ is calculated either using (6.3) or using field measurements at a distance of d_0.

The mean path loss $\overline{PL}(d)$ in (6.35) is not adequate to describe any particular setting or signal path because it does not consider the fact that the surrounding environment clutter may be greatly different at two different locations with the same distance between the transmitter and receiver. Cox et al. [10] and Bernhardt [11] showed that for any value of the distance d, the path loss $PL(d)$ is a random variable having a log-normal distribution about the mean distance-dependent value. Thus, the path loss $PL(d)$ in decibels can be expressed in terms of the mean path loss $\overline{PL}(d)$ plus a random variable X_σ as follows,

$$\begin{aligned} PL(d) &= \overline{PL}(d) + X_\sigma \\ &= \overline{PL}(d_0) + 10n \log_{10}\left(\frac{d}{d_0}\right) + X_\sigma \quad \text{(dB)}, \qquad (6.36) \end{aligned}$$

where X_σ is a zero-mean Gaussian distributed random variable with a standard deviation σ in decibels.

The log-normal distribution is that the path loss $PL(d)$ over the large-scale fading approaches a normal distribution when plotted on a logarithmic scale (in decibels). Thus, the log-normal distribution can be described by using the following probability-density function,

$$p(x) = \begin{cases} \frac{1}{\sqrt{\pi}\sigma} \exp\left[-\frac{(\log_{10} x - \mu)^2}{2\sigma^2}\right], & x > 0 \\ 0, & \text{otherwise,} \end{cases} \quad (6.37)$$

where x in decibels is a random variable representing the large-scale signal fading and μ and σ are the mean and standard deviation of x.

6.2.7 Small-Scale Fading

Small-scale fading can be referred to as the rapid fluctuation of the amplitude of a radio wave over a short period of time or travel distance. Multipath propagation in the wireless communication channel creates the small-scale fading effects. The three most important effects given by Rappaport [1] include: (1) signal strength rapidly changes over a small travel distance or time interval; (2) random frequency modulations happen because of varying Doppler shifts on different multipath signals; and (3) time dispersions are caused due to multipath propagation delays.

Rayleigh Fading

The small-scale fading is also known as *Rayleigh fading*. If the multipath reflective paths are large in number and are all NLOS, the envelope of the received signal can be statistically expressed by using a Rayleigh fading distribution. The Rayleigh fading distribution has a probability density function given by

$$p(r) = \begin{cases} \frac{r}{\sigma^2} \exp\left(-\frac{r^2}{2\sigma^2}\right), & r \geq 0 \\ 0, & \text{otherwise,} \end{cases} \quad (6.38)$$

where r is the envelope amplitude of the received signal, and σ^2 is the time-average power of the multipath signal. The corresponding cumulative distribution function, which is the probability of the envelope of the received signal not exceeding a specified value of R, is given by the expression

$$P(R) = Pr(r \leq R)$$

$$= \int_0^R p(r)dr$$
$$= 1 - \exp\left(-\frac{R^2}{2\sigma^2}\right). \quad (6.39)$$

The Rayleigh faded component is also called the *random, scatter*, or *diffuse component*.

Ricean Fading

The small-scale fading is called the *Ricean fading* when there is a dominant nonfading signal component present, such as the LOS propagation path. The Ricean fading is often observed in microcellular and mobile satellite applications

The Ricean fading distribution has a probability density function given by

$$p(r) = \begin{cases} \frac{r}{\sigma^2} \exp\left(-\frac{r^2+A^2}{2\sigma^2}\right) I_0\left(\frac{Ar}{\sigma^2}\right), & r \geq 0 \text{ and } A \geq 0 \\ 0, & r < 0, \end{cases} \quad (6.40)$$

where r and σ^2 are the envelope and average of the Ricean fading signal, respectively, A is the peak amplitude of the dominant signal, and $I_0(\cdot)$ is the modified Bessel function of zero order, which is given by Lee [7],

$$I_0(z) = \sum_{n=0}^{\infty} \frac{z^{2n}}{2^{2n} n! n!}. \quad (6.41)$$

Note that the Ricean fading distribution in (6.40) presents two extreme cases: (1) if the absence of a dominant signal $A = 0$, $p(r)$ becomes the Rayleigh fading distribution in (6.38); and (2) if the dominant signal A is large, $p(r)$ becomes a Gaussian distribution.

The Ricean fading distribution is usually expressed in terms of a parameter k that is defined as the ratio between the deterministic signal power and the variance of the multipath. The parameter k is given by

$$k = \frac{A^2}{2\sigma^2}, \quad (6.42)$$

or in decibels
$$K = 20\log_{10}\left(\frac{A}{\sqrt{2}\sigma}\right), \quad (6.43)$$

where k (or K) is known as the Ricean factor that completely specifies the Ricean fading distribution. Note that the Ricean fading distribution becomes the Rayleigh fading distribution as $A \to 0$, $K \to -\infty$ dB.

Frequency-Selective Fading

A multipath delay spread leads to *time dispersion* of the multipath channel, which causes a transmitted signal to undergo either frequency-selective or flat fading.

The transmitted signal experiences the *frequency-selective fading* if
$$B_s > F_0 \quad \text{or} \quad T_s < \sigma_\tau, \quad (6.44)$$

where B_s is the transmitted signal bandwidth, F_0 is the coherence bandwidth defined in (6.23), T_s is the symbol duration, and σ_τ is the rms delay spread given by (6.20). In other words, for frequency-selective fading, the transmitted signal has the bandwidth B_s that is greater than the coherence bandwidth F_0 of the multipath channel. In this case, the multipath channel becomes frequency selective where the channel gain is different for different frequency components. This is because the frequency-selective fading is caused by the multipath delay spread that exceeds the duration of the symbol time T_s. As time varies, the multipath channel varies in gain and phase across the spectrum of the transmitted signal, thereby resulting in time variation and ISI distortion in the received signal.

The transmitted signal experiences *flat fading* if
$$B_s \ll F_0 \quad \text{or} \quad T_s \gg \sigma_\tau. \quad (6.45)$$

This implies that if the multipath channel has a constant gain and linear phase response over a coherence bandwidth F_0, which is greater than the bandwidth of the transmitted signal B_s, the spectral characteristics of the transmitted signal through the multipath channel are then preserved at the receiver. Flat fading does not induce ISI distortion, but performance degradation can still be expected because of a loss in SNR whenever the transmitted signal is fading.

Time-Selective Fading

A Doppler spread leads to *frequency dispersion* of the multipath channel, which causes a transmitted signal to undergo either fast or slow fading. The *fast fading* is also referred to as *time-selective fading* because amplitude of the transmitted signal varies with time. Depending on how rapidly the transmitted signal changes as compared to the rate of change of the multipath channels, the multipath channels can be classified either as a fast fading or slow fading channel.

In a fast fading channel, the transmitted signal undergoes the fast fading (or time-selective fading) if

$$T_s > T_0 \quad \text{or} \quad B_s < D_s, \tag{6.46}$$

where T_s is the duration of symbol time, T_0 is the coherence time defined in (6.29), B_s is the transmitted signal bandwidth, and D_s is the Doppler spread that is given by (6.28). Equation (6.46) indicates that the coherence time of the multipath channel is smaller than the symbol duration of the transmitted signal. This causes frequency dispersion because of the Doppler spread, thereby leading to signal distortion at the receiver. The signal distortion of the fast fading increases if the Doppler spread relative to the bandwidth of the transmitted signal increases.

In a slow fading channel, the impulse response of the multipath channel changes much more slowly than the transmitted signal. This indicates that the Doppler spread of the multipath channel is much less than the bandwidth of the transmitted signal in the frequency domain. Thus, the transmitted signal experiences the slow fading if

$$T_s \ll T_0 \quad \text{or} \quad B_s \gg D_s. \tag{6.47}$$

As can be seen, (6.46) and (6.47) indicate that the velocity of mobile and the transmitted signal in the multipath channel determine whether the transmitted signal undergoes fast or slow fading.

6.3 Wired Channels

One of the most commonly used wired channels for data transmission has been the transmission line composed of a pair of wires (twisted wires) or a coaxial cable. The coaxial cable is traditionally used for digital communication inside a building, and for high capacity long-distance facilities in the telephone network. The pair of wires is used for connection of customer premises equipment (CPE) at home to a central office (CO). Broadband access approaches have been developed to provide a very high data rate over the pair of wires. These broadband access approaches are commonly known as DSL.

During the last decade, DSL technologies have been an attractive broadband access service for residential and small business areas. Several DSL standards, including HDSL, SHDSL, ADSL, and even VDSL, have been successfully established with tens of millions of customers throughout the world. Currently, the most popular asymmetric (ADSL) service reliably delivers up to 8 Mbps downstream and 800 Kbps upstream, depending on the distance of the CPE from the CO. However, the ADSL technology does not support long ranges extended to 18,000 feet and even longer, which cover up to 90% of telephone customers. The need for a long-range extended DSL (LDSL) with capabilities of transmitting the minimum data rate over 200 Kbps is already evident due to the increasing demands of the customers imposed by the proliferation of long-reach services [12]. However, in any case, to transmit higher data rates over a longer distance from CPE to CO, DSL technologies face challenges due to the wired channel of wire pairs with serious distortions.

6.3.1 Transmission Loop

A transmission line comprising in-line sections, bridge taps, a drop wire, and lumped elements can be analyzed by using a *two-port*

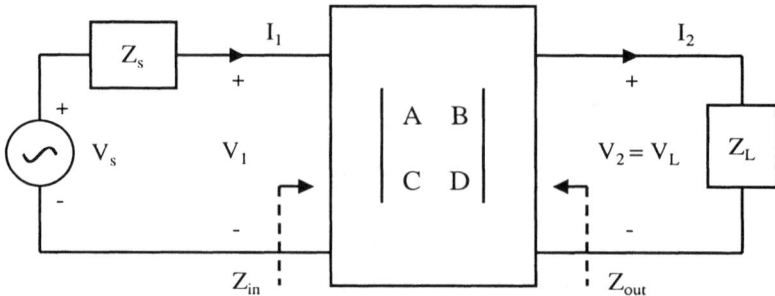

Figure 6.5 Block diagram of a two-port network with a definition of voltages and currents for the chain matrix.

network via its *chain matrix* [13–17]. The two-port network has an input port and output port for which the input and output currents are complementary, as shown in Figure 6.5. The matrix that relates the input voltage and current to the output voltage and current can be expressed by

$$\begin{bmatrix} V_1 \\ I_1 \end{bmatrix} = \begin{bmatrix} A & B \\ C & D \end{bmatrix} \begin{bmatrix} V_2 \\ I_2 \end{bmatrix}, \qquad (6.48)$$

where all quantities are complex functions of frequency, A denotes the open-circuit transfer funcion, B denotes the open-circuit transfer admittance, C denotes the short-circuit transfer impedance, and D denotes the short-circuit current ratio. Equation (6.48) illustrates the two-port transfer function completely and can be used to analyze cascade connections of the two-port network. Using the voltage source V_s and load voltage V_L as shown in Figure 6.5, the voltage transfer function $H(f)$ is obtained by

$$H(f) = \frac{V_L}{V_s} = \frac{V_2}{V_s}. \qquad (6.49)$$

Using (6.48) and the parameters in Figure 6.5, we obtain

$$\frac{V_2}{V_1} = \frac{Z_L}{B + AZ_L}, \qquad (6.50)$$

and
$$V_1 = I_1 Z_s + V_s = \frac{V_1}{Z_{in}} Z_s + V_s, \qquad (6.51)$$

where Z_{in} is the *input impedance* given by

$$Z_{in} = \frac{V_1}{I_1} = \frac{AZ_L + B}{CZ_L + D}. \qquad (6.52)$$

From (6.51), we then obtain,

$$V_1 = \left(\frac{Z_{in}}{Z_{in} - Z_s} \right) V_s. \qquad (6.53)$$

Substituting V_1 of (6.53) into (6.50) obtains the voltage transfer function,

$$H(f) = \frac{Z_{in} Z_L}{(B + AZ_L)(Z_{in} - Z_s)}. \qquad (6.54)$$

Substituting (6.52) into (6.54) yields the voltage transfer function as follows,

$$H(f) = \frac{Z_L}{AZ_L + B - (CZ_L + D)Z_s}. \qquad (6.55)$$

The output impedance is obtained by

$$Z_{out} = \frac{V_2}{I_2} = \frac{DZ_s + B}{CZ_s + A}. \qquad (6.56)$$

Generally, a pair of twisted wire loops of the transmission line contains several loop segments, each having a particular loop type or gauge. Each of the loop segments is a transmission line that can be viewed as a cascade combination of two-port networks. Figure 6.6 shows a typical end-to-end connection, which includes in-line sections, bridge taps, a drop wire, and lumped elements, with the information of the loop segment lengths and gauges. In case of the cascade combination of N two-port networks, the $ABCD$ matrix in (6.48) can be expressed as

$$\begin{bmatrix} A & B \\ C & D \end{bmatrix} = \prod_{i=1}^{N} \begin{bmatrix} A_i & B_i \\ C_i & D_i \end{bmatrix}. \qquad (6.57)$$

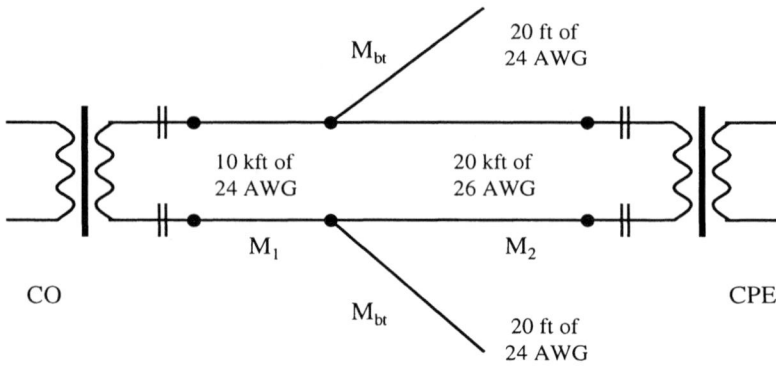

Figure 6.6 A typical end-to-end connection with in-line sections, bridge taps, and lumped elements.

In-Line Two-Port Networks

An in-line $ABCD$ matrix of the two-port network with the loop length L_i (feet) is given by Lee and Messerschmitt [15],

$$M_i = \begin{bmatrix} A_i & B_i \\ C_i & D_i \end{bmatrix}$$
$$= \begin{bmatrix} \cosh(\gamma_i L_i) & Z_{0i} \sinh(\gamma_i L_i) \\ Y_{0i} \sinh(\gamma_i L_i) & \cosh(\gamma_i L_i) \end{bmatrix}, \quad (6.58)$$

where Z_{0i} is the impedance given by

$$Z_{0i} = \sqrt{\frac{R_i + j\omega L_i}{G_i + j\omega C_i}}, \quad (6.59)$$

Y_{0i} is the *admittance*, which is the reciprocal of the impedance given by $Y_{0i} = \frac{1}{Z_{0i}}$, and

$$\gamma_i = \sqrt{(R_i + j\omega L_i)(G_i + j\omega C_i)}, \quad (6.60)$$

where R_i, L_i, G_i, and C_i, are the resistance, inductance, conductance, and capacitance for ith in-line loop segment, respectively.

Given the $RLGC$ parameters, the $ABCD$ matrix can be determined. Thus, the $RLGC$ is important to designers. Usually, the $RLGC$ parameters are found from field or laboratory tests, depending on various loop types and gauges.

Bridge Taps

Bridge taps present their open-circuit admittance in the shunt between two in-line two-port networks as shown in Figure 6.6. For this case, the $ABCD$ matrix of a bridge tap is given by Bingham [13],

$$M_{bt} = \begin{bmatrix} A_i & B_i \\ C_i & D_i \end{bmatrix}$$
$$= \begin{bmatrix} 1 & 0 \\ Y_{0bt}\tanh(\gamma_{bt}L_{bt}) & 1 \end{bmatrix}. \quad (6.61)$$

Note that the bridge tap represents an open circuit of the wires that has a minima of its input impedance to generate a notch in the end-to-end transfer function at a frequency for which the length segment is an odd number of quarter wavelengths. Werner [16] introduces the condition for the first notch as follows,

$$f_{notch} = \frac{v(f_{notch})}{4L_{bt}}, \quad (6.62)$$

where L_{bt} is the length of the bridge tap expressed in feet and $v(f_{notch})$ is the propagation velocity of wires over a DSL frequency range. Other notches occur at frequencies that are equal to $(2k+1)f_{notch}$, where $k = 1, 2, 3, \cdots$. For loops with several bridge taps, the location of the notches can be determined by using superposition heuristically.

In a case of unknown propagation velocity $v(f_{notch})$, the first notch in the loop's transfer function is approximately located at

$$f_{notch} \approx \frac{150}{L_{bt}/1,000} \text{ (kHz)}. \quad (6.63)$$

The signal propagates down the bridge tap and reflects back. The attenuation of the signal is approximately given by [13],

$$A_{att} \approx \alpha 2 L_{bt} \sqrt{f_{notch}}, \qquad (6.64)$$

where $\alpha \approx 6.2$ (dB/kft/$\sqrt{\text{MHz}}$) for a 24 AWG gauge and $\alpha \approx 7.8$ (dB/kft/$\sqrt{\text{MHz}}$) for a 26 AWG gauge.

Cascade Combination of the Loop

For the end-to-end loop as shown in Figure 6.6, the chain $ABCD$ matrix of cascade combination of the loop is given by (6.57)

$$M = \begin{bmatrix} A & B \\ C & D \end{bmatrix} = M_1 M_{bt} M_2. \qquad (6.65)$$

Thus, based on (6.65), a voltage transfer function for the end-to-end loop can be obtained by using (6.55).

6.3.2 Crosstalk

Crosstalk between pairs in a multipair cable as illustrated in Figure 6.7 is the dominant impairment in any type of DSL system. The cause of the crosstalk is capacitive and inductive coupling between the wires due to imbalance in the couplings. A precise knowledge of individual pair-to-pair crosstalk transfer functions will be needed in order to implement crosstalk cancellation.

NEXT

Assume that one pair j as shown in Figure 6.7 is considered as the interferer and the voltages and currents induced in the other pair i travel in both directions. Thus, the signals, which come back toward the source of the interferer, add up to form crosstalk. This crosstalk is referred to as *near-end crosstalk* (NEXT). The NEXT represents a crosstalk of a local transmitter into a local receiver and experiences attenuation. The NEXT power transfer function can be written as [13, 15, 16, 18, 20]

$$|H_{NEXT}(f, N)|^2 = K_{NEXT} |f|^{3/2} N^{0.6}, \qquad (6.66)$$

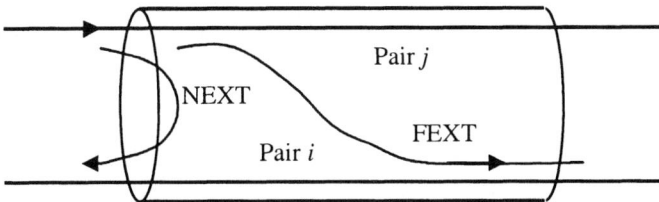

Figure 6.7 Crosstalk of the NEXT and FEXT generations in a multipair cable.

where K_{NEXT} is the aggregate coupling coefficient that will be different for various standards and N is the total number of NEXT disturbers.

FEXT

If one pair j is the interferer and the voltages and currents induced in the other pair i travel in both directions, the signals that continue in the same direction as the interfering signal add up to form crosstalk. The crosstalk is known as *far-end crosstalk* (FEXT). The FEXT represents a crosstalk of a local transmitter into a remote receiver, and also experiences attenuation. The FEXT power transfer function is given by

$$|H_{FEXT}(f, L, N)|^2 = K_{FEXT} L \, |L(f)|^2 \, |f|^2 N^{0.6}, \qquad (6.67)$$

where K_{FEXT} is the aggregate coupling coefficient that will be different for every pair of wire pairs, L is the length of the loop in feet, N is the total number of FEXT disturbers, and $L(f)$ is the insertion loss of the loop through which the interferer passes while the interferer and the signal are adjacent in the same binder.

6.3.3 Simulation Loop Model

A simulation loop model defines a function diagram of the end-to-end loop with composite impairment noise as shown in Figure 6.8.

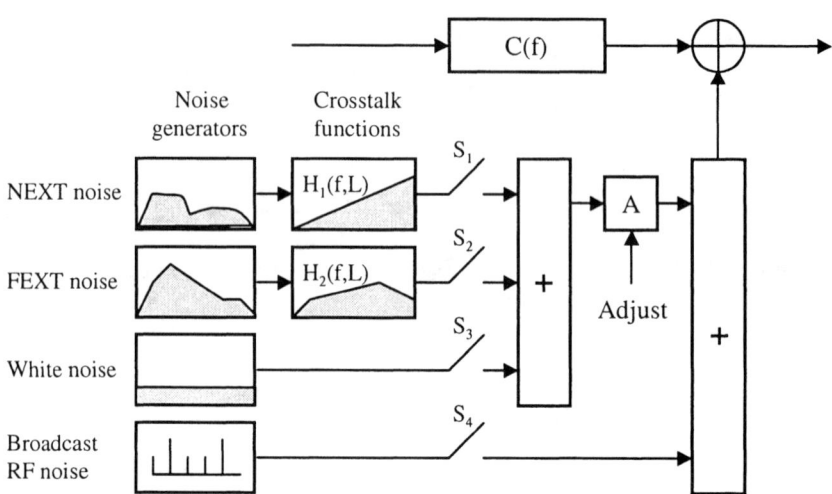

Figure 6.8 A simulation loop model with crosstalk, white, and broadcast RF noise.

The simulation loop model provides a functional description of the combined end-to-end loop with impairment noise that must be probed at the CPE receiver input of DSL transceiver.

The functional diagram of the simulation loop model in Figure 6.8 has the following elements. (1) The four impairment noise generators, including the NEXT, FEXT, white, and broadcast radio frequency (RF) noise, produce noise for the simulation loop model. (2) The transfer functions $H_1(f, L)$ and $H_2(f, L)$ describe the length and frequency depending on the NEXT and FEXT impairment, respectively. These transfer functions are independent of the simulation loop models, but can change with the electrical length of the simulation loop models. (3) Switches S_1 to S_4 are used to determine whether a specific impairment generator contributes to the total impairment during a simulation. (4) An amplifier A is adjustable and can model the property of increasing the level of the NEXT, FEXT, and white noise to perform the noise margin simulation. Thus, the simulation loop models allow designers to test performance of various DSL technologies based on loop topology.

Note that depending on different DSL standards, such as SHDSL, ADSL, and VDSL [18–20], the simulation loop models are not the same for different loop topology. In addition, the composite impairment noise, including NEXT, FEXT, broadcast RF, and background, which is injected into the simulation loop models, is also different from each other in the various DSL standards. Therefore, simulating the specific loop models requires referring to the DSL standards.

6.4 Channel Distortion

Many physical communications channels, including telephone and mobile radio multipath channels, accept a continuous-time waveform as the input signal. Consequently, a sequence of source bits that represent data or a digitized analog signal must be converted to a continuous-time waveform at the transmitter. Such channels may be generally characterized as bandlimited linear filters and are expressed by their frequency response $C(f)$ as

$$C(f) = A(f)e^{j\phi(f)}, \qquad (6.68)$$

where $A(f)$ is called *amplitude response* and $\phi(f)$ is called *phase response*. Thus, *envelope delay* or *group delay* is defined by

$$\tau(f) = -\frac{1}{2\pi}\left[\frac{d\phi(f)}{df}\right]. \qquad (6.69)$$

Notice that a physical channel is ideal or nondistortion if the amplitude response $A(f)$ is a constant in (6.68) and the envelope delay $\tau(f)$ is a constant in (6.69) or the phase response $\phi(f)$ is a linear function of frequency given channel bandwidth W occupied by the transmitted signal. On the other hand, the physical channel distorts the signal within the channel bandwidth W occupied by the transmitted signal if the amplitude response $A(f)$ and the envelope delay $\tau(f)$ are not constant. Thus, a distortion is often referred to as *amplitude distortion* if $A(f)$ is not a constant and the distortion is called *delay distortion* if $\tau(f)$ is not a constant. Therefore, a succession of pulses transmitted through the physical channel at rates comparable to the channel bandwidth is smeared to positions that are no longer distinguishable as well-defined transmitted pulses in the communication receiver if there are amplitude $A(f)$ and delay distortion $\phi(f)$ caused by a nonideal channel frequency response characteristic $C(f)$. Instead, those transmitted pulses overlap, resulting in ISI.

6.4.1 Intersymbol Interference

In this section, we consider the transmission of a pulse amplitude modulation (PAM) signal through a communication system including transmitter filter $H_T(f)$, channel $C(f)$, receiver filter $H_R(f)$, and sampling, as shown in Figure 6.9. The message bits $\{b_i\}$ are mapped to a sequence of amplitude level A_i as the input to transmitter pulse-shaping filter $h_T(t)$, where $h_T(t)$ represents the impulse response of the transmitter pulse-shaping filter $H_R(f)$. The input signal to the transmitter pulse-shaping filter $H_T(f)$ is the modulated sequence of delta functions that is expressed as

$$\{A_i\} = \sum_{i=0}^{N-1} A_i \delta(t - iT), \quad (6.70)$$

where $1/T$ is the symbol rate per second. Thus, the transmitted signal $q(t)$ can be obtained by

$$q(t) = \sum_{i=0}^{N-1} A_i h_T(t - iT). \quad (6.71)$$

Assume that $H(f)$ is the overall transfer function of the combined transmitter, channel, and receiver. Then, the overall transfer function $H(f)$ is expressed as

$$H(f) = H_T(f) C(f) H_R(f). \quad (6.72)$$

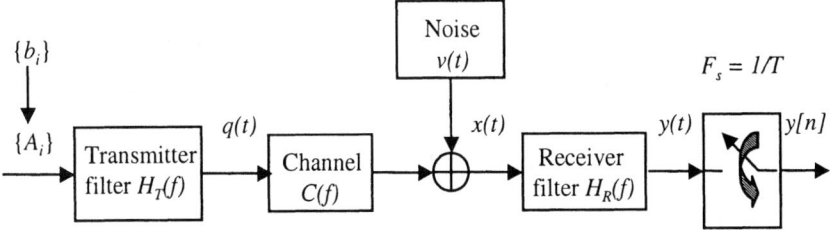

Figure 6.9 Block diagram of baseband model of a PAM system.

Corresponding of the impulse response in (6.72) is obtained by

$$h(t) = h_T(t) * C(t) * h_R(t). \qquad (6.73)$$

Then, the output of the receiver filter in Figure 6.9 is

$$y(t) = \sum_{i=0}^{N-1} A_i h(t - iT) + h_R(t) * v(t), \qquad (6.74)$$

where the second term of $h_R(t) * v(t)$ is a convolution output of the receiver filter $h_R(t)$ with an input noise $v(t)$. The discrete-time samples are produced at the output of the sampling function at the symbol rate of $1/T$. Thus, (6.74) can be rewritten at the kth sample of $y(t)$ as

$$\begin{aligned} y[n] &= y(kT) \\ &= \sum_{i=0}^{N-1} A_i h(kT - iT) + h_R(kT) * v(kT) \\ &= A_k h[0] + \sum_{i=1, i \neq k}^{N-1} A_i h(kT - iT) \\ &\quad + h_R(kT) * v(kT). \end{aligned} \qquad (6.75)$$

Notice that the first term on the right side of (6.75) is the kth transmitted symbol scaled by the overall system impulse response

$h[0]$ at $t = 0$. This term is the source bits without error by multiplying the received samples of $1/h[0]$ if no other terms are presented at the right side of (6.75). However, this is impossible in practical applications because other terms cannot be eliminated without a special device in the receiver. The second term in (6.75) is referred to as the ISI that represents the neighboring symbols interfering with the detection of each desired symbol. The third term in (6.75) denotes noise intensity because of the input noise interference with the channel.

6.4.2 Eye Diagrams

In a digital communications system, the amount of the ISI and noise intensity in (6.75) can be displayed on an oscilloscope. The received PAM signal $y(t)$ in (6.74) can be displayed on the vertical input with the horizontal sweep rate at $1/T$. The resulting display is called an *eye diagram*. For an illustration, Figure 6.10 shows the eye diagrams based on binary and quaternary PAM modulation. The effect of the ISI in the system is to cause the eye diagram to close, thereby reducing the margin for additive noise to cause errors. The effect of the ISI on reducing the opening of a binary eye diagram is graphically illustrated in Figure 6.11. There are four types of distortions including *zero crossing*, *noise margin*, *peak*, and sensitivity to *timing error*.

The vertical opening in Figure 6.11 is the separation between signal levels with worst-case ISI. In some cases, the vertical opening disappears altogether if the ISI and noise are large enough. In this case, the eye diagram is called *eye-closed*. Otherwise, the eye diagram is referred to as *eye-open*. An eye-closed means that the bit decisions are no longer sure and some fraction of these is wrong. In the limit, this leads to a probability of error close to 0.5. On the other hand, the wide eye-open in the vertical spacing between signal levels implies a large degree of immunity to additive noise. In general, the location of *optimum sampling* should be placed at the time instants $kT + t_0$, $k = 0, 1, 2, \cdots, N$, where t_0 is selected such that the maximum of the vertical eye-open can be obtained.

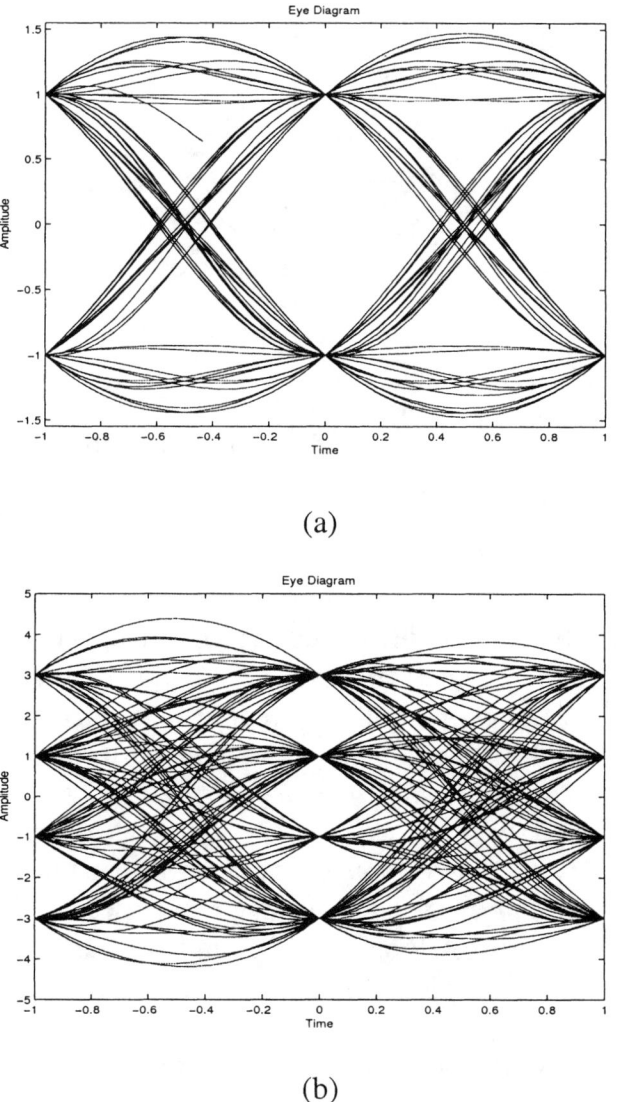

Figure 6.10 Eye diagrams: (a) a binary PAM and (b) a quaternary PAM.

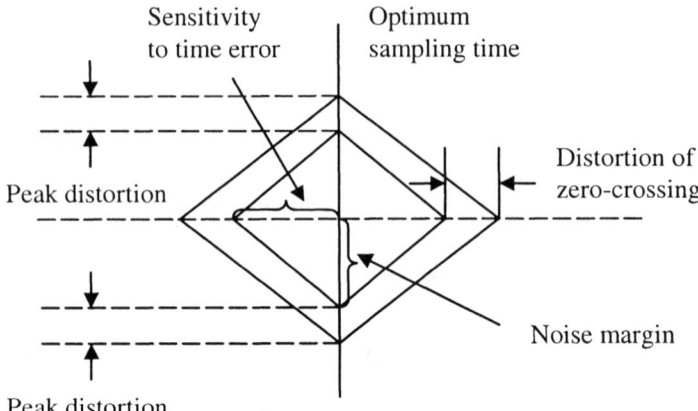

Figure 6.11 Diagram of the intersymbol interference effect on an eye pattern.

The width of the horizontal eye opening indicates how much a communication system is sensitive to the *timing offset*, as illustrated in Figure 6.11. A very narrow eye opening means that a small timing offset will result in sampling, while a wide horizontal eye opening means that a large timing offset can be tolerated.

The slope of the inner eye opening in Figure 6.11 also indicates the sensitivity to *timing jitter* or variance in the timing offset. A very steep slope means that the eye diagram closes rapidly when the timing offset increases. As a result, a large amount of timing jitter in the sampling times would significantly increase the probability of error in the receiver.

The ISI distorts the position of the zero crossing and causes a reduction in the eye opening. Therefore, the ISI causes the PAM communication system to be more sensitive to a synchronization error.

For phase-shift keying (PSK), quadrature amplitude modulation (QAM), or quadrature phase-shift keying (QPSK) signals, a useful

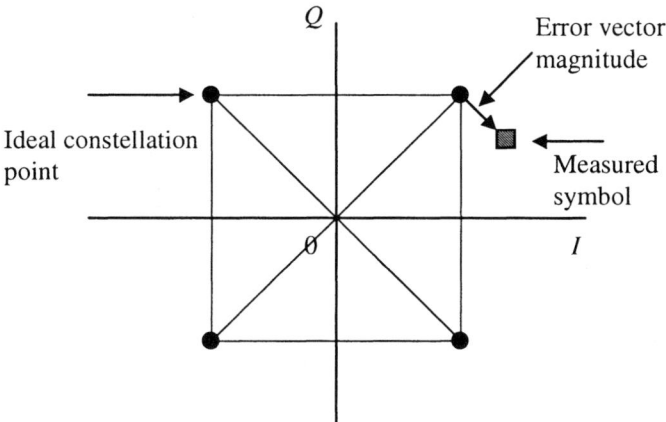

Figure 6.12 Eye diagram of a four-phase QPSK constellation error vector magnitude.

display is a two-dimensional scatter plot of the sampled values $y(kT)$, which represent the decision variables at the sampling instants. Figure 6.12 shows a constellation diagram for a four-phase QPSK constellation. Without the ISI and noise, the superimposed signals at the sampling instants would result in four distinct points corresponding to the four transmitted signal phases. In the case of the ISI and noise, the result is a deviation of the received samples $y(kT)$ from the desired four-phase QPSK signal. As a result, the larger the ISI and noise in a communication system, the larger the scattering of the received signal relative to the transmitted signal points.

Error vector magnitude (EVM) [21, 22] is used to measure the transmitted modulation accuracy between the difference of the actual transmitted waveform and the ideal signal waveform for PSK, QAM, or QPSK modulation. For example, the ideal complex I and Q constellation points associated with QPSK modulation shall be used as the reference as shown in Figure 6.12. The average magnitude of I and Q samples is given by

$$I_{mag} = \sum_{i=0}^{N-1} \frac{|I[n] - I_{mean}|}{N}, \qquad (6.76)$$

and

$$Q_{mag} = \sum_{i=0}^{N-1} \frac{|Q[n] - Q_{mean}|}{N}, \qquad (6.77)$$

where the DC offset for I samples is defined by

$$I_{mean} = \sum_{i=0}^{N-1} \frac{|I[n]|}{N}, \qquad (6.78)$$

and the DC offset for Q samples is defined as

$$Q_{mean} = \sum_{i=0}^{N-1} \frac{|Q[n]|}{N}. \qquad (6.79)$$

The normalized EVM (NEVM) for I and Q pairs is then obtained by

$$V_{NEVM}[n] = \sqrt{\frac{1}{2}\left[\left(\frac{I[n] - I_{mean}}{I_{mag}}\right)^2 + \left(\frac{Q[n] - Q_{mean}}{Q_{mag}}\right)^2\right]}. \qquad (6.80)$$

In practice, the ISI and noise based on a reference receiver system can be accepted for a QAM, PSK, or QPSK communication system if the following condition of the normalized EVM is satisfied $V_{NEVM}[n] < 0.35$.

6.4.3 Nyquist Criterion

In order to eliminate the ISI in (6.75), one of the possible solutions is to choose the transmitter filter $h_T(kT)$ and receiver filter $h_R(kT)$ such that

$$h(kT) = \begin{cases} 1, & k = 0 \\ 0, & k \neq 0. \end{cases} \qquad (6.81)$$

Then, the kth received sample in (6.75) is obtained as

$$y(kT) = A_k + h_R(kT) * v(kT). \tag{6.82}$$

Therefore, in this case, (6.82) indicates that the ISI has been completely eliminated. As a result, the selection of the transmitter filter $h_T(kT)$ and receiver filter $h_R(kT)$ in (6.81) is called a *zero-forcing* solution because it forces the ISI to zero. However, a zero-forcing solution may not be an optimal solution depending on the type of detection scheme used [6]. This is because the probability of error may also depend on the second term of noise intensity in (6.82) that generally increases when the ISI is completely eliminated.

Nyquist [23] stated that the ISI could be completely cancelled if the overall response of the communication system is designed such that the response due to all symbols except the current symbol is equal to zero at every sampling instant at the receiver. That is, if $H_{eff}(f)$ and $h_{eff}(t)$ are the transfer function and impulse response of the overall communication system given by (6.72) and (6.73), respectively, this condition is mathematically stated as

$$h_{eff}(kT) = \begin{cases} K, & k = 0 \\ 0, & k \neq 0, \end{cases} \tag{6.83}$$

where T is the symbol period, K is a nonzero constant, and k is the integer. The corresponding discrete Fourier transform of the impulse response $h_{eff}(kT)$ is expressed as $H_{eff}(e^{j2\pi fT})$. Thus, the condition in (6.83) is equivalent to the frequency-domain condition as follows,

$$H_{eff}(e^{j2\pi fT}) = K. \tag{6.84}$$

The relation of either (6.83) or (6.84) is known as the *Nyquist criterion*.

In order to satisfy the Nyquist criterion, the channel bandwidth W must be at least equal to $\frac{1}{2T}$. For the minimum channel bandwidth $W = \frac{1}{2T}$, (6.84) can be rewritten as

$$H_{eff}(e^{j2\pi fT}) = \begin{cases} T, & |f| < \frac{1}{2T} \\ 0, & \text{otherwise}. \end{cases} \tag{6.85}$$

Equation (6.85) can be solved to obtain the corresponding impulse response $h_{eff}(t)$ by using the inverse discrete Fourier transform given by Miao and Clements [24]

$$h_{eff}(t) = \frac{\sin(\pi t/T)}{\pi t/T}, \qquad (6.86)$$

which is called the minimum bandwidth Nyquist pulse. The frequency band of $|f| < \frac{1}{2T}$ in (6.85) is called the *Nyquist bandwidth*.

Note that the transfer function in (6.85) satisfies the zero ISI with a minimum of bandwidth $\frac{1}{2T}$, but it corresponds to a rectangular filter. The rectangular filter cannot be implemented in practice since it corresponds to a noncausal system, that is, $h_{eff}(t)$ exists for $t < 0$.

Nyquist also stated that the transfer function in (6.85) convolved with any arbitrary even function $Q(f)$ with zero magnitude outside the passband of the rectangular filter results in satisfying the zero ISI condition. This can be mathematically expressed as

$$H_{com}(e^{2j\pi fT}) = \frac{1}{T}\left[H_{eff}(e^{2j\pi fT})Q(e^{2j\pi fT})\right], \qquad (6.87)$$

where $Q(e^{2j\pi fT}) = Q(e^{-2j\pi fT})$ and $Q(e^{2j\pi fT}) = 0$ for $|f| \geq \frac{1}{2T}$. This implies that the corresponding impulse response of the transfer function in (6.87) is given by

$$h_{com}(t) = \left[\frac{\sin(\pi t/T)}{\pi t}\right]q(t). \qquad (6.88)$$

The Nyquist criterion states that any filter having the form in (6.88) can also eliminate ISI. A filter that satisfies the Nyquist criterion is referred to as the *Nyquist filter*.

6.5 Pulse Shaping

Communication systems, which operate with the minimum bandwidth, have not only ISI effects but also out-of-band radiation. Thus, it is highly desirable to reduce modulation bandwidth and suppress the out-of-band radiation while eliminating the ISI simultaneously. Pulse shaping is one of the techniques that is used to reduce the ISI effects and spectral width of a modulated signal.

6.5.1 Raised-Cosine Pulse

In practice, given the transmitter filter $H_T(f)$, receiver filter $H_R(f)$, and channel response $C(f)$, the cascade $H_T(f)C(f)H_R(f)$ is designed to yield zero ISI. Assume that the channel is an ideal. Thus, the cascade filter of $H_T(f)C(f)H_R(f)$ may be selected such that

$$H_T(f)C(f)H_R(f) = H_{RC}(f), \quad (6.89)$$

where $H_{RC}(f)$ is a raised-cosine frequency response given by

$$H_{RC}(f) = \begin{cases} T, & 0 \leq |f| \leq \frac{(1-\alpha)}{2T} \\ \frac{T}{2}\left\{1 - \sin\left[\frac{\pi T}{\alpha}\left(|f| - \frac{1}{2T}\right)\right]\right\}, & \frac{(1-\alpha)}{2T} \leq |f| \leq \frac{(1+\alpha)}{2T} \\ 0, & \text{otherwise,} \end{cases}$$
(6.90)

where α is called the *rolloff factor* and $0 \leq \alpha \leq 1$, and $1/T$ is the symbol rate. The frequency of $1/2T$ is known as the *Nyquist frequency*. The raised-cosine frequency response $H_{RC}(f)$ in (6.90) becomes an ideal brick wall filter with bandwidth frequency occupancy at $1/2T$ when $\alpha = 0$. When $\alpha > 0$, the frequency bandwidth of the raised-cosine frequency response $H_{RC}(f)$ beyond the Nyquist frequency $1/2T$ is referred to as the *excess bandwidth*. For example, the excess bandwidth of the raised-cosine frequency response $H_{RC}(f)$ is 50% when $\alpha = 1/2$. The excess bandwidth of the raised-cosine frequency response $H_{RC}(f)$ is 100% when $\alpha = 1$.

The corresponding time-domain raised-cosine pulse of the raised-cosine frequency response in (6.90) is given by

$$h_{RC}(t) = \left[\frac{\sin(\pi t/T)}{\pi t/T}\right]\left[\frac{\cos(\alpha \pi/T)}{1-(2\alpha t/T)^2}\right]. \quad (6.91)$$

Note that there is no ISI from adjacent symbols at the sampling instants $t = kT$ for $k \neq 0$ when the channel does not have distortion.

In this case, the receiver implements the same filter as

$$\begin{aligned} H_R(f) &= H_T(f) \\ &= \sqrt{H_{RC}(f)}. \end{aligned} \quad (6.92)$$

Equation (6.92) is referred to as a *square root raised cosine* (SRRC) pulse. The time-domain SRRC corresponding to $\sqrt{H_{RC}(f)}$ can be obtained by taking the inverse Fourier transform, thereby resulting in

$$h_T(t) = \frac{4\alpha \cos[(1+\alpha)\pi t/T] + \sin[(1-\alpha)\pi t/T](t/T)^{-1}}{\pi\sqrt{T}[1 - (4\alpha t/T)^2]}. \quad (6.93)$$

The SRRC in (6.93) is not a causal system. Therefore, in order to design the causal system of the SRRC transmit and receiver filter, a truncated time domain and time-shifted pulse is used [2]. On the other hand, the SRRC filter can also be designed in the frequency domain $\sqrt{H_{RC}(f)}$ based on sampling technology given by [25, 26]. This method results in the lengths of SRRC filter coefficients being significantly less than the SRRC filter coefficients obtained by the truncated time-domain method.

6.5.2 Gaussian Shaping Pulse

A Gaussian shaped pulse like the raised-cosine pulse has a transfer function, but with no zero crossings. It has the transfer function given by

$$H_G(f) = \exp\left[-(\alpha f)^2\right], \quad (6.94)$$

where

$$\alpha = \frac{\sqrt{2\ln 2}}{B}, \quad (6.95)$$

and B is the 3-dB bandwidth of the baseband shaping filter. The transfer function $H_G(f)$ in (6.94) is a bell shape and symmetric at $f = 0$.

The corresponding impulse response is given by [1]

$$h_G(t) = \frac{\sqrt{\pi}}{\alpha} \exp\left[-\left(\frac{\pi t}{\alpha}\right)^2\right]. \quad (6.96)$$

The spectral occupancy of the Gaussian shaping pulse decreases and its time dispersion of the applied signal increases when α increases.

The Gaussian shape pulse has a narrow absolute bandwidth, sharp cut off, low overshoot, and a preservation property of pulse area, but it does not satisfy the Nyquist criterion for ISI cancellation. Reducing its spectral occupancy may create degradation in performance because of increased ISI. This leads to a trade-off design in the choice of α for mobile communication systems given by [27].

6.6 Matched Filtering

Consider the transmission of an isolated pulse $A_0\delta(t)$ through the communication system shown in Figure 6.9. Also assume that the noise $v(t)$ is white with a spectral density of $N_0/2$. Thus, the input signal to the receiver filter $H_R(f)$ can be expressed as

$$x(t) = A_0 h_{TC}(t) + v(t), \qquad (6.97)$$

where $h_{TC}(t)$ is the inverse Fourier transform of the overall transfer function of the combined transmitter filter and channel,

$$H_{TC}(f) = H_T(f) H_C(f). \qquad (6.98)$$

Then, the output of the receiver filter is

$$\begin{aligned} y(t) &= h_R(t) * x(t) \\ &= A_0[h_R(t) * h_{TC}(t)] + h_R(t) * v(t). \end{aligned} \qquad (6.99)$$

It should be noted that the first term and the second term on the right side of (6.99) are the desired signal and the noise, respectively. The SNR at the sampling instant is therefore obtained by

$$\text{SNR} = \frac{E\{|A_0|^2\} \left| \int_{-\infty}^{\infty} h_R(-t) * h_{TC}(t) dt \right|^2}{\frac{N_0}{2} \int_{-\infty}^{\infty} |h_R(t)|^2 dt}. \qquad (6.100)$$

The receiver filter $h_R(t)$ is called an *optimal matched filter* if the maximum of SNR in (6.100) can be achieved. In this case, to maximize this SNR expression, the receiver impulse response is $h_R(t) = h_{TC}^*(-t)$ and its corresponding transfer function is $H_R(f) = H_{TC}^*(f)$.

The matched filter that is derived for the isolated pulse case does not take into account ISI. In the presence of channel distortion due to the ISI and noise, the ISI given in the second term of (6.75) is no longer zero even by using the ideal SRRC transmitter and receiver filter. Thus, an appropriate front end of the communication receiver needs to consist of the matched filter along with further signal processing technology, such as a channel equalizer, for improved performance. Hence, the matched filter preceded by a channel equalizer in the communication receiver is usually needed to minimize its effect on communication system performance. However, designing the channel equalizer usually requires one to know the characterization of channel distortion. Therefore, the channel distortion needs to be estimated either by sending a training sequence or by using a blind identification method without the training sequence.

6.7 Summary

In this chapter, channel characterization and distortion were introduced. We first presented the characteristics of wireless channels with focusing on multipath propagation and fading, and the characteristics of wired channels with emphases on the topology of transmission loop and crosstalk. Second, we described the fundamental theory of channel distortion, which generates ISI effects because of bandlimited channel bandwidth, and introduced the Nyquist criterion for ISI cancellation. We then brought in pulse shaping techniques for eliminating the ISI and suppressing out-of-band spectral width at the transmitter, and derived the optimal matched filter for the receiver.

Wireless channels experience multipath propagation due to reflection, diffraction, and/or scattering of radiated energy off of objects located in the environment. Signals at the receiver are much feebler than transmitted signals because of propagation path loss. In addition, received signals may display fading over traveling distance from the transmitter. The fading includes large-scale fading and small-scale fading. The large-scale fading represents the average signal power attenuation or the propagation path loss because of motion over large areas, while the small-scale fading represents the dramatic changes of the amplitude and phase because of the multipath propagation over a spatial separation between the receiver and transmitter. The small-scale fading is Rayleigh fading if multipath reflective paths are large in number and there is an NLOS signal component. When there is a dominant nonfading signal component present, such as an LOS propagation path, the small-scale fading is referred to as Ricean fading. In any case, a mobile radio roaming over a large area has to process the received signals, which experience both types of fading with the small-scale fading superimposed on the large-scale fading.

Delay spread leads to time dispersion of the multipath channel, which makes a transmitted signal undergo either frequency-selective or flat fading. The frequency-selective fading is caused by the delay spread, which exceeds the duration of symbol time. In the frequency domain, the bandwidth of the transmitted signal is greater than the coherence bandwidth of the multipath channel. This results in the transmitted signal becoming time varying and hence ISI distortion at the receiver. Flat fading happens if the multipath channel has a constant gain and linear phase response over the coherence bandwidth, which is greater than the bandwidth of the transmitted signal. In this case, the spectral characteristics of the transmitted signal through the multipath channel are preserved at the receiver. Flat fading does not create ISI distortion, but performance degradation is expected because of loss in SNR whenever the transmitted signal is fading.

Doppler spread leads to frequency dispersion of the multipath channel that causes the transmitted signal to experience either fast fading (time-selective fading) or slow fading. The transmitted signal undergoes fast fading if coherence time of the multipath channel is smaller than the symbol duration of the transmitted signal. On the other hand, the transmitted signal experiences the slow fading if the coherence time of the multipath channel changes much slower than the symbol duration of the transmitted signal. Hence, the velocity of mobile radio and the transmitted signal in the multipath channel determine whether the transmitted signal undergoes the fast or slow fading.

For data communications, a commonly used wired channel is a transmission line based on a pair of wires. They have been used for broadband access to carry a very high data rate. The transmission line usually comprises loop segments, bridge taps, a drop wire, and lumped elements, each of them having a particular loop gauge. The bridge taps have an open circuit of the pair of wires, which generate notches in an end-to-end transfer function.

The two-port network can be used to analyze the transmission line via a chain matrix. The chain matrix establishes a relationship between input voltage and current to output voltage and current through an $ABCD$ matrix. This leads to a simulation loop model for the transmission line.

A dominant impairment for the transmission line is crosstalk including NEXT and FEXT. The NEXT represents the crosstalk of a local transmitter into a local receiver, while the FEXT represents the crosstalk of the local transmitter into a remote receiver. The crosstalk is caused by capacitive and inductive coupling between the wires due to imbalance in the couplings. However, to cancel crosstalk, a precise knowledge of crosstalk transfer functions is necessary.

Digital communications through both bandlimited wireless and wired channels are subject to ISI, which can cause performance degradation at the receiver. An eye diagram, which is a simple method to study the ISI effects and other channel impairments, is widely used for analyzing the performance degradation. In addition, the eye diagram can be used to determine the decision point at the widest opening of an "eye." On the other hand, in order to eliminate the ISI and out-of-band radiation, we have to design a combined transfer function of a transmitter filter, channel, and receiver filter to be a raised-cosine pulse, which meets the Nyquist criteria for ISI cancellation. In practice, given the transmitter filter and channel characteristics, we can design the optimal matched filter to maximize SNR at the receiver. However, this SNR is derived for an isolated pulse case, which does not take into account ISI. Therefore, in the presence of the ISI and noise, the receiver should consist of the optimal matched filter preceded by a channel equalizer for further signal processing.

References

[1] Rappaport, T. S., *Wireless Communications: Principles and Practice*, Prentice Hall, Upper Saddle River, New Jersey, 1996.

[2] Stüber, G. L., *Principles of Mobile Communication*, 2nd ed., Kluwer Academic Publishers, Boston, Massachusetts, 2001.

[3] Seidel, S. Y., et al., "Path Loss, Scattering and Multipath Delay Statistician Four European Cities for Digital Cellular and Microcellular Radiotelephone," *IEEE Trans. on Vehicular Technology*, Vol. 40, No. 4, pp. 721–730, November 1991.

[4] Hernando, J. M., and F. Pérez-Fontàn, *Introduction to Mobile Communications Engineering*, Artech House, Norwood, Massachusetts, 1999.

[5] Bernard, S. "Rayleigh Fading Channels in Mobile Digital Communication Systems, Part I: Characterization," *IEEE Communications Magazine*, pp. 136–146, September 1997.

[6] Gibson, J. D., (ed.), *The Mobile Communications Handbook*, 2nd ed., CRC Press LLC, Danvers, Massachusetts, 1999.

[7] Lee, W. C. Y., *Mobile Communications Engineering: Theory and Applications*, 2nd ed., McGraw-Hill, New York, 1998.

[8] Gans, M. J., "A Power Spectral Theory of Propagation in the Mobile Radio Environment," *IEEE Trans. on Vehicular Technology*, Vol. VT-21, pp. 27–38, February 1972.

[9] Clarke, R. H., "A Statistical Theory of Mobile Radio Reception," *Bell Systems Technical Journal*, Vol. 47, pp. 957–1000, July–August 1968.

[10] Cox, D. C., R. Murray, and A. Norris, "800 MHz Attenuation Measured in and around Suburban Houses," *AT&T Bell Laboratory Technical Journal*, Vol. 673, No. 6, pp. 921–954, July–August 1984.

[11] Bernhardt, R. C., "Macroscopic Diversity in Frequency Reuse Systems," *IEEE Journal on Selected Areas in Communications*, Vol-SAC 5, pp. 862–878, June 1987.

[12] Miao, G. J., "Introduction to Long-Range Extended DSL Systems: LDSL," *White Paper*, pp. 1-21, December 2002.

[13] Bingham, J. A. C., *ADSL, VDSL, and Multicarrier Modulation*, John Wiley & Sons, New York, 2000.

[14] Su, K. L., *Fundamentals of Circuits, Electronics, and Signal Analysis*, Waveland Press, Prospect Heights, Illinois, 1978.

[15] Lee, E. A., and D. G. Messerschmitt, *Digital Communications*, Kluwer Academic Publishers, Norwell, Massachusetts, 1988.

[16] Werner, J. J., "The HDSL Environment," *IEEE Journal on Selected Areas in Communications*, Vol. 9, No. 6, pp. 785–800, August 1991.

[17] Erickson, B., "Cascade ABCD Two-Port Networks," *EDN*, pp. 107–108, December 23, 1999.

[18] ITU-T, "Single-Pair High-Speed Digital Subscriber Line (SHDSL) Transceivers," G.991.2, February 2001.

[19] ANSI, "Asymmetric Digital Subscriber Line (ADSL) Metallic Interface," ANSI Standard T1.413, 1995.

[20] T1E1.4, "Very-High-Bit-Rate Digital Subscriber Line (VDSL) Metallic Interface, Part I: Functional Requirement and Common Specification," November 2000.

[21] IEEE Std 802.11b, "Part 11: Wireless LAN Medium Access Control (MAC) and Physical Layer (PHY) Specifications: Higher-Speed Physical Layer Extension in the 2.4 GHz Band," *IEEE-SA Standards Board*, 1999.

[22] IEEE Std 802.11a, "Part 11: Wireless LAN Medium Access Control (MAC) and Physical Layer (PHY) Specifications: Higher-Speed Physical Layer Extension in the 5 GHz Band," *IEEE-SA Standards Board*, 1999.

[23] Nyquist, H., "Certain Topics in Telegraph Transmission Theory," *Transactions of AIEE*, Vol. 47, pp. 617–644, February 1928.

[24] Miao, G. J., and M. A. Clements, *Digital Signal Processing and Statistical Classification*, Artech House, Norwood, Massachusetts, 2002.

[25] Acharya, T., and G. J. Miao, "Square Root Raised Cosine Symmetric Filter for Mobile Telecommunications," U.S. Patent No. 6,731,706, pp. 1–15, May 2004.

[26] Acharya, T., and G. J. Miao, "Chip Rate Selection Square Root Raised Cosine Filter for Mobile Telecommunications," U.S. Patent No. 6,650,688, pp. 1–18, November 2003.

[27] Murota, K., and K. Hirade, "GMSK Modulation for Digital Mobile Radio," *IEEE Trans. on Communications*, Vol. 29, pp. 1044–1050, July 1981.

7

Channel Estimation and Blind Identification

7.1 Introduction

Channel estimation is an important step in signal detection for a digital communication system, especially when there is little or no knowledge about a wireless or wired transmission channel. The channel estimation can be classified as two categories: *nonblind* and *blind*. The essential ingredient in nonblind techniques is the use of a known training sequence embedded in the transmission data, which then determines the channel impulse response. We call this method *nonblind channel estimation* or *channel estimation*. On the other hand, if the training sequence is not available, then it is referred to as *blind identification*.

The channel estimation or blind identification is a necessary step for wireless and wired digital communication system designs. For the wireless digital communications, in the reverse link, the channel estimation or blind identification is used to provide the channel information for equalization, Rake receiver, and/or orthogonal frequency division multiplexing (OFDM), which provide diversity and reduce *cochannel interference* (CCI). The CCI arises from cellular frequency reuse, and thus limits the quality and capacity of wireless networks. In the forward link, the channel estimation or blind identification is needed to design weight vectors to deliver energy to a selected user without causing significant CCI to other users. For wired digital communications, the channel estimation or blind identification is required for using the equalization either in a time domain or in a frequency domain and discrete multitone (DMT). Generally, the channel estimation method is to measure a channel impulse response and noise power spectral density (PSD) or their equivalents, such as direct measurement of signal-to-noise ratio (SNR) without measuring impulse response and noise separately. In a multitone communication channel, the channel estimation directly estimates signal and noise parameters for each of the subchannels. However, these estimates can always be converted to an aggregate channel and noise estimate for a single-channel transmission.

Traditionally, channel estimation is achieved either by sending the training sequence or by designing a channel estimator based on a priori knowledge of the channel. This approach is often not suitable for a wireless channel since little knowledge about such a wireless channel can be assumed a priori. Even adaptive approaches, which are attractive in handling time-varying channels, have to waste a fraction of transmission time for the training sequence. Recently, in contrast to the adaptive approaches, blind channel identification attracts much attention in research and practice since this method does not require the training sequence. The channel identification is obtained by using only channel output sequences along with certain assumptions of statistical information on input sources. Blind identification methods demonstrate the potential to increase transmission capability because of elimination of the training sequence in both wireless and wired high-speed digital communications.

This chapter is organized as follows. In this section, background, overview, and types of communication channels along with the methods of channel estimations and blind identifications are briefly presented. In Section 7.2, we introduce characteristics of discrete-time channel models. In Section 7.3, we describe the channel estimators based on the technique methods of maximum likelihood, least squares, generalized least squares, and minimum mean-square error (MMSE). Subsequently, fundamentals of adaptive channel estimation and algorithms are given in Section 7.4, with emphases on least mean square (LMS) and recursive least squares (RLS) and their corresponding convergence and excess mean-squared error (EMSE) analysis. Section 7.5 discusses channel models of single-input single-output (SISO), single-input multiple-output (SIMO), multiple-input multiple-output (MIMO), and their estimations based on the higher-order statistics (HOS). Blind identifications for the discrete-time SISO, SIMO, and MIMO channels are given in Section 7.6. A brief summary of this chapter is given in Section 7.7.

7.2 Discrete-Time Channel Models

In digital communications, the transmitter sends discrete-time symbols at a symbol rate of $1/T$ per second, and the sampled output of the matched filter at the receiver is also a discrete-time signal with sampling rate of $1/T$ per second. In this case, the cascade of the analog transmitter $H_T(f)$, the channel $C(f)$, the receiver matched filter $H_R(f)$, and the sampler (or A/D converter) can be described by using the equivalent discrete-time transversal finite impulse response (FIR) filter with the tap coefficient $h[n]$, where n is some arbitrary positive integer.

Assume that the discrete-time transversal FIR tap coefficient is

$$h[n] = \begin{cases} b_n, & n = -N, ..., -1, 0, 1, ..., N \\ 0, & \text{otherwise}, \end{cases} \quad (7.1)$$

and the channel is a bandlimited frequency. Figure 7.1 shows an equivalent discrete-time channel model, where z^{-1} is the delay of T. Its input is the sequence of information symbols $s[n]$. Thus, its output is the discrete-time sequence $y[n]$ given by

$$y[n] = \sum_{k=-N}^{N} h[k]s[n-k] + v[n], \quad (7.2)$$

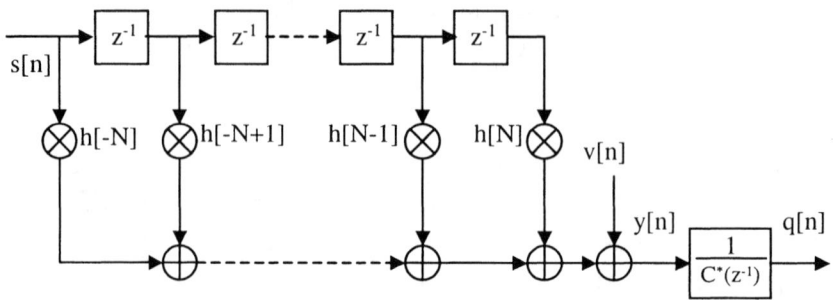

Figure 7.1 Equivalent discrete-time FIR channel model.

where $v[n]$ is the additive noise sequence. The corresponding z-domain transfer function of (7.2) is then obtained by

$$Y(z) = H(z)S(z) + V(z), \qquad (7.3)$$

where the z-domain transfer function of the channel $H(z)$ is expressed as

$$H(z) = \sum_{n=-N}^{N} h[n]z^{-n}. \qquad (7.4)$$

Let the discrete-time channel $h[n]$ be symmetric, that is, $h[n] = h^*[-n]$. Then it follows that the z-domain transfer function of the discrete-time channel is

$$H(z) = H^*(z^{-1}). \qquad (7.5)$$

Consequently, the $H(z)$ has the $2N$ symmetric roots, where $(1/\rho)^*$ is a root if ρ is a root. Therefore, the z-domain transfer function of the discrete-time channel $H(z)$ can be factored as

$$H(z) = C(z)C^*(z^{-1}), \qquad (7.6)$$

where $C(z)$ is a polynomial of degree N with the roots ρ_i and $C^*(z^{-1})$ is a polynomial of degree N having the roots $1/\rho_i$, where

$i = 1, 2, \cdots, N$. Proakis [1] suggested that one choose a unique $C^*(z^{-1})$ with minimum phase, that is, the polynomial function has all its roots inside the unit circle. This is to say that $1/C^*(z^{-1})$ is a physically realizable, stable, and discrete-time infinite impulse response (IIR) filter. Thus, by using the discrete-time IIR filter $1/C^*(z^{-1})$, the output sequence $y[n]$ in Figure 7.1 results in an output sequence $q[n]$ in the z-domain that can be expressed as

$$\begin{aligned} Q(z) &= Y(z)\left[\frac{1}{C^*(z^{-1})}\right] \\ &= [H(z)S(z) + V(z)]\left[\frac{1}{C^*(z^{-1})}\right] \\ &= C(z)S(z) + G(z), \end{aligned} \quad (7.7)$$

where $G(z) = V(z)/C^*(z^{-1})$ is a z-domain white Gaussian noise. Hence, corresponding the impulse response $q[n]$ in (7.7) is obtained by

$$q[n] = \sum_{k=0}^{N} c[k]s[n-k] + g[n], \quad (7.8)$$

where the sequence $g[n]$ is a white Gaussian noise sequence and the sequence $c[n]$ is a set of tap coefficients of an equivalent discrete-time FIR filter with a transfer function $C(z)$.

Note that the matched filter at the communication receiver becomes a time-varying filter if the channel impulse response is changing slowly with time. As a result, the time variations of the channel and matched filter pair require a discrete-time filter with time varying filter coefficients. Consequently, intersymbol interference (ISI) effects are time-varying, where the tap coefficient is slowly varying with time. In general, the compensation methods for the time-varying ISI, which will be discussed in Chapter 8, are called *adaptive equalization techniques*. However, in designing a good equalizer, one usually needs to know the characteristics of a channel distortion.

7.3 Channel Estimators

One of the design objectives for receivers is to minimize the detection error. Generally, the design of an optimal detector requires the knowledge of the communication channel. In practice, channel parameters need to be estimated, preferably using only a limited amount of training data samples. The efficiency of the *channel estimator* is particularly important, because the efficiency is a statistical measure of how effectively an algorithm utilizes the available training data samples.

The channel estimator is defined as any statistic (known function of observable random variables that is itself a random variable) whose values are used to estimate $\tau(\theta)$, where $\tau(\cdot)$ is some function of the parameter θ. In other words, the channel estimator is always a statistic, which is both a random variable and a function.

Channel estimation admits two steps: (1) to devise some means of obtaining a statistic to use as a channel estimator, and (2) to select criteria and techniques to define and find a "best" estimator among possible estimators. In this section, we consider three types of channel estimators based on the framework of maximizing the likelihood function.

7.3.1 Maximum Likelihood Estimator

There are several methods of finding the channel estimator. One of these, and probably the most important, is the *method of maximum likelihood*. The *maximum likelihood estimator* can be derived in a systematic way and has been proved to be asymptotically optimal [2].

Let us reconsider the discrete-time channel model given by (7.2), where we now assume that the discrete-time channel vector has a finite impulse response of order L,

$$\mathbf{h} = \{h[1], h[2], \cdots, h[L]\}^T, \tag{7.9}$$

where $\{\cdot\}^T$ is the transpose of a vector. Suppose that we have received N samples of the observations,

$$\mathbf{y} = \{y[0], y[1], \cdots, y[N-1]\}^T. \tag{7.10}$$

We then have the following linear model given by

$$\mathbf{y} = \mathbf{sh} + \mathbf{v}, \tag{7.11}$$

where $\mathbf{s}$ is an $N \times L$ Toeplitz matrix consisting of samples of the input sequence $\{s[n], n = 0, 1, \cdots, N-1\}$ given by

$$\mathbf{s} = \begin{bmatrix} s[0] & s[N-1] & \cdots & s[N-L+1] \\ s[1] & s[0] & \cdots & s[N-L+2] \\ \vdots & \vdots & \ddots & \vdots \\ s[N-1] & s[N-2] & \cdots & s[0] \end{bmatrix}, \tag{7.12}$$

and $\mathbf{v}$ is a noise vector given by

$$\mathbf{v} = \{v[0], v[1], \cdots, v[N-1]\}^T. \tag{7.13}$$

Let θ be the vector of unknown parameters that may contain the channel vector $\mathbf{h}$ and possibly the entire or part of the input vector $\mathbf{s}$. Assume that the probability space, which describes jointly the noise vector $\mathbf{v}$ and the input vector $\mathbf{s}$, is known. In principle, we can then obtain the probability density function of the observation vector $\mathbf{y}$ (assuming it exists). The joint pdf of the observation $f_\mathbf{y}(\mathbf{y}; \theta)$ is referred to as the *likelihood function*, which is considered to be a function of θ.

The most important case is that a random sample is used to form the joint pdf $f_\mathbf{y}(\mathbf{y}; \theta)$, where θ is a single parameter. The likelihood function is then given by

$$f_\mathbf{y}(\mathbf{y}; \theta) = f(y[0], \theta) f(y[1], \theta) \cdots f(y[N-1], \theta). \tag{7.14}$$

Thus, the maximum likelihood estimator is the solution of the equation

$$\frac{df_\mathbf{y}(\mathbf{y}; \theta)}{d\theta} = 0. \tag{7.15}$$

Note that $f_\mathbf{y}(\mathbf{y};\theta)$ and $\log\left[f_\mathbf{y}(\mathbf{y};\theta)\right]$ have their maximums at the same value of θ. It is sometimes easier to find the maximum of the logarithm of the likelihood.

If the likelihood function contains the vector of θ, $\theta = [\theta_1, \theta_2, \cdots, \theta_k]$,

$$f_\mathbf{y}(\mathbf{y};\theta_1,\theta_2,\cdots,\theta_k) = \prod_{n=0}^{N-1} f_{y[n]}(y[n];\theta_1,\theta_2,\cdots,\theta_k), \quad (7.16)$$

then the maximum likelihood estimators of the parameters $\theta_1, \theta_2, \cdots, \theta_k$ are the solution of the k equations,

$$\frac{\partial\left[\prod_{n=0}^{N-1} f_{y[n]}(y[n];\theta_1,\theta_2,\cdots,\theta_k)\right]}{\partial \theta_i} = 0, \quad i = 1, 2, \cdots, k. \quad (7.17)$$

Equation (7.17) may also be easier to work with the logarithm of the likelihood.

Maximum likelihood estimators usually have good performance when the sample size of the observation is sufficiently large, but their implementation is sometimes computationally intensive. Furthermore, the optimization of the likelihood function is sometimes hampered because the equation $\partial f_\mathbf{y}(\mathbf{y};\theta)/\partial\theta$ locates minimums, local maximums, and maximums. Hence, one must avoid using a root that actually locates a minimum. In addition, the likelihood function $f_\mathbf{y}(\mathbf{y};\theta)$ may be represented by a curve where the actual maximum is at $\hat{\theta}$, but the derivative set equal to 0 would locate θ' as the maximum. Therefore, it is highly desirable that effective initialization techniques be used in conjunction with the method of maximum likelihood.

Example 7.1

Assume that a random sample of size N of observations from a channel output has the normal distribution given by

$$f_{x[n]}(x[n];\mu,\sigma) = \frac{1}{\sqrt{2\pi}\sigma}\exp\left[-\frac{1}{2\sigma^2}(x[n]-\mu)^2\right], \quad (7.18)$$

where $-\infty < \mu < \infty$ and $\sigma > 0$. Determine the maximum likelihood estimators of $\hat{\mu}$ and $\hat{\sigma}$.

Using (7.16), the likelihood function $L(x[n]; \mu, \sigma)$ is given by

$$L(x[n]; \mu, \sigma) = \prod_{n=1}^{N} \frac{1}{\sqrt{2\pi}\sigma} \exp\left[-\frac{1}{2\sigma^2}(x[n] - \mu)^2\right]$$

$$= \left(\frac{1}{2\pi\sigma^2}\right)^{N/2} \exp\left[-\frac{1}{2\sigma^2} \sum_{n=1}^{N} (x[n] - \mu)^2\right]. \quad (7.19)$$

The logarithm of the likelihood function in (7.19) is then obtained by

$$L^* = \ln L(x[n]; \mu, \sigma)$$

$$= -\frac{N}{2} \ln(2\pi) - \frac{N}{2} \ln \sigma^2 - \frac{1}{2\sigma^2} \sum_{n=1}^{N} (x[n] - \mu)^2. \quad (7.20)$$

In order to determine the location of the maximum likelihood in (7.20), we calculate

$$\frac{\partial L^*}{\partial \mu} = \frac{1}{\sigma^2} \sum_{n=1}^{N} (x[n] - \mu), \quad (7.21)$$

and

$$\frac{\partial L^*}{\partial \sigma^2} = -\left(\frac{N}{2}\right)\left(\frac{1}{\sigma^2}\right) + \frac{1}{2\sigma^4} \sum_{n=1}^{N} (x[n] - \mu)^2. \quad (7.22)$$

Setting these derivatives of (7.21) and (7.22) equal to 0 and solving the resulting equations for μ and σ^2, we find the maximum likelihood estimators

$$\hat{\mu} = \frac{1}{N} \sum_{n=1}^{N} x[n], \quad (7.23)$$

and
$$\hat{\sigma}^2 = \frac{1}{N} \sum_{n=1}^{N} (x[n] - \hat{\mu})^2. \tag{7.24}$$

Note that (7.23) and (7.24) turn out to be the sample moments corresponding to μ and σ^2.

7.3.2 Least Squares Estimator

Consider the discrete-time channel model given by (7.11)

$$\mathbf{y} = \mathbf{sh} + \mathbf{v}, \tag{7.25}$$

where $\mathbf{s}$ is now a known $N \times L$ matrix of rank L (assume that training sequences are available) and $\mathbf{v}$ is the noise vector of disturbances. Also assume merely that the noise vector of disturbances $\mathbf{v}$ is not normally distributed and has the mean $E\{\mathbf{v}\} = 0$ and the covariance $V\{\mathbf{v}\} = \Omega$. In the discrete-time channel model when the joint probability density function of the observable random variables is not given, then the estimation of $\mathbf{h}$ can be approached by using the *method of least squares*.

Choosing the values of $\mathbf{h}$ minimizes the residual sum of squares

$$L(\mathbf{y}; \mathbf{h}) = (\mathbf{y} - \mathbf{sh})^T (\mathbf{y} - \mathbf{sh}), \tag{7.26}$$

where $\mathbf{h}$ is defined as the *least squares estimator*. We then compute

$$\begin{aligned} \frac{\partial L(\mathbf{y}; \mathbf{h})}{\partial \mathbf{h}} &= -\mathbf{s}^T (\mathbf{y} - \mathbf{sh}) - (\mathbf{y} - \mathbf{sh})^T \mathbf{s} \\ &= -2\mathbf{s}^T (\mathbf{y} - \mathbf{sh}). \end{aligned} \tag{7.27}$$

Setting the derivative in (7.27) equal to 0 and solving the resulting equation for $\mathbf{h}$, we find the least squares estimator

$$\hat{\mathbf{h}} = \left(\mathbf{s}^T \mathbf{s}\right)^{-1} \mathbf{s}^T \mathbf{y}. \tag{7.28}$$

Note that the mean of the least squares estimator $\hat{\mathbf{h}}$ is obtained by

$$E\{\hat{\mathbf{h}}\} = \left(\mathbf{s}^T \mathbf{s}\right)^{-1} \mathbf{s}^T E\{\mathbf{y}\}$$

Channel Estimation and Blind Identification

$$= \left(\mathbf{s}^T\mathbf{s}\right)^{-1}\mathbf{s}^T\left(\mathbf{s}\mathbf{h} + E\{\mathbf{v}\}\right)$$

$$= \left(\mathbf{s}^T\mathbf{s}\right)^{-1}\left(\mathbf{s}^T\mathbf{s}\right)\mathbf{h}. \tag{7.29}$$

Since

$$\left(\mathbf{s}^T\mathbf{s}\right)^{-1}\left(\mathbf{s}^T\mathbf{s}\right) = \mathbf{I}, \tag{7.30}$$

the mean of the least squares estimator $\hat{\mathbf{h}}$ is then

$$E\{\hat{\mathbf{h}}\} = \mathbf{h}. \tag{7.31}$$

Thus, $\hat{\mathbf{h}}$ is the *unbiased estimate* of $\mathbf{h}$. Also, the covariance of the least squares estimator $\hat{\mathbf{h}}$ is given by (see Example 7.2 for proof),

$$V\{\hat{\mathbf{h}}\} = \left(\mathbf{s}^T\mathbf{s}\right)^{-1}\left(\mathbf{s}^T\Omega\mathbf{s}\right)\left(\mathbf{s}^T\mathbf{s}\right)^{-1}, \tag{7.32}$$

where Ω is the covariance matrix. In particular, assuming $\Omega = \sigma^2\mathbf{I}$, (7.32) simplifies to

$$V\{\hat{\mathbf{h}}\} = \sigma^2\left(\mathbf{s}^T\mathbf{s}\right)^{-1}. \tag{7.33}$$

In this case, (7.33) leads to the *optimal property* for the least squares estimator $\hat{\mathbf{h}}$.

The optimal property is referred to as the Gauss-Markov theorem in multivariate statistics [3]. We restate the optimal property for the channel estimator as follows. Consider the discrete-time channel model in (7.25) and assume the disturbance terms are uncorrelated with each other $\Omega = \sigma^2\mathbf{I}$. The least squares estimator for the channel estimation given by (7.28) has a covariance matrix that is smaller than any other linear estimator. In other words, the least squares estimator is the best linear unbiased estimator.

Example 7.2

The discrete-time channel model is given by (7.25), where the noise vector of disturbances $\mathbf{v}$ is not normally distributed and has the mean $E\{\mathbf{v}\} = 0$ and the covariance $V\{\mathbf{v}\} = \Omega$. Applying the method of least squares obtains the least squares estimator

$$\hat{\mathbf{h}} = \left(\mathbf{s}^T\mathbf{s}\right)^{-1}\mathbf{s}^T\mathbf{y}. \tag{7.34}$$

Show that the covariance matrix of the least squares estimator $\hat{\mathbf{h}}$ is
$$V\{\hat{\mathbf{h}}\} = \left(\mathbf{s}^T\mathbf{s}\right)^{-1} \left(\mathbf{s}^T\mathbf{\Omega}\mathbf{s}\right) \left(\mathbf{s}^T\mathbf{s}\right)^{-1}. \tag{7.35}$$

We know that the covariance is given by
$$\begin{aligned}
V\{\hat{\mathbf{h}}\} &= E\{(\hat{\mathbf{h}} - \mathbf{h})(\hat{\mathbf{h}} - \mathbf{h})^T\} \\
&= E\left\{\left[\left(\mathbf{s}^T\mathbf{s}\right)^{-1}\mathbf{s}^T\mathbf{y} - \mathbf{h}\right]\left[\left(\mathbf{s}^T\mathbf{s}\right)^{-1}\mathbf{s}^T\mathbf{y} - \mathbf{h}\right]^T\right\} \\
&= E\left\{\left(\mathbf{s}^T\mathbf{s}\right)^{-1}\mathbf{s}^T\mathbf{y}\mathbf{y}^T\mathbf{s}\left(\mathbf{s}^T\mathbf{s}\right)^{-1} - 2\mathbf{h}\left(\mathbf{s}^T\mathbf{s}\right)^{-1}\mathbf{s}^T\mathbf{y} + \mathbf{h}\mathbf{h}^T\right\}.
\end{aligned} \tag{7.36}$$

Since
$$\begin{aligned}
E\left\{2\mathbf{h}\left(\mathbf{s}^T\mathbf{s}\right)^{-1}\mathbf{s}^T\mathbf{y}\right\} &= 2\mathbf{h}\left(\mathbf{s}^T\mathbf{s}\right)^{-1}\mathbf{s}^T E\{\mathbf{y}\} \\
&= 2\mathbf{h}\left(\mathbf{s}^T\mathbf{s}\right)^{-1}\mathbf{s}^T\left(\mathbf{s}\mathbf{h} + E\{\mathbf{v}\}\right) \\
&= 2\mathbf{h}\left(\mathbf{s}^T\mathbf{s}\right)^{-1}\left(\mathbf{s}^T\mathbf{s}\right)\mathbf{h} \\
&= 2\mathbf{h}\mathbf{h}^T,
\end{aligned} \tag{7.37}$$

then (7.36) can be rewritten as
$$V\{\hat{\mathbf{h}}\} = \left(\mathbf{s}^T\mathbf{s}\right)^{-1}\mathbf{s}^T E\{\mathbf{y}\mathbf{y}^T\}\mathbf{s}\left(\mathbf{s}^T\mathbf{s}\right)^{-1} - \mathbf{h}\mathbf{h}^T. \tag{7.38}$$

Note that
$$\begin{aligned}
E\{\mathbf{y}\mathbf{y}^T\} &= E\{(\mathbf{s}\mathbf{h} + \mathbf{v})(\mathbf{s}\mathbf{h} + \mathbf{v})^T\} \\
&= \mathbf{s}\mathbf{h}\mathbf{h}^T\mathbf{s}^T + 2\mathbf{s}\mathbf{h}E\{\mathbf{v}\} + E\{\mathbf{v}^2\}.
\end{aligned} \tag{7.39}$$

Since $E\{\mathbf{v}\} = 0$, the covariance $V\{\mathbf{v}\} = E\{\mathbf{v}^2\} = \mathbf{\Omega}$. Now, (7.39) simplifies to
$$E\{\mathbf{y}\mathbf{y}^T\} = \mathbf{s}\mathbf{h}\mathbf{h}^T\mathbf{s}^T + \mathbf{\Omega}. \tag{7.40}$$

Substituting (7.40) into (7.38) yields the desired result of the covariance matrix of the least squares estimator $\hat{\mathbf{h}}$,
$$V\{\hat{\mathbf{h}}\} = \left(\mathbf{s}^T\mathbf{s}\right)^{-1} \left(\mathbf{s}^T\mathbf{\Omega}\mathbf{s}\right) \left(\mathbf{s}^T\mathbf{s}\right)^{-1}. \tag{7.41}$$

7.3.3 Generalized Least Squares Estimator

The least squares estimator for the channel estimation discussed above is the best linear unbiased estimator when the covariance matrix of $\mathbf{v}$ equals $\sigma^2 \mathbf{I}$. But, this is a special case for the least squares estimator. In general, the covariance matrix $V(\mathbf{v}) = \Omega \neq \sigma^2 \mathbf{I}$. In this case, the least squares estimator is not the best linear unbiased estimator. In this section, we introduce a *generalized least squares estimator*.

Reconsider the discrete-time channel model in (7.25) if the noise vector of disturbances $\mathbf{v}$ is not normally distributed. Let the mean $E\{\mathbf{v}\} = 0$ and the covariance matrix $V\{\mathbf{v}\} = \Omega$. Also assume that the covariance matrix Ω is known or can be estimated. Multiplying the matrix $\frac{1}{\sqrt{\Omega}}$ on both sides of (7.25) yields a new transformed channel model,

$$\mathbf{z} = \frac{1}{\sqrt{\Omega}} \mathbf{sh} + \mathbf{u}, \qquad (7.42)$$

where

$$\mathbf{z} = \frac{1}{\sqrt{\Omega}} \mathbf{y}, \qquad (7.43)$$

and

$$\mathbf{u} = \frac{1}{\sqrt{\Omega}} \mathbf{v}. \qquad (7.44)$$

Since $E\{\mathbf{v}\} = 0$, then the mean $E\{\mathbf{u}\} = \frac{1}{\sqrt{\Omega}} E\{\mathbf{v}\} = 0$, and the covariance matrix,

$$\begin{aligned} V(\mathbf{u}) &= E\{\mathbf{u}^2\} - E^2\{\mathbf{u}\} \\ &= \frac{1}{\Omega} E\{\mathbf{v}^2\} \\ &= \frac{1}{\Omega} V(\mathbf{v}) \\ &= \mathbf{I}. \end{aligned} \qquad (7.45)$$

Thus, (7.45) indicates that the transformed channel model in (7.42) satisfies the assumptions of the optimal property (Gauss-Markov theorem). Hence, using the method of least squares, the best linear unbiased estimate of **h** is given by

$$\tilde{\mathbf{h}} = \left(\mathbf{s}^T \frac{1}{\Omega}\mathbf{s}\right)^{-1} \mathbf{s}^T \frac{1}{\sqrt{\Omega}}\mathbf{z}$$

$$= \left(\mathbf{s}^T \frac{1}{\Omega}\mathbf{s}\right)^{-1} \mathbf{s}^T \frac{1}{\Omega}\mathbf{y}. \qquad (7.46)$$

Also, the covariance matrix of $\tilde{\mathbf{h}}$ is given by

$$V(\tilde{\mathbf{h}}) = \left(\mathbf{s}^T \frac{1}{\Omega}\mathbf{s}\right)^{-1}. \qquad (7.47)$$

The estimator given by (7.46) is called the generalized least squares estimator for channel estimation. In this case, the covariance matrix in (7.47) is the same as the best minimum mean-square error (MMSE) since the generalized least squares estimator is the best linear unbiased estimator.

7.3.4 MMSE Estimator

In this section, we present a linear optimum discrete-time solution that is known as an *MMSE estimator* based on an FIR Wiener filter [4–6]. The main result is to derive the discrete-time Wiener-Hopf equations that provide the FIR filter coefficients of the optimum filter in the sense of MMSE for the channel estimation.

The MMSE estimator based on the FIR Wiener filter is shown in Figure 7.2. It is used to produce the MMSE estimate, $d[n]$ for $y[n]$, given two wide-sense stationary processes jointly, $s[n]$ and $y[n]$, which are statistically related to each other. It is also assumed that the autocorrelation functions, $r_s(k)$ and $r_y(k)$, and the cross-correlation function, $r_{ys}(k)$, are known or can be estimated.

In order to develop the MMSE estimator for the channel estimation, we need to determine a set of coefficients for the FIR Wiener filter, $w[n]$, which minimizes the mean square error (MSE) of the filter output as compared to a channel model output, $y[n]$. The error signal $e[n]$ is given by

$$e[n] = y[n] - d[n]$$

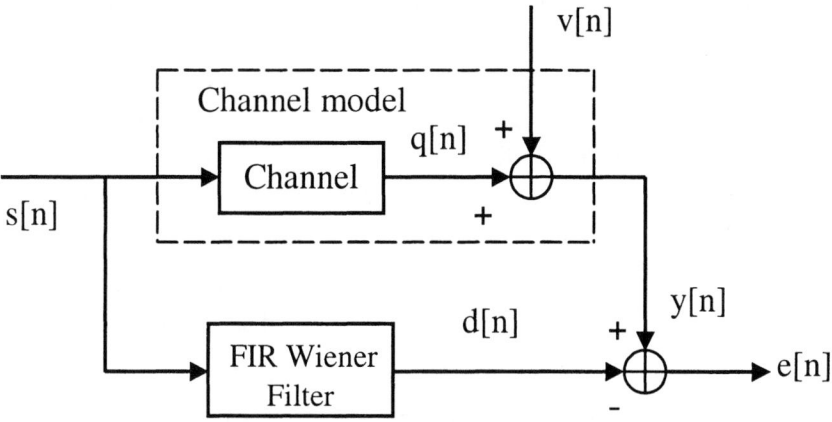

Figure 7.2 Illustration of the MMSE estimator based on the Wiener-Hopf optimum solution for the channel estimation.

$$= y[n] - \sum_{k=0}^{M-1} w[k]s[n-k], \qquad (7.48)$$

where M is the number of coefficients of the FIR Wiener filter. In vector form, the error signal in (7.48) is written as

$$e[n] = y[n] - \mathbf{w}^T[n]\mathbf{s}[n]. \qquad (7.49)$$

The MSE is a function of the coefficient weight vector $\mathbf{w}[n]$ chosen and is obtained by

$$\begin{aligned}
\xi_{MSE} &= E\{|e[n]|^2\} \\
&= E\{e[n]e^T[n]\} \\
&= E\{(y[n] - \mathbf{w}^T[n]\mathbf{s}[n])(y[n] - \mathbf{w}^T[n]\mathbf{s}[n])^T\} \\
&= E\{y[n]y^T[n] - y[n]\mathbf{s}^T[n]\mathbf{w}[n] \\
&\quad - \mathbf{w}^T[n]\mathbf{s}[n]y^T[n] + \mathbf{w}^T[n]\mathbf{s}[n]\mathbf{s}^T[n]\mathbf{w}[n]\} \\
&= \text{Var}\{y[n]\} - \mathbf{r}_{ys}[n]\mathbf{w}[n] - \mathbf{w}^T[n]\mathbf{r}_{ys}[n] \\
&\quad + \mathbf{w}^T[n]\mathbf{R}_s[n]\mathbf{w}[n], \qquad (7.50)
\end{aligned}$$

where $\mathbf{r}_{ys}[n] = E\{\mathbf{s}[n]y^T[n]\}$ is the product of the cross-correlation function, and $\mathbf{R}_s[n] = E\{\mathbf{s}[n]\mathbf{s}^T[n]\}$ is the autocorrelation function.

To minimize ξ_{MSE}, it is necessary and sufficient that the derivative of ξ_{MSE} with respect to $\mathbf{w}[n]$ be equal to zero,

$$\mathbf{R}_s[n]\mathbf{w}[n] = \mathbf{r}_{ys}[n], \qquad (7.51)$$

where $\mathbf{R}_s[n]$ is an $M \times M$ Hermitian-Toeplitz matrix of autocorrelation,

$$\mathbf{R}_s[n] = \begin{bmatrix} r_s[0] & r_s[1] & \cdots & r_s[M-1] \\ r_s[1] & r_s[0] & \cdots & r_s[M-2] \\ \vdots & \vdots & \ddots & \vdots \\ r_s[M-1] & r_s[M-2] & \cdots & r_s[0] \end{bmatrix}, \qquad (7.52)$$

$\mathbf{w}[n]$ is the vector of the FIR Wiener filter coefficients,

$$\mathbf{w}[n] = \begin{pmatrix} w[0], & w[1], & \cdots, & w[M-1] \end{pmatrix}^T, \qquad (7.53)$$

and $\mathbf{r}_{ys}[n]$ is the vector of cross-correlation between the channel model output $y[n]$ and the input signal $s[n]$,

$$\mathbf{r}_{ys}[n] = \begin{pmatrix} r_{ys}[0], & r_{ys}[1], & \cdots, & r_{ys}[M-1] \end{pmatrix}^T. \qquad (7.54)$$

The matrix form in (7.51) is called the *Wiener-Hopf equation*.

If the matrix $\mathbf{R}_s[n]$ is invertible, then $\mathbf{w}[n]$ can be solved by

$$\mathbf{w}[n] = \mathbf{R}_s^{-1}[n]\mathbf{r}_{ys}[n]. \qquad (7.55)$$

Equation (7.55) is called the MMSE estimator for the channel estimation or *normal equation*, since the error signal is orthogonal to each element of the input vector given by

$$E\{\mathbf{s}[n]e^T[n]\} = 0. \qquad (7.56)$$

The MMSE of the estimate of the channel model output $y[n]$ can be then computed by using (7.50) as follows,

$$\begin{aligned} \xi_{MMSE} &= E\{|e[n]|^2\} \\ &= E\{e[n](y[n] - \mathbf{w}^T[n]\mathbf{s}[n])\} \\ &= E\{e[n]y[n]\} - \mathbf{w}^T[n]E\{\mathbf{s}[n]e^T[n]\}. \end{aligned} \qquad (7.57)$$

Since $E\{\mathbf{s}[n]e^T[n]\} = 0$ given by (7.56), the second term in (7.57) is equal to zero. Hence,

$$\begin{aligned}\xi_{MMSE} &= E\{e[n]y[n]\} \\ &= E\{(y[n] - \mathbf{w}^T[n]\mathbf{s}[n])y[n]\} \\ &= E\{y^2[n]\} - \mathbf{w}^T[n]E\{\mathbf{s}[n]y[n]\} \\ &= R_y(0) - \mathbf{r}_{ys}^T[n]\mathbf{w}[n]. \end{aligned} \quad (7.58)$$

The MMSE can also be expressed in terms of the autocorrelation matrix $\mathbf{R}_s[n]$ and the cross-correlation vector $\mathbf{r}_{ys}[n]$ in the following:

$$\xi_{MMSE} = R_y(0) - \mathbf{r}_{ys}^T[n]\mathbf{R}_s^{-1}[n]\mathbf{r}_{ys}[n]. \quad (7.59)$$

Now, it is useful to consider what must be known to obtain the MMSE estimator and to estimate ξ_{MMSE} in the following steps:

1. Calculate the autocorrelation matrix $\mathbf{R}_s[n]$ of the input vector, $\mathbf{s}[n]$.
2. Determine the cross-correlation vector $\mathbf{r}_{ys}[n]$ between the input vector, $\mathbf{s}[n]$, and the channel model output, $y[n]$.
3. Estimate the variance $R_y(0)$ of the channel model output, $y[n]$.
4. Compute the optimal coefficients $\mathbf{w}[n]$ and the ξ_{MMSE} for the channel estimation.

Further note that, in Figure 7.2, a desired channel output $q[n]$ is to be estimated from the noise corrupted observation of the channel model output $y[n]$

$$y[n] = q[n] + v[n]. \quad (7.60)$$

If we assume that the noise $v[n]$ has zero mean and σ^2 variance, and that it is uncorrelated with the desired channel output $q[n]$, then we obtain

$$E\{q[n]v[n-k]\} = 0, \quad (7.61)$$

and the cross-correlation function between the desired channel output $q[n]$ and the channel model output $y[n]$

$$r_{qy}(k) = E\{q[n]y[n-k]\}$$

$$\begin{aligned} &= E\{q[n](q[n-k]+v[n-k])\} \\ &= E\{q[n]q[n-k]\} + E\{q[n]v[n-k]\} \\ &= r_q(k). \end{aligned} \qquad (7.62)$$

In the matrix form, (7.62) can be written as

$$\mathbf{r}_{qy}[n] = \mathbf{r}_q[n]. \qquad (7.63)$$

Since the noise $v[n]$ and the desired channel output $q[n]$ are uncorrelated, it follows that

$$\begin{aligned} r_y(k) &= E\{y[n+k]y[n]\} \\ &= E\{(q[n+k]+v[n+k])(q[n]+v[n])\} \\ &= E\{q[n+k]q[n]\} + E\{v[n+k]v[n]\} \\ &= r_q(k) + r_v(k). \end{aligned} \qquad (7.64)$$

Therefore, making the autocorrelation matrix $\mathbf{R}_q[n]$ for the desired channel output $q[n]$, and the autocorrelation matrix $\mathbf{R}_v[n]$ for the noise $v[n]$, the MMSE estimator based on the Wiener-Hopf equation is given by

$$\left(\mathbf{R}_q[n] + \mathbf{R}_v[n]\right)\mathbf{w}[n] = \mathbf{r}_q. \qquad (7.65)$$

Equation (7.65) can be further simplified if specific information of the statistic of the signal and noise are available.

The MMSE estimator based on the Wiener-Hopf solution is a single-step optimal algorithm and can be used to solve the channel estimation. However, from the point of view of implementation, it is not an efficient algorithm in terms of computation complexities, especially in a real-time digital signal processing operation. Furthermore, the optimal coefficient vector $\mathbf{w}[n]$ needs to be computed again if the input signals $x[n]$ and the output response $y[n]$ are nonstationary. In other words, the optimal coefficient vector $\mathbf{w}[n]$ must be recomputed if the channel is time-varying. This leads to a very high computation load for each iteration. Therefore, the direct implementation of the MMSE estimator based on the Wiener-Hopf solution is not recommended for the real-time digital channel estimation in the environment of wireless mobile communications.

Example 7.3

In this example, we introduce the channel estimation by using the MMSE estimator based on the FIR Wiener filter solution as shown in Figure 7.2. Suppose that the channel output $q[n]$ is observed in the presence of uncorrelated white noise $v[n]$

$$y[n] = q[n] + v[n], \qquad (7.66)$$

where $v[n]$ is the zero mean and a variance of σ^2. Assume that the discrete-time channel can be expressed in the difference equation

$$q[n] = \sum_{k=0}^{L-1} b[k]s[n-k], \qquad (7.67)$$

and in the z-transfer function

$$H(z) = \sum_{k=0}^{L-1} b[k]z^{-k}. \qquad (7.68)$$

The goal of this example is to estimate the channel output $q[n]$, reduce the noise in $y[n]$, and to obtain the optimal coefficients of the FIR Wiener filter for the MMSE estimator.

Consider a first-order discrete-time channel ($L = 2$) in this example. By using (7.65), the MMSE estimator based on the Wiener-Hopf equation is obtained by

$$\begin{pmatrix} r_q[0] + r_v[0] & r_q[1] + r_v[1] \\ r_q[1] + r_v[1] & r_q[0] + r_v[0] \end{pmatrix} \begin{pmatrix} w[0] \\ w[1] \end{pmatrix} = \begin{pmatrix} r_q[0] \\ r_q[1] \end{pmatrix}. \qquad (7.69)$$

The correlation function of the channel output $q[n]$ is given by [7]

$$r_q(k) = \sum_{l=0}^{L-k-1} b[l+k]b[l], \qquad (7.70)$$

and the correlation function of the noise is $r_v(k) = \sigma^2 \delta(k)$. Thus the MMSE estimator based on the Wiener-Hopf equation becomes

$$\begin{pmatrix} b^2[0] + b^2[1] + \sigma^2 & b[0]b[1] \\ b[0]b[1] & b^2[0] + b^2[1] + \sigma^2 \end{pmatrix} \begin{pmatrix} w[0] \\ w[1] \end{pmatrix}$$
$$= \begin{pmatrix} b^2[0] + b^2[1] \\ b[0]b[1] \end{pmatrix}. \qquad (7.71)$$

Let $b[0] = 1$, $b[1] = -0.5$ and $\sigma^2 = 1$ in (7.71). We have

$$\begin{pmatrix} 2.25 & -0.5 \\ -0.5 & 2.25 \end{pmatrix} \begin{pmatrix} w[0] \\ w[1] \end{pmatrix} = \begin{pmatrix} 1.25 \\ -0.5 \end{pmatrix}. \quad (7.72)$$

Solving for $w[0]$ and $w[1]$ yields

$$\begin{pmatrix} w[0] \\ w[1] \end{pmatrix} = \begin{pmatrix} 0.5325 \\ -0.1039 \end{pmatrix}. \quad (7.73)$$

In this case, the MMSE estimator in the z-transfer domain is

$$W(z) = 0.5325 - 0.1039 z^{-1}. \quad (7.74)$$

For the MMSE, we obtain the result as follows:

$$\begin{aligned} \xi_{MMSE} &= E\{|e[n]|^2\} \\ &= r_q[0] - (w[0]r_q[0] + w - [1]r_q[1]) \\ &= 0.5324. \end{aligned} \quad (7.75)$$

7.4 Adaptive Channel Estimation and Algorithms

The objective of adaptive channel estimation is to estimate the channel impulse response $h[n]$ in (7.2) given the input signal source $x[n]$ and the noise $v[n]$. The structure of adaptive channel estimation is shown in Figure 7.3, which is also referred to as the *system identification* given by Miao and Clements [6].

In the channel estimation, the adaptive filter model is used to provide a linear model that represents the best fit to an unknown channel impulse response $h[n]$. The unknown channel impulse response $h[n]$ and the adaptive filter model use the same input signal source $s[n]$. Then output of the unknown channel impulse response $h[n]$ provides the desired signal response for the adaptive filter model. When the unknown channel impulse response $h[n]$ is dynamic, the linear model will be time-varying in the channel estimation.

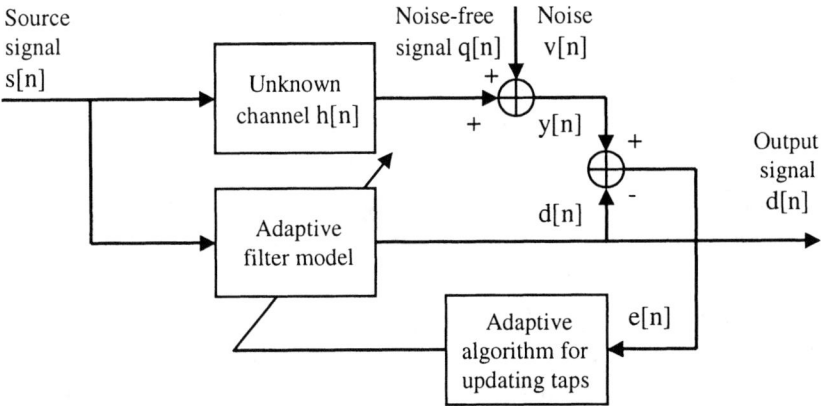

Figure 7.3 Block diagram of adaptive channel estimation.

The adaptive filter model can be assumed to be an FIR filter model,

$$d[n] = \sum_{k=0}^{M-1} b_n[k]s[n-k], \qquad (7.76)$$

where $b_n[k]$ is the time-varying set of the tap coefficients for the FIR filter model at the index of n. In a vector form, (7.76) can be expressed as

$$d[n] = \mathbf{b}_n^T[n]\mathbf{s}[n], \qquad (7.77)$$

where $\mathbf{b}_n^T[n] = [b_n[0], b_n[1], ..., b_n[M-1]]$. The adaptive process updates the filter tap coefficients so that they move from an initial position toward the MMSE solution. The MSE of $\xi_{MSE}[n]$ is obtained by

$$\xi_{MSE}[n] = E\{|e[n]|^2\}, \qquad (7.78)$$

where the error sequence $e[n]$ as shown in Figure 7.3 is written as

$$\begin{aligned} e[n] &= y[n] - d[n] \\ &= \sum_{k=0}^{M-1} h[k]s[n-k] + v[n] - \sum_{k=0}^{M-1} b_n[k]s[n-k] \end{aligned}$$

$$= (\mathbf{h}[n] - \mathbf{b}_n)^T \mathbf{s}[n] + v[n], \quad (7.79)$$

where the channel matrix $\mathbf{h}^T[n] = [h[0], h[1], ..., h[M-1]]$.

In order to minimize the MSE function given by (7.78), we need to update the tap coefficients of the adaptive FIR filter model to estimate the channel impulse response $h[n]$ at each iteration so that the MSE in (7.78) can be achieved in the sense of MMSE. The adaptive algorithm is controlled by the error signal $e[n]$ in (7.79). The adaptive algorithm uses the error signal $e[n]$ to minimize the MSE function in (7.78) and updates the tap coefficients of the adaptive filter in a manner that iteratively reduces the MSE. The commonly used adaptive algorithms for the channel estimation include *least mean squares* (LMS) and *recursive least squares* (RLS).

7.4.1 The LMS Algorithms

In this section, we introduce a set of widely used adaptive techniques called LMS algorithms for the channel estimation. The LMS algorithms are the simplest and the most used algorithm that serves as a benchmark standard against other adaptive algorithms.

Consider an adaptive FIR filter used for estimating a desired signal $d[n]$ from an input signal $x[n]$ as shown in Figure 7.4. The structure of the adaptive FIR filter is called a *transversal filter*, which has $N-1$ delay elements, N taps, and N tunable complex multipliers (or tap weights). The tap weights of the adaptive FIR filter are described by their physical location in the delay line structure and have a subscript, n, to explicitly show that they vary with time. The LMS algorithm, either on a sample-by-sample basis or on a block-by-block basis, continuously updates these tap weights of the adaptive FIR filter.

The transfer function of the adaptive FIR filter is described by

$$\begin{aligned} y[n] &= \sum_{q=0}^{M-1} b_n[q] x[n-q] \\ &= \mathbf{B}_n^T \mathbf{x}[n], \end{aligned} \quad (7.80)$$

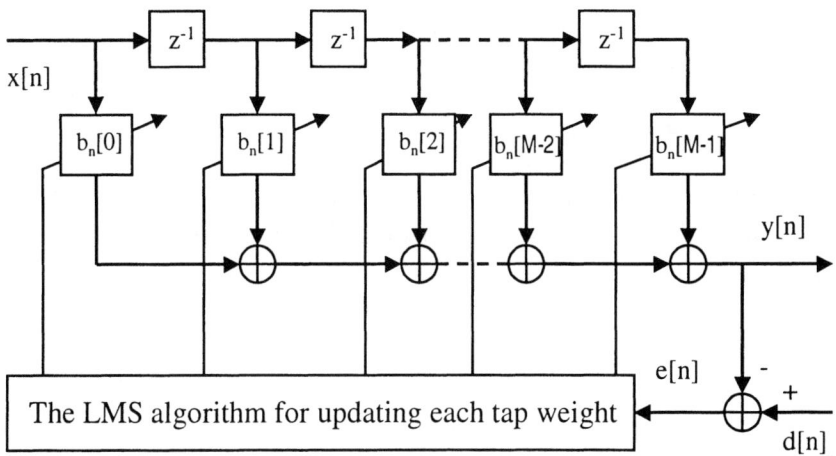

Figure 7.4 The LMS algorithm for updating the adaptive FIR filter.

where $b_n[q]$ is a time-varying set of coefficients for the FIR filter and $\mathbf{B}_n^T = \{b_n[0], b_n[1], \cdots, b_n[M-1]\}$. The objective of the adaptive process is to adjust the filter tap weights so that they move from their current position toward the MMSE solution. The MSE of $\xi_{MSE}[n]$ is defined by

$$\xi_{MSE}[n] = E\{|e[n]|^2\}, \tag{7.81}$$

where the error signal $e[n]$ in Figure 7.4 is given by

$$\begin{aligned} e[n] &= d[n] - y[n] \\ &= d[n] - \mathbf{B}_n^T \mathbf{x}[n], \end{aligned} \tag{7.82}$$

where $d[n]$ is either an exact scaled replica of the transmitted signal or a known property of the transmitted signal.

The LMS algorithm updates the tap weights based on the method of *steepest descent*. The steepest descent algorithm is an iterative procedure, which minimizes the MMSE at time n using an estimate of the tap-weight vector $\mathbf{B}_n$. At time $n + 1$, a new estimate of the tap-weight vector $\mathbf{B}_{n+1}$ is formed by adding a small correction with a *step size* μ to the tap-weight vector $\mathbf{B}_n$, which is used to move the tap-weight vector $\mathbf{B}_n$ closer to the desired optimal solution. This process can be expressed as follows:

$$\mathbf{B}_{n+1} = \mathbf{B}_n - \frac{1}{2}\mu \nabla \xi[n], \qquad (7.83)$$

where $\mathbf{B}_{n+1}$ is the tap-weight vector at time $n+1$, and the step size μ is used to control the rate of convergence. The gradient, which is the derivative of $E\{|e(n)|^2\}$ with respect to the tap-weight vector $\mathbf{B}_n$, is given by

$$\begin{aligned}\nabla \xi[n] &= \frac{\partial E\{|e(n)|^2\}}{\partial \mathbf{B}_n} \\ &= -2E\{e[n]\mathbf{x}[n]\}, \end{aligned} \qquad (7.84)$$

where $\mathbf{x}[n]$ is the input signal vector and $e[n]$ is the error signal as shown in Figure 7.4. Thus, the update equation for tap-weight vector $\mathbf{B}_n$ in the steepest descent algorithm in (7.83) becomes

$$\mathbf{B}_{n+1} = \mathbf{B}_n + \mu E\{e[n]\mathbf{x}[n]\}. \qquad (7.85)$$

Note that if the input signal $x[n]$ and the desired signal $d[n]$ are jointly wide-sense stationary, then we obtain

$$\begin{aligned} E\{e[n]\mathbf{x}[n]\} &= E\{(d[n] - \mathbf{B}_n^T\mathbf{x}[n])\mathbf{x}[n]\} \\ &= E\{d[n]\mathbf{x}[n]\} - E\{\mathbf{x}[n]\mathbf{x}[n]\mathbf{B}_n^T\} \\ &= \mathbf{r}_{dx}[n] - \mathbf{R}_x[n]\mathbf{B}_n. \end{aligned} \qquad (7.86)$$

Thus, in this case, the steepest descent algorithm in (7.85) becomes

$$\mathbf{B}_{n+1} = \mathbf{B}_n + \mu(\mathbf{r}_{dx}[n] - \mathbf{R}_x[n]\mathbf{B}_n). \qquad (7.87)$$

When $\mathbf{B}_n = \mathbf{R}_x^{-1}[n]\mathbf{r}_{dx}[n]$, the second term in (7.87) is zero. Therefore, (7.87) means that the steepest descent of the adaptive filter tap-weight vector $\mathbf{B}_n$ converges to the solution of the Wiener-Hopf equation when

$$\lim_{n \to \infty} \mathbf{B}_n = \mathbf{R}_x^{-1}[n]\mathbf{r}_{dx}[n]. \qquad (7.88)$$

The step size μ satisfies the condition [5]

$$0 < \mu < \frac{2}{\lambda_{max}}, \tag{7.89}$$

where λ_{max} is the maximum eigenvalue of the autocorrelation matrix $\mathbf{R}_x[n]$.

In general, the steepest descent algorithm has a practical limitation because the expectation $E\{e[n]\mathbf{x}[n]\}$ is unknown. However, we can estimate it by using the sample mean,

$$E\{e[n]\mathbf{x}[n]\} = \frac{1}{N}\sum_{k=0}^{N-1} e[n-k]\mathbf{x}[n-k], \tag{7.90}$$

where N is the number of samples. Substituting (7.90) into (7.85), the update equation for the tap-weight vector $\mathbf{B}_n$ becomes

$$\mathbf{B}_{n+1} = \mathbf{B}_n + \frac{\mu}{N}\sum_{k=0}^{N-1} e[n-k]\mathbf{x}[n-k]. \tag{7.91}$$

In a special case, if we use only one sample for estimating the sample mean, the update equation for the tap-weight vector $\mathbf{B}_n$ in (7.91) has a particularly simple form

$$\mathbf{B}_{n+1} = \mathbf{B}_n + \mu e[n]\mathbf{x}[n]. \tag{7.92}$$

Equation (7.92) is called the *LMS algorithm*. Equivalently, we summarize the adaptive channel estimation using the LMS algorithm as follows:

1. Determine the parameters, including the filter length M and the step size μ.
2. Set the initialization of the filter tap-weight vector, $\mathbf{B}_0 = \mathbf{0}$.
3. Compute the LMS algorithm: For $n = 0, 1, 2, ..., N$,

    (a) Calculate the filter output:

    $$y[n] = \mathbf{B}_n^T \mathbf{x}[n]. \tag{7.93}$$

(b) Estimate the error signal:

$$e[n] = d[n] - y[n]. \qquad (7.94)$$

(c) Update the adaptation of the tap-weight vector:

$$\mathbf{B}_{n+1} = \mathbf{B}_n + \mu e[n]\mathbf{x}[n]. \qquad (7.95)$$

In practice, the adaptive filter system with M tap weights using the LMS algorithm requires M multiplications and M additions to update the filter tap weights. Moreover, one addition is needed to compute the error signal $e[n]$, and one multiplication is needed to form the product $\mu e[n]\mathbf{x}[n]$. Thus, M multiplications and $M-1$ additions are needed to calculate the output signal of the adaptive filter. A total of $2(M+1)$ MACs (multiplier-accumulator) per sample output are therefore required.

7.4.2 The LMS Algorithm Convergence

The selection of the step size μ in (7.92) must be done carefully. If the step size μ is too small, the tap-weight vector $\mathbf{B}_n$ will adapt very slowly and may not react to changes in the input signal vector. However, if the step size μ is too large, the adaptive filter system using the LMS algorithm will unduly respond to noise in the signal and may not converge to the MMSE solution.

The LMS algorithm convergence is derived by using a statistical framework where the tap weight vector $\mathbf{B}_n$ is treated as a vector of random variables. Substituting (7.82) into (7.92) gives the update equation of the tap weight as follows:

$$\mathbf{B}_{n+1} = \mathbf{B}_n + \mu \left(d[n] - \mathbf{B}_n^T \mathbf{x}[n] \right) \mathbf{x}[n]. \qquad (7.96)$$

Taking the expectation of both sides of (7.96) obtains

$$E\{\mathbf{B}_{n+1}\} = E\{\mathbf{B}_n\} + \mu E\{d[n]\mathbf{x}[n]\} - \mu E\{\mathbf{x}[n]\mathbf{x}^T[n]\mathbf{B}_n\}. \qquad (7.97)$$

To perform the convergence analysis for (7.97), we assume that the input signal $x[n]$ and the desired signal $d[n]$ are jointly wide-sense stationary. Furthermore, the input signal $x[n]$ and the tap-weight vector $\mathbf{B}_n$ of the LMS adaptation are statistically independent. Note that this independence assumption for statistical analysis of the LMS algorithm is often approximately true, and its use is attributed to Haykin [8]. With this assumption, (7.97) can be rewritten as

$$\begin{aligned} E\{\mathbf{B}_{n+1}\} &= E\{\mathbf{B}_n\} + \mu \mathbf{r}_{dx}[n] - \mu E\{\mathbf{x}[n]\mathbf{x}^T[n]\}E\{\mathbf{B}_n\} \\ &= (\mathbf{I} - \mu \mathbf{R}_x[n])E\{\mathbf{B}_n\} + \mu \mathbf{r}_{dx}[n]. \end{aligned} \qquad (7.98)$$

Further note that the LMS adaptive filter converges to the solution of the Wiener-Hopf equation when $\mathbf{r}_{dx}[n] = \mathbf{R}_x[n]\mathbf{B}$ in (7.87), where $\mathbf{B}$ is the Wiener-Hopf solution. Thus, (7.98) becomes

$$E\{\mathbf{B}_{n+1}\} = (\mathbf{I} - \mu \mathbf{R}_x[n])E\{\mathbf{B}_n\} + \mu \mathbf{R}_x[n]\mathbf{B}. \qquad (7.99)$$

Subtracting $\mathbf{B}$ from both sides in (7.99) gives

$$E\{\mathbf{B}_{n+1}\} - \mathbf{B} = (\mathbf{I} - \mu \mathbf{R}_x[n])(E\{\mathbf{B}_n\} - \mathbf{B}). \qquad (7.100)$$

Since $\mathbf{R}_x[n]$ is an autocorrelation matrix that is symmetric and a positive definite, by using the spectral decomposition theorem in Appendix B, $\mathbf{R}_x[n]$ can be decomposed into

$$\mathbf{R}_x[n] = \mathbf{V}[n]\Lambda\mathbf{V}^T[n], \qquad (7.101)$$

where Λ is a diagonal matrix consisting of the eigenvalues of the autocorrelation matrix $\mathbf{R}_x[n]$, and

$$\mathbf{V}[n]\mathbf{V}^T[n] = \mathbf{I}. \qquad (7.102)$$

Then, substituting (7.101) into (7.100) and using the fact in (7.102), we obtain

$$\mathbf{V}^T[n](E\{\mathbf{B}_{n+1}\} - \mathbf{B}) = (\mathbf{I} - \mu\Lambda)\mathbf{V}^T[n](E\{\mathbf{B}_n\} - \mathbf{B}). \qquad (7.103)$$

Let a new vector be

$$\mathbf{u}_n = \mathbf{V}^T[n](E\{\mathbf{B}_n\} - \mathbf{B}). \qquad (7.104)$$

Substituting (7.104) into (7.103) yields

$$\mathbf{u}_{n+1} = (\mathbf{I} - \mu \Lambda)\mathbf{u}_n. \qquad (7.105)$$

Given an initial vector $\mathbf{u}_0$, (7.105) becomes

$$\mathbf{u}_{n+1} = (\mathbf{I} - \mu \Lambda)^{n+1}\mathbf{u}_0. \qquad (7.106)$$

Note that $(\mathbf{I} - \mu \Lambda)$ in (7.106) is a diagonal matrix. Thus, the ith components of the vector $\mathbf{u}_{n+1}$ may be expressed as follows:

$$u_{n+1}(i) = (1 - \mu \lambda_i)^{n+1} u_0(i). \qquad (7.107)$$

If we want the filter tap-weight vector $\mathbf{B}_{n+1}$ to converge to the mean vector $\mathbf{B}$, then the vector $\mathbf{u}_{n+1}$ must converge to zero. This requires

$$|1 - \mu \lambda_i| < 1, \qquad i = 0, 1, 2, ..., M, \qquad (7.108)$$

where M is the number of filter tap weights. Therefore, in order to prevent the LMS adaptation from becoming unstable, the value of the step size μ must be chosen such that

$$0 < \mu < \frac{2}{\lambda_{max}}, \qquad (7.109)$$

where λ_{max} is the maximum eigenvalue of the autocorrelation matrix $\mathbf{R}_x[n]$.

In practice, knowledge of λ_{max} is not available for the application of the adaptive filter system using the LMS algorithm. The trace of $\mathbf{R}_x[n]$ may be taken as a conservative estimate for λ_{max}. In this case, (7.109) can be rewritten as

$$0 < \mu < \frac{2}{\text{tr}\{\mathbf{R}_x[n]\}}, \qquad (7.110)$$

where $\text{tr}\{\mathbf{R}_x[n]\}$ denotes the trace of autocorrelation matrix $\mathbf{R}_x[n]$. If the input signal $x[n]$ is wide-sense stationary, then $\mathbf{R}_x[n]$ is a Toeplitz matrix. Thus, the trace $\text{tr}\{\mathbf{R}_x[n]\}$ can be rewritten as follows:

$$\text{tr}\{\mathbf{R}_x[n]\} = (M+1)E\{|x[n]|^2\}, \qquad (7.111)$$

where $E\{|x[n]|^2\}$ can be estimated by

$$E\{|x[n]|^2\} = \frac{1}{M} \sum_{k=0}^{M-1} |x(n-k)|^2. \quad (7.112)$$

Therefore, in this case, (7.109) can be reexpressed as follows:

$$0 < \mu < \frac{2}{\frac{M+1}{M} \sum_{k=0}^{M-1} |x(n-k)|^2}. \quad (7.113)$$

7.4.3 The LMS EMSE Analysis and Misadjustment

In the application of the channel estimation, the LMS tap-weight vector $\mathbf{B}_n$ tends to fluctuate about its optimum value when the tap-weight vector $\mathbf{B}_n$ begins to converge to the mean $\mathbf{B}$. This is because of the noisy gradient vectors that are used to produce the small corrections to the tap-weight vector $\mathbf{B}_n$. Thus, the tap-weight vector $\mathbf{B}_n$ does not converge to the mean vector $\mathbf{B}$ exactly. If the MSE is larger than the MMSE by an amount, then the amount is referred to as the *excess mean-squared error* (EMSE).

Using (7.82) and letting $\Psi = \mathbf{B}_n - \mathbf{B}$, we rewrite the error signal as follows:

$$\begin{aligned} e[n] &= d[n] - \mathbf{B}_n^T \mathbf{x}[n] \\ &= d[n] - (\mathbf{B} + \Psi)^T \mathbf{x}[n] \\ &= e_{min}[n] + \Psi^T \mathbf{x}[n], \end{aligned} \quad (7.114)$$

where $e_{min}[n] = d[n] - \mathbf{B}^T \mathbf{x}[n]$ is the error if the solution of the Wiener-Hopf equation were used. If the LMS filter is in steady-state such that $E\{\Psi\} = 0$, then the MSE can be expressed as

$$\begin{aligned} \xi[n] &= E\{|e[n]|^2\} \\ &= \xi_{MMSE} + \xi_{EMSE}[n], \end{aligned} \quad (7.115)$$

where $\xi_{MMSE} = E\{|e_{MMSE}[n]|^2\}$ is the MMSE and $\xi_{EMSE}[n]$ is the EMSE.

Haykin [8] showed that if the step size μ satisfies (7.109) and the condition

$$\sum_{i=1}^{M} \frac{2\lambda_i}{2 - \mu\lambda_i} < 1, \qquad (7.116)$$

where λ_i is the ith eigenvalue of the autocorrelation matrix $\mathbf{R}_x[n]$, then the EMSE for the LMS algorithm is less than the MMSE as follows:

$$\xi_{EMSE}[\infty] = \xi_{MMSE}\left[\mu \sum_{i=1}^{M} \frac{\lambda_i}{2 - \mu\lambda_i}\right]. \qquad (7.117)$$

Further note that if $\mu \ll \frac{2}{\lambda_{max}}$, then $\mu\lambda_i \ll 2$, and

$$\mu \sum_{i=1}^{M} \frac{\lambda_i}{2 - \mu\lambda_i} \approx \frac{\mu}{2} \sum_{i=1}^{M} \lambda_i$$

$$= \frac{\mu}{2}\text{tr}\{\mathbf{R}_x[n]\}. \qquad (7.118)$$

Substituting (7.118) into (7.117) yields the EMSE as follows:

$$\xi_{EMSE}[\infty] \approx \xi_{MMSE}\left[\frac{\mu}{2}\text{tr}\{\mathbf{R}_x\}\right]. \qquad (7.119)$$

Equation (7.119) indicates that the $\xi_{EMSE}[\infty]$ is proportional to the step size μ when $n \to \infty$.

The ratio of the steady-state EMSE $\xi_{EMSE}[\infty]$ to the MMSE ξ_{MMSE} is called as the *LMS misadjustment* $\mathcal{M}$. Using (7.117), the LMS misadjustment $\mathcal{M}$ is defined by

$$\begin{aligned}\mathcal{M} &= \frac{\xi_{EMSE}[\infty]}{\xi_{MMSE}} \\ &= \mu \sum_{i=1}^{M} \frac{\lambda_i}{2 - \mu\lambda_i}.\end{aligned} \qquad (7.120)$$

The LMS misadjustment $\mathcal{M}$ is less than unity if the step-size μ holds the condition in (7.116). If the step size is small such that $\mu \ll \frac{2}{\lambda_{max}}$, then the LMS misadjustment is approximately obtained by

$$\mathcal{M} \approx \frac{\mu}{2}\text{tr}\{\mathbf{R}_x[n]\}. \qquad (7.121)$$

The trade-off analysis among the rate of convergence, the amount of EMSE, and the ability of the adaptive filter to track the signal are important. Since the filter tap-weight vectors $\mathbf{B}_n$ are far from the optimal solution when the LMS adaptation algorithm begins, the step size μ can be large in order to move the tap-weight vectors $\mathbf{B}_n$ rapidly toward the desired solution. However, the step size μ should be decreased to reduce the EMSE $\xi_{EMSE}[\infty]$ when the LMS filter begins to converge in the MMSE solution. Therefore, using the LMS algorithm with a time-varying step size $\mu[n]$ is desirable in the application of channel estimation.

7.4.4 The RLS Algorithms

The RLS algorithm is derived from the least squares solution of the adaptive filter system for the channel estimation. The RLS algorithm utilizes information contained in the particular processed input sample data. This is to say that we can get different adaptive filters for different input sample data. As a result, the adaptive filter tap weights that minimize the least squares error will be optimal for the given sample data rather than statistically optimal over a particular class of processes.

To derive the RLS algorithm for the adaptive filter system in the channel estimation, we need to minimize the total sum of squared errors for all input sample data at discrete-time n. The least squares error function is defined by

$$\xi[n] = \sum_{i=0}^{n} \lambda^{n-i} |e[i]|^2, \qquad (7.122)$$

where $0 < \lambda \leq 1$ is an exponential weighting factor and $e[i]$ is the error signal given by

$$\begin{aligned} e[i] &= d[i] - y[i] \\ &= d[i] - \mathbf{w}^T[n]\mathbf{x}[i], \end{aligned} \qquad (7.123)$$

where $\mathbf{w}[n]$ is the tap-weight vector of the adaptive filter system at discrete-time n, and $\mathbf{x}[i]$ is the input vector at discrete-time i.

Setting the derivative of $\xi[n]$ in (7.122) with respect to $w_k[n]$ equal to zero for $k = 0, 1, 2, \cdots, M$, we generate the optimum value of the tap-weight vector $\mathbf{w}[n]$ defined by the normal equations written in matrix form:

$$\mathbf{R}_x[n]\mathbf{w}[n] = \mathbf{r}_{dx}[n], \qquad (7.124)$$

where $\mathbf{R}_x[n]$ is an $(M + 1) \times (M + 1)$ exponentially weighted deterministic autocorrelation matrix for $x[n]$ as follows,

$$\mathbf{R}_x[n] = \sum_{i=0}^{n} \lambda^{n-i} \mathbf{x}^*[i] \mathbf{x}^T[i], \qquad (7.125)$$

and where $\mathbf{r}_{dx}[n]$, the deterministic cross-correlation between $d[n]$ and $x[n]$ is expressed by

$$\mathbf{r}_{dx}[n] = \sum_{i=0}^{n} \lambda^{n-i} d[i] \mathbf{x}^*[i]. \qquad (7.126)$$

To calculate the least squares estimate of the tap-weight vector $\mathbf{w}[n]$ in (7.124), we need to compute the inverse of the autocorrelation matrix $\mathbf{R}_x[n]$,

$$\mathbf{w}[n] = \mathbf{R}_x^{-1}[n]\mathbf{r}_{dx}[n]. \qquad (7.127)$$

Equation (7.127) is referred to as the least squares estimate $\mathbf{w}[n]$ for the tap-weight vector. However, performing such an implementation directly is too time consuming if the number of tap weights, M, is relatively high. Therefore, in practice, we compute the least squares estimate of the tap-weight vector $\mathbf{w}[n]$ recursively for $n = 1, 2, 3, \cdots, M$.

Since $\mathbf{R}_x[n]$ and $\mathbf{r}_{dx}[n]$ both depend on n, we obtain the recursive equation for updating the value of the autocorrelation matrix as follows:

$$\mathbf{R}_x[n] = \lambda \mathbf{R}_x[n-1] + \mathbf{x}^*[n]\mathbf{x}^T[n], \qquad (7.128)$$

where $\mathbf{R}_x[n-1]$ is the value of the autocorrelation matrix at discrete-time $n - 1$, and the matrix product $\mathbf{x}^*[n]\mathbf{x}^T[n]$ can be considered as the "correction" term in the updating operation.

In a similar way, the cross-correlation matrix may be updated recursively,
$$\mathbf{r}_{dx}[n] = \lambda \mathbf{r}_{dx}[n-1] + d[n]\mathbf{x}^*[n]. \tag{7.129}$$

In order to invert the autocorrelation matrix $\mathbf{R}_x[n]$, we use the matrix inversion lemma as follows:
$$\begin{aligned}\mathbf{A}^{-1} &= (\mathbf{B}^{-1} + \mathbf{C}\mathbf{D}^{-1}\mathbf{C}^T)^{-1} \\ &= \mathbf{B} - \mathbf{BC}(\mathbf{D} + \mathbf{C}^T\mathbf{BC})^{-1}\mathbf{C}^T\mathbf{B}.\end{aligned} \tag{7.130}$$

Let $\mathbf{A} = \mathbf{R}_x[n]$, $\mathbf{B}^{-1} = \lambda \mathbf{R}_x[n-1]$, $\mathbf{C} = \mathbf{x}[n]$, and $\mathbf{D} = 1$. Then, substituting the above definitions into (7.130) obtains the recursive expression of the inverse autocorrelation matrix as follows:

$$\mathbf{R}_x^{-1}[n] = \lambda^{-1}\mathbf{R}_x^{-1}[n-1] - \frac{\lambda^{-2}\mathbf{R}_x^{-1}[n-1]\mathbf{x}[n]\mathbf{x}^T\mathbf{R}_x^{-1}[n-1]}{1 + \lambda^{-1}\mathbf{x}[n]\mathbf{R}_x^{-1}[n-1]\mathbf{x}[n]}. \tag{7.131}$$

To simplify this formula, we further let
$$\mathbf{P}[n] = \mathbf{R}_x^{-1}[n], \tag{7.132}$$

where $\mathbf{P}[n]$ is known as the *inverse autocorrelation matrix*, and
$$\mathbf{k}[n] = \frac{\mathbf{P}[n-1]\mathbf{x}[n]}{\lambda + \mathbf{x}^T[n]\mathbf{P}_x[n-1]\mathbf{x}[n]}, \tag{7.133}$$

where $\mathbf{k}[n]$ is referred to as the *gain vector*. Using the definitions in (7.132) and (7.133), we rewrite (7.131) as follows:
$$\mathbf{P}[n] = \frac{1}{\lambda}\{\mathbf{I} - \mathbf{k}[n]\mathbf{x}^T[n]\}\mathbf{P}[n-1]. \tag{7.134}$$

Equation (7.134) is referred to as the *Riccati* formula for the RLS algorithm. As a result, the desired recursive equation for updating the coefficient-weight vector is given by
$$\begin{aligned}\mathbf{w}[n] &= \mathbf{w}[n-1] + \triangle \mathbf{w}[n-1] \\ &= \mathbf{w}[n-1] + \mathbf{k}[n]\xi[n],\end{aligned} \tag{7.135}$$

where $\xi[n]$ is an *a priori estimation error* defined by

$$\xi[n] = d[n] - \mathbf{w}^T[n-1]\mathbf{x}[n]. \tag{7.136}$$

As we can see, the implementation of the RLS algorithm requires the initialization of the recursion formula of (7.134) with a starting value $\mathbf{P}[0]$, which is the nonsingularity of the autocorrelation matrix $\mathbf{R}_x[n]$. To simplify this procedure of obtaining the initialization value, we change the expression of (7.128) slightly for the autocorrelation matrix $\mathbf{R}_x[n]$ and write as follows:

$$\mathbf{R}_x[n] = \lambda \mathbf{R}_x[n-1] + \delta \lambda^n \mathbf{I}, \tag{7.137}$$

where $\mathbf{R}_x[n-1]$ is given by (7.125), $\mathbf{I}$ is the $(M+1) \times (M+1)$ identity matrix, and δ is a small positive constant. Note that if $n = 0$, (7.137) becomes

$$\mathbf{R}_x[0] = \delta \mathbf{I}. \tag{7.138}$$

Therefore, the initial value of $\mathbf{P}[n]$ is obtained by

$$\begin{aligned} \mathbf{P}[0] &= \mathbf{R}_x^{-1}[0] \\ &= \frac{1}{\delta} \mathbf{I}. \end{aligned} \tag{7.139}$$

Finally, we summarize all of the steps of the RLS algorithm for the channel estimation as follows:

1. Select the filter order of the tap weights M, the exponential weighting factor, λ, and a small positive constant, δ, to initialize the matrix $\mathbf{P}[0]$ given in (7.139).
2. Initialize the RLS algorithm by choosing $\mathbf{P}[0] = \frac{1}{\delta}\mathbf{I}$, where δ is a small positive constant and $\mathbf{w}[0] = \mathbf{0}$.
3. For each instant of time, $n = 1, 2, \cdots, N$, we calculate as follows:

    (a) Using (7.136) finds the a priori estimation error.
    (b) Using (7.133) calculates the gain vector.

Channel Estimation and Blind Identification 223

(c) Using (7.135) updates the tap-weight vector.
(d) Using (7.134) computes the *Riccati* formula.

In application for the channel estimation, the RLS algorithm requires $[3(N + 1)(N + 2)]$ MACs (multiplier-accumulator) [6]. Thus, the computational complexity of the RLS algorithm is much higher than for the LMS algorithm. However, the RLS algorithm generally converges faster than the LMS algorithm.

For theoretic treatment of the RLS algorithm, we suggest Haykin [5]. For other RLS methods based on the slide window and state-space approaches, we refer the reader to Zhao et al. [9] and Sayed and Kailath [10], respectively.

7.4.5 The RLS Algorithm Convergence

In order to perform the analysis of the RLS algorithm convergence, we let $\lambda = 1$ for mathematical simplification. The autocorrelation matrix $\mathbf{R}_x[n]$ in (7.125) can be then reexpressed as follows:

$$\mathbf{R}_x[n] = \sum_{i=0}^{n} \mathbf{x}^*[i]\mathbf{x}^T[i], \qquad (7.140)$$

and the cross-correlation matrix (7.126) becomes

$$\mathbf{r}_{dx}[n] = \sum_{i=0}^{n} d[i]\mathbf{x}^*[i]. \qquad (7.141)$$

Furthermore, we rewrite (7.123) as follows:

$$d[i] = \mathbf{w}_c^T \mathbf{x}[i] + e_w[i], \qquad (7.142)$$

where $e_w[i]$ is white noise with zero mean and variance σ^2, and the weight vector $\mathbf{w}_c$ is constant. Then, substituting (7.142) into (7.141) yields the cross-correlation matrix as follows:

$$\begin{aligned}
\mathbf{r}_{dx}[n] &= \sum_{i=0}^{n}(\mathbf{w}_c^T \mathbf{x}[i] + e_w[i])\mathbf{x}^*[i] \\
&= \sum_{i=0}^{n} \mathbf{x}^*[i]\mathbf{x}^T[i]\mathbf{w}_c + \sum_{i=0}^{n} e_w[i]\mathbf{x}^*[i] \\
&= \mathbf{R}_x[n]\mathbf{w}_c + \sum_{i=0}^{n} e_w[i]\mathbf{x}^*[i].
\end{aligned} \qquad (7.143)$$

Substituting (7.143) into (7.127), we can rewrite (7.127) as follows:

$$\begin{aligned}\mathbf{w}[n] &= \mathbf{R}_x^{-1}[n](\mathbf{R}_x[n]\mathbf{w}_c + \sum_{i=0}^{n} e_w[i]\mathbf{x}^*[i]) \\ &= \mathbf{w}_c + \mathbf{R}_x^{-1}[n]\sum_{i=0}^{n} e_w[i]\mathbf{x}^*[i].\end{aligned} \qquad (7.144)$$

In order to analyze the RLS algorithm convergence further, we now need to make two assumptions as follows:

1. The input vectors, $\mathbf{x}[1]$, $\mathbf{x}[2]$, ..., $\mathbf{x}[n]$, are statistically independent.
2. The measurement error $e_w[i]$ is independent of the input vector $\mathbf{x}[i]$ for $n = 1, 2, ..., n$.

Given these two assumptions, we take the expectation value for both sides of (7.144) with $n \geq M$ and obtain

$$\begin{aligned}E\{\mathbf{w}[n]\} &= E\{\mathbf{w}_c\} + E\left\{\mathbf{R}_x^{-1}[n]\sum_{i=0}^{n} e_w[i]\mathbf{x}^*[i]\right\} \\ &= \mathbf{w}_c,\end{aligned} \qquad (7.145)$$

where M is the number of tap weights in the adaptive filter system and the measurement error $e_w[i]$ has zero mean. As a result, (7.145) indicates that the RLS algorithm is convergent in the mean value, $\mathbf{w}_c$, for $n \geq M$. In other words, this means that the RLS algorithm does not wait for convergence for $n \to \infty$.

7.4.6 The RLS EMSE Analysis and Misadjustment

The MSE of the RLS algorithm with setting $\lambda = 1$ is given by [6, 8]

$$\xi[n] = \sigma^2 + \frac{M\sigma^2}{n - M - 1}, \quad n > M + 1 \qquad (7.146)$$

where σ^2 is the variance of the measurement error $e_w[n]$ and M is the number of filter tap weights in the adaptive filter system. When $n \to \infty$, the second term in (7.146) is approximately zero. Under a stationary environment, (7.146) indicates that the RLS algorithm generates zero EMSE when $n \to \infty$. Therefore, using (7.120), we obtain that the RLS misadjustment $\mathcal{M} \approx 0$.

7.4.7 Comparison of the Adaptive Algorithms

The RLS algorithm convergence rate is usually an order of magnitude faster than the LMS algorithm convergence. The reason is that there are fundamental differences between the RLS algorithm and the LMS algorithm. The LMS algorithm uses the step size μ, while the RLS algorithm uses the inverse of the autocorrelation matrix $\mathbf{R}_x^{-1}[n]$ of the input signal vector. This results in a profound impact on the convergence behavior of the RLS algorithm. Thus, the RLS algorithm convergence does not wait for $n \to \infty$.

The RLS algorithm has approximately zero EMSE. On the other hand, the EMSE of the LMS algorithm approximately equals $\xi_{MMSE}\left[\frac{\mu}{2}\text{tr}\{\mathbf{R}_x\}\right]$. This is because the EMSE of the LMS algorithm is proportional to the step size μ when $n \to \infty$. Thus, the misadjustment of the LMS algorithm is much greater than the RLS algorithm.

The LMS algorithm has much fewer computation complexities compared with the RLS algorithm. In addition, the LMS algorithm does not have a stability issue, while the stablity of RLS algorithms needs to be considered during operation. This leads to the LMS algorithm as the most often used adaptive algorithm for adaptive channel estimation.

Example 7.4

In this example, we show an adaptive channel estimation using the exponentially weighted RLS algorithm based on the block diagram of adaptive channel estimation as shown in Figure 7.3. Assume that the transfer function of unknown channel $H(z)$ has an expression form as follows:

$$H(z) = 0.5 + 0.95z^{-1} - 0.5z^{-2} - 0.25z^{-3}, \qquad (7.147)$$

and $v[n]$ is white noise with zero mean and unit variance. The input signal $s[n]$ is the random sequence with 1 or -1.

The goal of the example is to estimate the unknown tap weights of the fourth-order adaptive FIR filter given by

$$y[n] = \sum_{k=0}^{3} w_n[k]x[n-k], \qquad (7.148)$$

where $w_n[k]$ ($k = 0, 1, 2, 3$) are the tap weights to be estimated using the exponentially weighted RLS algorithm. Then the error signal is obtained by

$$e[n] = d[n] - y[n], \qquad (7.149)$$

and the estimated average mean square error is given by:

$$\xi_{MMSE}[n] = \frac{1}{N} \sum_{k=0}^{N-1} |e[k]|^2, \qquad (7.150)$$

where N is the number of input samples.

For initialization of the RLS algorithm, we let the exponential weight $\lambda = 0.999$ and $\delta = 0.0001$. The simulation results are shown in Figure 7.5. After 1,000 iterations, we see that the tap weights $w_n[0]$, $w_n[1]$, $w_n[2]$, and $w_n[3]$ are converged to 0.5, 0.95, -0.55, and -0.25, respectively.

7.5 Channel Models and HOS Estimations

In this section, we introduce three commonly used types of the discrete-time channel structures and their estimation methods based on statistic assumptions of the *higher-order statistics* (HOS) (the order ≥ 3). Depending on the number of channel inputs and outputs, these discrete-time channel structures are classified as either SISO, SIMO, or MIMO.

7.5.1 SISO Channel Model and Estimation

Consider a discrete-time channel model consisting of a linear time-invariant (LTI) impulse response $h[n]$ followed by additive noise $v[n]$ as shown in Figure 7.6. This discrete-time channel is called the SISO channel. The output sequence $y[n]$ is obtained by

$$y[n] = q[n] + v[n], \qquad (7.151)$$

where the noise-free signal sequence $q[n]$ can be expressed as

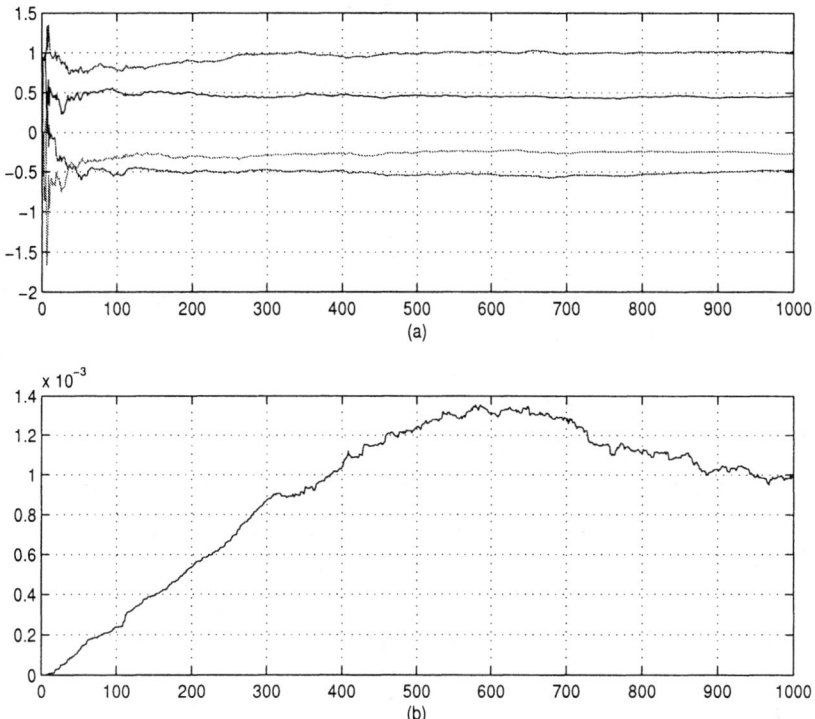

Figure 7.5 Adaptive channel estimation using the exponentially weighted RLS algorithm with 1,000 iterations: (a) estimated coefficients $w_n[1]$, $w_n[2]$, $w_n[3]$, and $w_n[4]$; and (b) estimated average mean square errors.

$$q[n] = \sum_{k=-N}^{N} h[k]s[n-k]. \qquad (7.152)$$

Thus, (7.152) is the noise-free signal distorted by an unknown discrete-time LTI SISO channel $h[n]$ and a noise sequence $v[n]$. The receiver signal-to-noise (SNR) associated with the received signal $y[n]$ for the discrete-time SISO channel is defined as

$$\text{SNR}_{SISO} = \frac{E\{|q[n]|^2\}}{E\{|v[n]|^2\}}. \qquad (7.153)$$

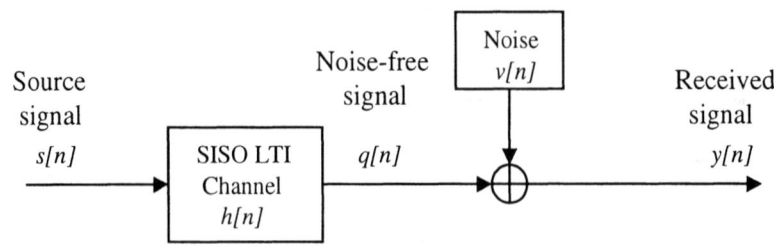

Figure 7.6 The discrete-time single-input single-output (SISO) channel.

The discrete-time SISO channel can be estimated either by sending a training sequence or by using a blind estimator. For a method of the HOS (the order ≥ 3), the blind channel estimation method is generally based on the following assumptions for the received signal $y[n]$ modeled by (7.151) and (7.152):

1. The discrete-time LTI SISO channel $h[n]$ is stable.
2. The source signal $s[n]$ is a zero-mean, independent, and identically distributed non-Gaussian random process with a variance $\sigma_s^2 = E\{|s[n]|^2\}$.
3. The noise sequence $v[n]$ is a zero-mean Gaussian random process that can be colored with the correlation function $R_v(k) = E\{v[n]v^*[n-k]\}$ and the $(p+q)$th-order cumulant $C_{p,q}\{s[n]\} \neq 0$.
4. The source signal $s[n]$ is statistically independent of the noise $v[n]$.

Figure 7.7 shows a block diagram of a discrete-time SISO channel followed by a linear equalizer $h_{le}[n]$. The equalization approach is equivalent to finding the linear equalizer $h_{le}[n]$ such that $H_{le}(z) = 1/H(z)$ or an inverse system estimate, that is, $z[n] = \beta s[n-\tau]$. This leads to the name "inverse filtering approach."

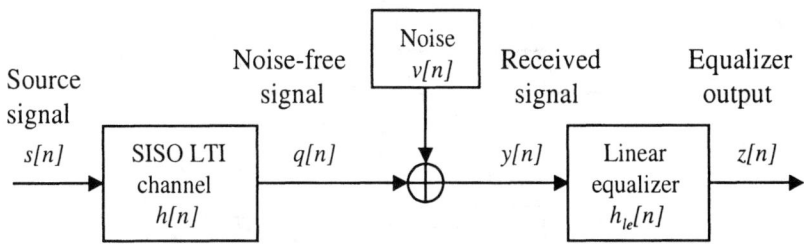

Figure 7.7 Block diagram of a discrete-time channel SISO and linear equalizer.

Assume that the output sequence $y[n]$ is generated by using (7.151) and (7.152) under the above assumptions. By using blind channel estimation with the estimate of the input signal source $\hat{s}[n] = z[n]$, the discrete-time SISO channel can then be estimated via

$$\hat{h}[n] = \frac{E\{y[n+1]\hat{s}[n]\}}{E\{|\hat{s}[n]|^2\}}. \qquad (7.154)$$

If $\text{SNR}_{SISO} = \infty$ in (7.153), the output of the linear equalizer is indeed equal to $z[n] = \beta s[n - \tau]$. In this case, (7.154) leads to the discrete-time SISO channel estimation $\hat{h}[n] = h[n]$. However, in practice, SNR_{SISO} in (7.153) has finite value. Hence, the resultant estimator $\hat{h}[n]$ has bias because of the noise in the output sequence $y[n]$. This results in the input signal estimate $\hat{s}[n]$ consisting of the estimation error from the output of the linear equalizer.

7.5.2 SIMO Channel Model and Estimation

Consider a channel model consisting of L subchannels fed from a common input source $s[n]$ as depicted in Figure 7.8. Assume that the discrete-time channel is modeled as an FIR filter, and the received signal $y[n]$ is oversampled at $t = iT/L$, with their individual outputs $y_i[n]$, $i = 0, 1, 2, \cdots, L - 1$, sampled at the symbol rate of $1/T$. This channel model is referred to as the discrete-time SIMO channel.

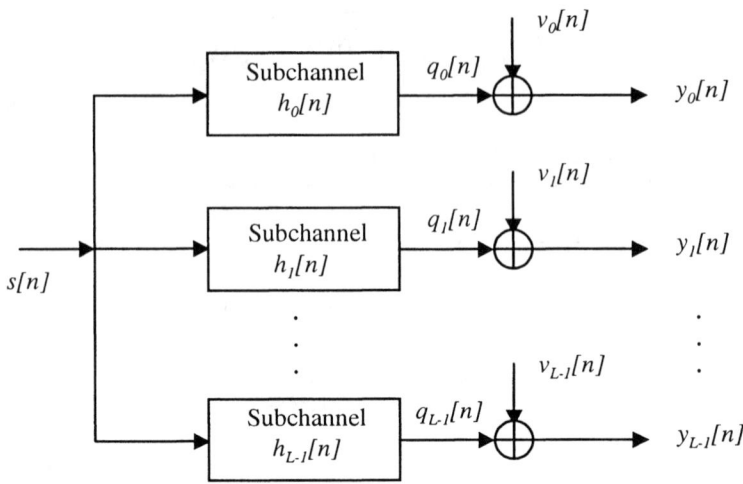

Figure 7.8 The single-input multiple-output (SIMO) discrete-time channel.

Each of the subchannels $h_i[n]$ has the same time index and its noise contribution is $v_i[n]$, where $i = 0, 1, 2, \cdots, L-1$. Thus, we may describe the oversampled discrete-time SIMO channel in the simplified vector form

$$\mathbf{y}_i[n] = \mathbf{h}_i[n]\mathbf{s}[n] + \mathbf{v}_i[n], \quad i = 0, 1, \cdots, L-1, \tag{7.155}$$

where the transmitted input signal source vector $\mathbf{s}[n]$ consisting of (M+N) symbols is defined as

$$\mathbf{s}[n] = [x[n], x[n-1], \cdots, x[n-M-N+1]]^T, \tag{7.156}$$

the $N \times 1$ noise vector is

$$\mathbf{v}_i[n] = [v_i[n], v_i[n-1], \cdots, v_i[n-N+1]]^T, \tag{7.157}$$

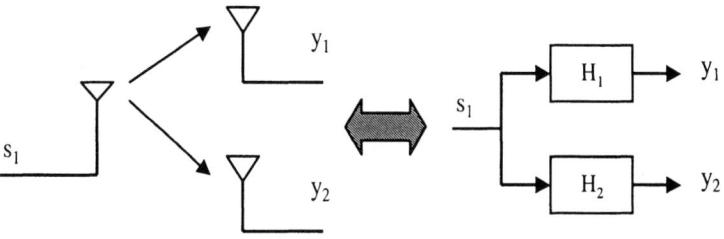

Figure 7.9 The relationship of a transmitter antenna and two receiver antennas for the SIMO channels.

the $N \times (M + N)$ filtering matrix $\mathbf{h}_i[n]$ (also referred to as a multichannel filtering matrix) is a Toeplitz structure given by

$$\mathbf{h}_i[n] = \begin{bmatrix} h_i[0] & h_i[1] & \cdots & h_i[M] & 0 & \cdots & 0 \\ 0 & h_i[0] & \cdots & h_i[M-1] & h_i[M] & \cdots & 0 \\ \vdots & \vdots & \ddots & \vdots & \vdots & \ddots & \vdots \\ 0 & 0 & \cdots & h_i[0] & h_i[1] & \cdots & h_i[M] \end{bmatrix},$$
(7.158)

and the received signal vector is

$$\mathbf{y}_i[n] = [y_i[n], y_i[n-1], \cdots, y_i[n-N+1]]^T. \quad (7.159)$$

At the receiver, each received signal $y_i[n]$ consists of NL samples.

It has been noted by Paulraj et al. [11] that the discrete-time sequence $y[n]$ in the discrete-time SISO channel, obtained by sampling at the symbol rate at 1/T, is wide-sense stationary, while the discrete-time sequence $y[n]$ in the discrete-time SIMO channel obtained by temporal oversampling (at a rate higher than $1/T$) or spatial oversampling (multiple antenna elements) is *cyclostationary*. Figure 7.9 shows the relationship between a transmitter antenna and two receiver antennas for the SIMO channels. The cyclostationary signal consists of a number of

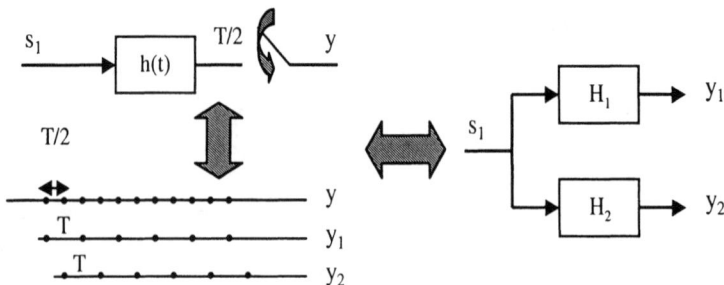

Figure 7.10 The relationship of temporal and spatial oversampling results in a polyphase-based discrete-time SIMO channel.

phases, each of which is stationary. A phase corresponds to a different sampling point in temporal oversampling and a different antenna element in spatial oversampling. For example, the duality relationship between temporal and spatial oversampling is illustrated in Figure 7.10 for polyphase discrete-time SIMO channels.

The cyclostationary property of oversampled signals in the discrete-time SIMO channel can carry important information about the channel phase that can be exploited in several ways for blind identification of the channel. This is because the oversampling of the discrete-time SIMO channel increases the number of samples in the signal sequence $y[n]$ and phases in the channel $\mathbf{h}_i[n]$, but does not change the value of the data for the duration of the symbol period. This allows the channel $\mathbf{h}_i[n]$ to become more rows than columns (tall) and to have a full-column rank. In addition, the stationarity of the discrete-time channel makes $\mathbf{h}_i[n]$ block Toeplitz. Tallness and Toeplitz properties are keys to the blind identification of the channel $\mathbf{h}_i[n]$. Therefore, exploiting the cyclostationarity of the signal output $y[n]$ in the discrete-time SIMO channel can lead to second-order statistics-based blind algorithms to identify the channels $\mathbf{h}_i[n]$, which are more attractive than HOS techniques.

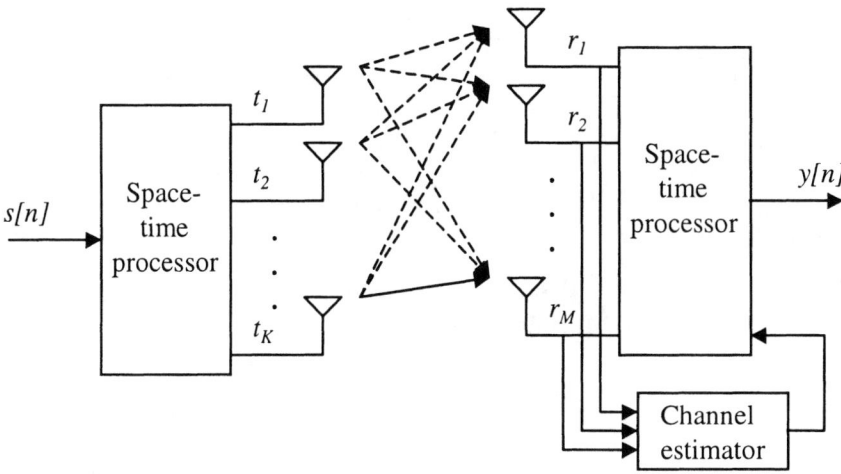

Figure 7.11 Block diagram of MIMO system with multiple antenna elements.

The discrete-time SIMO channel model has been extensively used in the fractionally spaced equalizer, the polyphase filter bank–based equalizer, and subspace-based multichannel identification given by Moulines et al. [12].

7.5.3 MIMO Channel Model and Estimation

MIMO channels have recently emerged as one of the most significant technical breakthroughs in modern digital communication. The MIMO channel is sometimes referred to as a "volume-to-volume" wireless link [13]. In an arbitrary wireless communication system, we consider a wireless link in which the transmitter end and the receiver end are connected with multiple antenna elements as shown in Figure 7.11. The rationale behind MIMO channel is to improve the bit error rate (BER) or the data rate (bits per second) of the communication for each MIMO user when the signal sequences in the transmit antennas at one end and the receive antennas at the other end are

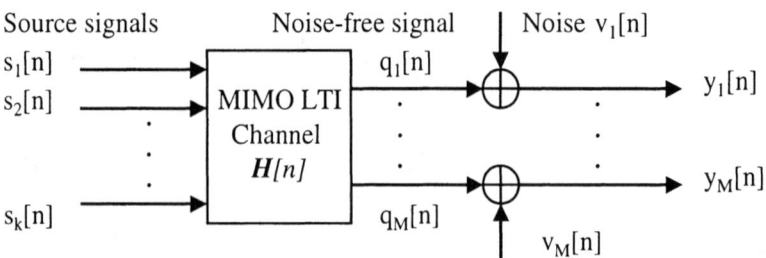

Figure 7.12 The discrete-time multiple-input multiple-output (MIMO) channel.

combined together. This is because MIMO channels and systems are able to turn multipath propagation into a benefit for the user and effectively take advantage of random fading as well as multipath delay spread to multiply transfer rates. Inherently, a key idea of MIMO channels and systems is *space-time signal processing*, in which time (or the natural dimensional of digital communication data) is complemented with the spatial dimension by the use of multiple spatially distributed antennas. Hence, a MIMO channel can be viewed as an extension of *smart antennas*, which is a popular technology using antenna arrays for improving wireless transmission.

Consider K different source signals $s_i[n]$, where $i = 1, 2, \cdots, K$, simultaneously transmitted through a multipath channel. At the receiver, all the source signals are received by an M-element antenna array in the presence of additive noise. The structure of discrete-time MIMO channel model is depicted in Figure 7.12.

Let the $K \times 1$ source signal vector be

$$\mathbf{s}[n] = [s_1[n], s_2[n], \cdots, s_k[n]]^T, \qquad (7.160)$$

the $M \times 1$ noise-free signal vector be

$$\mathbf{q}[n] = [q_1[n], q_2[n], \cdots, q_M[n]]^T, \qquad (7.161)$$

the $M \times 1$ noise vector be

$$\mathbf{v}[n] = [v_1[n], v_2[n], \cdots, v_M[n]]^T, \quad (7.162)$$

and the $M \times 1$ received signal vector be

$$\mathbf{y}[n] = [y_1[n], y_2[n], \cdots, y_M[n]]^T. \quad (7.163)$$

Then, the $M \times 1$ received signal vector can be expressed as

$$\mathbf{y}[n] = \mathbf{q}[n] + \mathbf{v}[n], \quad (7.164)$$

where the noise-free signal vector is

$$\mathbf{q}[n] = \sum_{k=-N}^{N} \mathbf{H}[k]\mathbf{s}[n-k], \quad (7.165)$$

where $\mathbf{H}[n] = \{h_{ji}[n]\}$ denotes the channel response from the transmitted antenna i ($i = 1, 2, \cdots, K$) to the received antenna j ($j = 1, 2, \cdots, M$). Thus, (7.165) is distorted by the $M \times K$ discrete-time LTI MIMO channel $\mathbf{H}[n]$, the $K \times 1$ source signal vector $\mathbf{s}[n]$, and the $M \times 1$ noise vector $\mathbf{v}[n]$.

Note that the discrete-time MIMO channel in (7.164) and (7.165) not only has the intersymbol interference (ISI) but also involves *multiple access interference* (MAI). This is because each component of $\mathbf{q}[n]$ is a mixture of all the signal $s_i[n]$, $i = 1, 2, \cdots, K$. When there is only one source signal for the transmitter, the discrete-time MIMO channel becomes the discrete-time SIMO channel.

Given the received signal $\mathbf{y}[n]$ in (7.164), the receiver SNR associated with the received signal $\mathbf{y}[n]$ is obtained by

$$\begin{aligned} \text{SNR}_{MIMO} &= \frac{E\{|\mathbf{q}[n]|^2\}}{E\{|\mathbf{v}[n]|^2\}} \\ &= \frac{E\{|\mathbf{y}[n] - \mathbf{v}[n]|^2\}}{E\{|\mathbf{v}[n]|^2\}}. \end{aligned} \quad (7.166)$$

The discrete-time MIMO channel can be estimated either by using an adaptive approach with sending a training sequence or by using blind identification. The performance of the adaptive estimation and/or blind identification for the discrete-time MIMO channel is mainly dependent on the SNR_{MIMO} in (7.166). In the past decade, the blind identification of the discrete-time MIMO channel using HOS has been extensively studied by Tugnait [14], Papadias and Paulraj [15], Li and Liu [16], Chi and Chen [17], and Chi et al. [18]. This is generally based on the following assumptions:

1. The discrete-time MIMO channel $\mathbf{H}[n]$ is stable.
2. The source signal $\mathbf{s}[n]$ is a zero-mean, independent, and identically distributed non-Gaussian random process with the variance $\sigma_\mathbf{s}^2[k] = E\{|\mathbf{s}[n]|^2\}$ and the $(p+q)$th-order cumulant $C_{p,q}\{s_i[n]\} \neq 0$.
3. The source signal $s_i[n]$, $i = 1, 2, \cdots, K$, is statistically independent of $s_j[n]$ for all $j \neq i$.
4. The noise sequence $\mathbf{v}[n]$ is a zero-mean Gaussian random process that can be spatially correlated and temporally colored with covariance matrix $R_\mathbf{v}(k) = E\{\mathbf{v}[n]\mathbf{v}^*[n-k]\}$.
5. The source signal $\mathbf{s}[n]$ is statistically independent of the noise $\mathbf{v}[n]$.

Figure 7.13 shows a block diagram of a discrete-time MIMO channel followed by a set of linear equalizers $h_{le,i}[n]$, $i = 1, 2, \cdots, M$, in parallel. Let

$$\mathbf{h}_{le,i}[n] = [h_{le,1}[n], h_{le,2}[n], \cdots, h_{le,M}[n]] \qquad (7.167)$$

be a multiple-input single-output (MISO) equalizer that consists of a linear FIR filter bank with an infinite length. This equalization approach is to find the linear equalizer $\mathbf{h}_{le}[n]$ such that the equalizer output for a perfect equalization is

$$z[n] = \beta s_i[n - \tau], \qquad (7.168)$$

where β is a real or complex constant and τ is an integer.

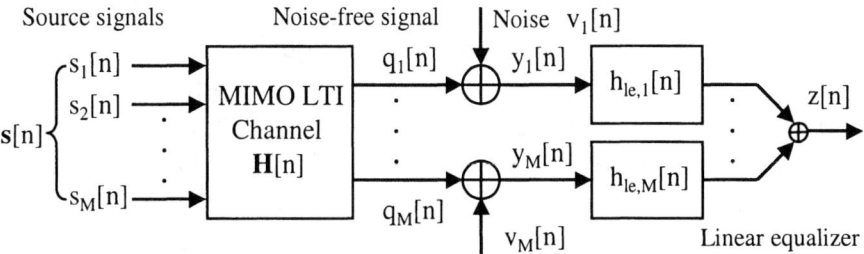

Figure 7.13 Block diagram of a discrete-time MIMO channel cascaded with a set of linear equalizers in parallel.

Also define the $M \times 1$ $\mathbf{h}_k[n]$ as the kth column of the $M \times K$ channel impulse response matrix of $\mathbf{H}[n]$. Thus, the received signal in (7.164) can be rewritten as

$$\mathbf{y}[n] = [\mathbf{h}_1[n], \mathbf{h}_2[n], \cdots, \mathbf{h}_K[n]] * \mathbf{s}[n] + \mathbf{v}[n]$$
$$= \sum_{i=1}^{K} \sum_{k=-N}^{N} \mathbf{h}_i[k] s_i[n-k] + \mathbf{v}[n]. \qquad (7.169)$$

Assume that the estimate of the input source signal is $\hat{s}_i[n] = z[n]$ by using (7.168). Then, the discrete-time MIMO channel of $\mathbf{h}_k[n]$ can be estimated by

$$\hat{\mathbf{h}}_k[n] = \frac{E\{\mathbf{y}[n+1]\hat{s}_k[n]\}}{E\{|\hat{s}_k[n]|^2\}}. \qquad (7.170)$$

Updating the received signal $\mathbf{y}[n+1]$ is obtained by

$$\mathbf{y}[n+1] = \mathbf{y}[n] - \sum_{i=-N}^{N} \hat{\mathbf{h}}_k[i] * s_k[n-i]. \qquad (7.171)$$

Equation (7.171) cancels the component of $s_k[n]$ from the received signal matrix $\mathbf{y}[n]$. As a result, one column of the MIMO channel matrix $\mathbf{H}[n]$ can be estimated by using (7.170) and (7.171) for each stage. Therefore, the entire estimate of discrete-time MIMO channel $\mathbf{H}[n]$ can be obtained after running K stages.

The output of the filter bank of linear equalizers is indeed equal to (7.168) when $\text{SNR}_{MIMO} = \infty$ in (7.166). In this case, (7.170) and (7.171) lead to the discrete-time MIMO channel estimation $\hat{\mathbf{H}}[n] = \mathbf{H}[n]$. However, in reality, SNR_{MIMO} in (7.166) is finite valued. As a result, the resultant estimator $\hat{\mathbf{H}}[n]$ has bias because of the noise in the received matrix $\mathbf{y}[n]$. In addition, the estimate of a column of the MIMO channel matrix $\mathbf{H}[n]$ obtained at the kth stage may suffer from the estimate error for a larger k because of error propagation of the estimate procedure. This results from the input signal estimate $\hat{s}[n]$ consisting of the estimation errors from the output of the filter bank of the linear equalizers $\mathbf{h}_{le}[n]$.

7.6 Blind Channel Identification

Wireless and wired digital communications often require the identification of the channel impulse response that can facilitate channel equalization and maximum likelihood sequence detection. A blind channel identification is used to estimate channel without using a training sequence. Instead, the channel identification is achieved by using only the channel output along with certain a priori statistical information on the input source. The blind channel identification methods attributed to Tong et al. [19, 20] have attracted research attention using the second-order cyclostationary statistics. In fact, the blind channel identification based on the method of the second-order cyclostationary statistics has been considered a major technical contribution. In this section, we introduce the time-domain approach of blind identification for a SISO channel based on the second-order cyclostationary statistics, subspace-based blind identification for a SIMO channel, and the frequency-domain approach of blind identification for a MIMO channel.

7.6.1 Blind Identification for SISO Channel

Consider the SISO channel as shown in Figure 7.6 with the received signal vector $\mathbf{y}[n]$ in (7.151) satisfying

$$\mathbf{y}[n] = \mathbf{h}\mathbf{s}[n] + \mathbf{v}[n], \qquad (7.172)$$

where $\mathbf{h}$ is the channel matrix, $\mathbf{s}[n]$ is the source input vector, and $\mathbf{v}[n]$ is the noise vector. If the received signal in (7.172) is sampled at the baud rate $1/T$, then $\mathbf{y}[n]$ is a wide-sense stationary process. In this case, only minimum-phase channels can be identified from the second-order statistics. This is because the phase information of the channel is lost in the second-order statistics when its output is sampled at the baud rate of $1/T$. If the sampling rate is higher than the baud rate of $1/T$, the received signal vector $\mathbf{y}[n]$ is then widesense cyclostationary. As a result, the second-order statistics of the observation with oversampling rates include the phase information of the SISO channel.

The objective of the blind channel identification in (7.172) is to identify the channel matrix $\mathbf{h}$ from the received signal $\mathbf{y}[n]$. In order to identify the channel matrix, $\mathbf{h}$ must be a matrix with an $M \times d$ full column rank condition. Oversampling can satisfy this. The input signal source $\mathbf{s}[n]$ is a zero mean stationary process with the autocorrelation function

$$\mathbf{R_S}(k) = E\{\mathbf{s}[n]\mathbf{s}^H[n-k]\} \\ = \begin{cases} \mathbf{J}^k, & k \geq 0 \\ \left(\mathbf{J}^H\right)^{|k|}, & k < 0, \end{cases} \qquad (7.173)$$

where $(\cdot)^H$ is the Hermitian transpose and $\mathbf{J}$ is a $d \times d$ shifting matrix as

$$\mathbf{J} = \begin{bmatrix} 0 & 0 & \cdots & 0 & 0 \\ 1 & 0 & \cdots & 0 & 0 \\ 0 & 1 & \cdots & 0 & 0 \\ \vdots & \vdots & \ddots & \vdots & \vdots \\ 0 & 0 & \cdots & 1 & 0 \end{bmatrix}. \qquad (7.174)$$

Then the correlation matrix of the received signal $\mathbf{R_y}(k)$ is obtained by

$$\mathbf{R_y}(k) = \mathbf{h R_S}(k)\mathbf{h}^H + \mathbf{R_V}(k). \tag{7.175}$$

With the assumption of white noise, the noise correlation matrix $\mathbf{R_V}(k)$ can be expressed as

$$\begin{aligned}\mathbf{R_V}(k) &= E\{\mathbf{v}[n]\mathbf{v}^H[n-k]\} \\ &= \sigma^2 \mathbf{J}^{kT_s},\end{aligned} \tag{7.176}$$

where σ^2 is the unknown noise variance, $\mathbf{J}$ is the shift matrix given in (7.174), and T_s is an integer such that $T = T_s \Delta$.

There is no a priori information either on the noise covariance or on the signal space dimension d. However, they can be estimated from the data covariance matrix $\mathbf{R_y}(k)$ in (7.175) when $k = 0$. It can be shown by Miao and Clements [6] that the spectral decomposition or the singular value decomposition (SVD) of $\mathbf{R_y}(0)$ must have the expression form as

$$\mathbf{U}^H \mathbf{R_y}(0)\mathbf{U} = \Lambda, \tag{7.177}$$

where $\mathbf{U}$ is an orthogonal matrix whose columns are normalized eigenvectors, and

$$\Lambda = \text{diag}(\lambda_1 + \sigma^2, \cdots, \lambda_d + \sigma^2, \sigma^2, \cdots, \sigma^2), \tag{7.178}$$

where $\lambda_1 \geq \lambda_2 \geq \cdots \geq \lambda_d > 0$. Therefore, both the noise variance σ^2 and the signal space d can be determined by observing the most significant singular values of $\mathbf{R_y}(0)$.

Tong et al. [19] have shown that if the channel matrix $\mathbf{h}$ and the input signal source $\mathbf{s}[n]$ satisfy the linear equation in (7.172) and its constraints, then the channel matrix $\mathbf{h}$ is uniquely identified by using the second-order cyclostationary statistics of $\mathbf{R_y}(0)$ and $\mathbf{R_y}(1)$.

The correlation matrices $\mathbf{R_y}(0)$ and $\mathbf{R_y}(1)$ can be estimated from the observation data $\mathbf{y}[n]$ via time-index averaging,

$$\hat{\mathbf{R}}_\mathbf{y}(0) = \frac{1}{N}\left[\sum_{k=0}^{N-1} \mathbf{y}[n]\mathbf{y}^H[n]\right], \tag{7.179}$$

and
$$\hat{\mathbf{R}}_{\mathbf{y}}(1) = \frac{1}{N}\left[\sum_{k=0}^{N-1} \mathbf{y}[n]\mathbf{y}^H[n-1]\right]. \qquad (7.180)$$

Thus, the noise covariance σ^2 and the signal space d can be estimated by using (7.177) along with the estimate of (7.179). Subtracting the corresponding noise covariance σ^2 from the observation correlation matrix $\hat{\mathbf{R}}_{\mathbf{y}}(0)$ yields

$$\tilde{\mathbf{R}}_0 = \hat{\mathbf{R}}_{\mathbf{y}}(0) - \hat{\sigma}^2 \mathbf{I}. \qquad (7.181)$$

The SVD of $\tilde{\mathbf{R}}_0$ in (7.181) has the form

$$\mathbf{U}_s^H \left[\hat{\mathbf{R}}_{\mathbf{y}}(0) - \hat{\sigma}^2 \mathbf{I}\right] \mathbf{U}_s = \Lambda_{sr}, \qquad (7.182)$$

where $\mathbf{U}_s$ is the singular vector associated with the largest singular value, and Λ_{sr} is the positive square-root of the d largest singular value.

Let the matrix $\mathbf{F} = \Lambda_{sr}^{-1}\mathbf{U}_s^H$ and $\mathbf{z}_d$ denote the smallest singular value. Then the SVD of the second-order $\tilde{\mathbf{R}}_1$ can be formed by subtracting the corresponding noise correlation $\mathbf{R}_{\mathbf{V}}(1)$ from the estimated observation correlation matrix $\hat{\mathbf{R}}_{\mathbf{y}}(1)$,

$$\tilde{\mathbf{R}}_1 = \mathbf{F}\left[\hat{\mathbf{R}}_{\mathbf{y}}(1) - \mathbf{R}_{\mathbf{V}}(1)\right]\mathbf{F}^H, \qquad (7.183)$$

where the second-order noise correlation matrix $\mathbf{R}_{\mathbf{V}}(1)$ is given by

$$\mathbf{R}_{\mathbf{V}}(1) = \hat{\sigma}^2 \mathbf{J}^{T_s}. \qquad (7.184)$$

The blind channel identification based on the second-order cyclostationary statistics is obtained as

$$\hat{\mathbf{H}} = \mathbf{U}_s \Lambda_{sr} \mathbf{Q}, \qquad (7.185)$$

where

$$\mathbf{Q} = [1, \tilde{\mathbf{R}}_1, \tilde{\mathbf{R}}_1^2, \cdots, \tilde{\mathbf{R}}_1^{d-1}]\mathbf{z}_d. \qquad (7.186)$$

The blind channel identification method of (7.186) can provide the exact identification for nonminimum-phase channels if the correlation matrix $\mathbf{R_y}(k)$ of the received signal is known exactly. In addition, this method is usually insensitive to the timing recovery. This is because sampling of the received signal $y[n]$ is at a rate higher than the baud rate. Moreover, the full rank condition of channel matrix $\mathbf{h}$ can be achieved if the sampling frequency F_s satisfies the following condition:

$$F_s > \frac{1+d}{L}. \qquad (7.187)$$

That is to say that the sampling frequency F_s must be at least $(d+k)/k$ times faster than the baud rate if $L = kT$ and there are d symbols having contributions to the received signal $y[n]$. Thus, the blind channel identification method can identify possible nonminimum-phase channels based on the second-order statistics by exploiting the cyclostationary of the received signal via oversampling.

In order to obtain a performance measure of the blind channel identification, the normalized MSE (NMSE) is used. For an M Monte Carlo trial, the NMSE is defined as

$$\text{NMSE} = \frac{1}{M \sum_{n=0}^{N-1} |h[n]|^2} \left[\sum_{i=0}^{M-1} \sum_{n=0}^{N-1} \left| \hat{h}_i[n] - h[n] \right|^2 \right], \qquad (7.188)$$

where $\hat{h}_i[n]$ is the estimate of the channel from the ith trial.

7.6.2 Subspace Blind Identification for SIMO Channel

Consider the oversampled discrete-time SIMO channel given in (7.155). The set of L equations can be combined into a single matrix form as

$$\mathbf{Y}[n] = \mathbf{H}[n]\mathbf{s}[n] + \mathbf{V}[n], \qquad (7.189)$$

where $\mathbf{Y}[n]$ is the $LN \times 1$ multichannel received signal vector

$$\mathbf{Y}[n] = [\mathbf{y}_0[n], \mathbf{y}_1[n], \cdots, \mathbf{y}_{L-1}[n]]^T, \qquad (7.190)$$

and $\mathbf{H}[n]$ is the $LN \times (M+N)$ multichannel filtering matrix

$$\mathbf{H}[n] = [\mathbf{h}_0[n], \mathbf{h}_1[n], \cdots, \mathbf{h}_{L-1}[n]]^T, \qquad (7.191)$$

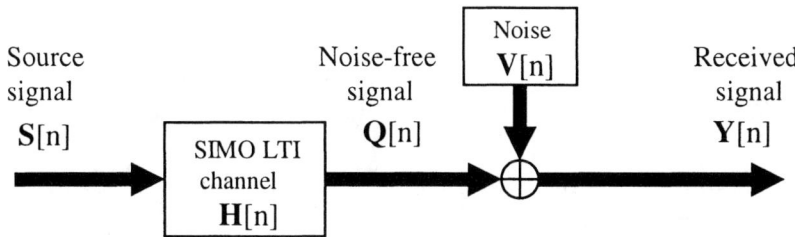

Figure 7.14 The matrix representation of an oversampled discrete-time single-input and multiple-output (SIMO) channel.

where the individual filtering matrix that provides a matrix description of an oversampled channel is given in (7.158) and $\mathbf{V}[n]$ is the $LN \times 1$ multichannel noise vector

$$\mathbf{V}[n] = [\mathbf{v}_0[n], \mathbf{v}_1[n], \cdots, \mathbf{v}_{L-1}[n]]^T. \quad (7.192)$$

Figure 7.14 shows a block diagram representation of this equation that may be viewed as a condensed version of the discrete-time SIMO channel. In order to proceed with blind channel identification via statistical channel characterization, the following assumptions are made:

1. The transmitted signal vector $\mathbf{s}[n]$ and multichannel noise vector $\mathbf{v}[n]$ are wide-sense stationary processes that are statistically independent.
2. The transmitted signal vector $\mathbf{s}[n]$ has a zero mean and a correlation matrix given by

$$\mathbf{R_s}[n] = E\{\mathbf{s}[n]\mathbf{s}^H[n]\} \quad (7.193)$$

where the $\mathbf{R_s}$ has an $(M+N) \times (M+N)$ full column rank.

3. The $N \times 1$ noise vector $\mathbf{v}[n]$ has a zero mean and a correlation matrix

$$\begin{aligned}\mathbf{R_V}[n] &= E\{\mathbf{v}[n]\mathbf{v}^H[n]\} \\ &= \sigma^2 \mathbf{I},\end{aligned} \quad (7.194)$$

where the noise variance σ^2 is assumed to be known.

Accordingly, the received signal vector $\mathbf{Y}[n]$ has an $LN \times 1$ zero mean and a correlation matrix defined by

$$\begin{aligned}\mathbf{R_Y}[n] &= E\{\mathbf{Y}[n]\mathbf{Y}[n]^H\} \\ &= E\{(\mathbf{H}[n]\mathbf{s}[n] + \mathbf{V}[n])(\mathbf{H}[n]\mathbf{s}[n] + \mathbf{V}[n])\} \\ &= E\{\mathbf{H}[n]\mathbf{s}[n]\mathbf{s}^H[n]\mathbf{H}^H[n]\} + E\{\mathbf{V}[n]\mathbf{V}^H[n]\} \\ &= \mathbf{H}[n]\mathbf{R_s}[n]\mathbf{H}^H[n] + \mathbf{R_V}[n].\end{aligned} \quad (7.195)$$

Invoking the theorem of spectral decomposition given by Miao and Clements [6], the $LN \times LN$ correlation matrix $\mathbf{R_Y}[n]$ in (7.195) can be expressed in terms of its eigenvalues and associated eigenvectors

$$\mathbf{R_Y}[n] = \sum_{k=1}^{LN} \lambda_k \mathbf{q}_k \mathbf{q}_k^H, \quad (7.196)$$

where the eigenvalues decrease in order as

$$\lambda_0 \leq \lambda_1 \leq \cdots \leq \lambda_{LN-1}. \quad (7.197)$$

Then, the eigenvalues in (7.197) can be classified into two groups based on the filtering matrix rank theorem

$$\lambda_k > \sigma^2, \quad k = 0, 1, \cdots, M + N - 1, \quad (7.198)$$

and

$$\lambda_k = \sigma^2, \quad k = M + N, N + M + 1, \cdots, LN - 1. \quad (7.199)$$

As a result, the corresponding space spanned by the eigenvectors of the received signal vector $\mathbf{R_Y}[n]$ can be decomposed into two subspaces:

1. Signal subspace spanned by the eigenvectors associated with the eigenvalues $\lambda_0, \lambda_1, \cdots, \lambda_{M+N-1}$. These eigenvectors are denoted as

$$\mathbf{s}_k = \mathbf{q}_k, \quad k = 0, 1, \cdots, M+N-1. \qquad (7.200)$$

2. Noise space spanned by the eigenvectors associated with the remaining eigenvalues $\lambda_{M+N}, \lambda_{M+N+1}, \cdots, \lambda_{LN-1}$. These eigenvectors are written as

$$\mathbf{n}_k = \mathbf{q}_{M+N+k}, \quad k = 0, 1, \cdots, LN - (M+N+1). \qquad (7.201)$$

Note that the signal subspace is the orthogonal complement of the noise subspace. Therefore, using the matrix definition, the noise equation is obtained by

$$\mathbf{R}_\mathbf{Y}[n]\mathbf{n}_k = \sigma^2 \mathbf{n}_k, \quad k = 0, 1, \cdots, LN - (M+N+1). \qquad (7.202)$$

Substituting (7.195) along with $\mathbf{R}_\mathbf{V}[n] = \sigma^2 \mathbf{I}$ into (7.202) and simplifying yields

$$\mathbf{H}[n]\mathbf{R}_\mathbf{S}[n]\mathbf{H}^H[n]\mathbf{n}_k = 0, \quad k = 0, 1, \cdots, LN - (M+N+1). \qquad (7.203)$$

Equation (7.203) can be rewritten as

$$\mathbf{H}^H[n]\mathbf{n}_k = 0, \quad k = 0, 1, \cdots, LN - (M+N+1), \qquad (7.204)$$

because both matrices $\mathbf{H}[n]$ and $\mathbf{R}_\mathbf{S}[n]$ are all full column rank. We need to emphasize that the result of subspace blind identification in (7.204) is based on the following three assumptions:

1. The received signal is oversampled to ensure that the multichannel filter matrix $\mathbf{H}[n]$ has a full column rank.
2. There is the knowledge of the eigenvectors associated with $LN - (M+N)$ smallest eigenvalues of the received signal correlation matrix $\mathbf{R}_\mathbf{Y}[n]$.

3. The noise subspace is orthogonal to the columns of the unknown multichannel filter matrix $\mathbf{H}[n]$.

In addition, a successful use of the subspace decomposition for blind identification depends on the premise that the transfer functions of the discrete-time SIMO channel have no common zeros. This requires the exact knowledge of the channel model order. Therefore, given these requirements, (7.204) indicates that the cyclostationary second-order statistics of the received signal $\mathbf{Y}[n]$ associated with the correlation matrix $\mathbf{R_Y}[n]$ are indeed sufficient for blind identification of the discrete-time SIMO channel.

Equations (7.202) through (7.204) provide the fundamental theory of using subspace decomposition for blind identification of the discrete-time SIMO channel. This noise subspace procedure for blind identification was contributed by Moulines et al. [12] and later introduced by Haykin [5].

7.6.3 Blind Identification for MIMO Channel

Blind identifications of a discrete-time MIMO channel have been extensively reported either based on the second-order cyclostationary statistics of the output data long with corresponding identifiable conditions given by Gorokhov and Loubaton [21], Abed-Meraim and Hua [22], and Hua and Tugnait [23] or based on the higher-order statistics of the system outputs by Mendel [24], Shamsunder and Giannakis [25], Tugnait [14], and Chi et al. [18]. Diamantaras et al. [26] and Bradaric et al. [27] have researched the blind identification of FIR MIMO channels with cyclostationary input using second-order statistics, for which we mainly introduce results in this section.

The objective of blind identification of an $N \times N$ discrete-time MIMO channel is to identify an unknown convolution FIR system driven by N unknown input sources by using the second-order cycostationary input statistics. By exploiting the second-order input statistics with cyclostationarity, a cross-wise $N \times N$ convolution filtering mixture can be uniquely identified based on the second-order statistics of the discrete-time MIMO channel output.

Consider a general $N \times N$ discrete-time MIMO channel with N input signal sources $s_0[k], s_1[k], \cdots, s_{N-1}[k]$ and N stochastic output sequences $y_0[k], y_2[k], \cdots, y_{N-1}[k]$. Assume that each discrete-time channel is a FIR filter with length of L. Thus, the input and output relationship of this discrete-time MIMO channel can be written as

$$y_i[k] = \sum_{j=0}^{N-1} \sum_{l=0}^{L-1} h_{ij}[l] s_j[k-l], \quad i = 0, 1, \cdots, N-1. \quad (7.205)$$

Let $\mathbf{s}[n] = [s_0[n], s_1[n], \cdots, s_{N-1}[n]]^T$ be a vector of N input sources, $\mathbf{H}[n]$ be the $N \times N$ impulse response filtering matrix with elements $\{h_{ij}[n]\}$, and $\mathbf{y}[n] = [y_0[n], y_1[n], \cdots, y_{N-1}[n]]^T$ be the vector of system output observations. Thus, (7.205) can be expressed into matrix form as

$$\mathbf{y}[k] = \sum_{k=0}^{L-1} \mathbf{H}[k] \mathbf{s}[n-k]. \quad (7.206)$$

Taking the discrete-time Fourier transform (DTFT) of both sides of (7.206) yields

$$\mathbf{Y}(e^{j\omega}) = \mathbf{H}(e^{j\omega}) \mathbf{S}(e^{j\omega}), \quad (7.207)$$

where $\omega \in [0, 2\pi)$.

Note that the frequency-domain filter matrix $\mathbf{H}(e^{j\omega})$ in (7.207) is not identifiable unless some constraints are provided based on its structure. In other words, the problem as defined in (7.207) is ill posed and admits an infinite number of solutions. In order to identify the discrete-time MIMO channel $\mathbf{H}(e^{j\omega})$ in (7.207) blindly, we need to provide the following assumptions:

1. The input source $\{s_i[n]\}$, $i = 0, \cdots, N-1$ are uncorrelated, zero-mean, cyclostationary, and their statistics are unknown. In addition, the input sources are colors and unidentical, that is, $R_{s_1}(k) \neq R_{s_2}(k)$ for all k.
2. The frequency-domain MIMO filtering matrix $\mathbf{H}(e^{j\omega})$ is a full column rank for all $\omega \in [0, 2\pi)$.

3. $H_{ii}(e^{j\omega}) = 1$ for all ω when $i = 0, 1, \cdots, N - 1$. This indicates that the diagonal channels $h_{ii}[n]$ are unit impulse responses while the cross-channels $h_{ij}[n]$, $i \neq j$, have at least length 2.
4. The cross-channel $h_{ij}[n]$ are real or complex FIR filters without common zeros and zeros on the unit circle. Moreover, the cross-channel $h_{ij}[n]$ and $h_{ik}[n]$, $i \neq j, i \neq k$, do not have zeros in conjugate reciprocal pairs.

Under the assumptions of 1 and 2, the relationship of the identified MIMO channel filtering matrix $\hat{\mathbf{H}}(e^{j\omega})$ and the channel filter matrix $\mathbf{H}(e^{j\omega})$ can be obtained by

$$\hat{\mathbf{H}}(e^{j\omega}) = \mathbf{H}(e^{j\omega})\mathbf{P}(e^{j\omega})\Lambda(e^{j\omega}), \tag{7.208}$$

where $\mathbf{P}(e^{j\omega})$ is a column permutation matrix, and $\Lambda(e^{j\omega})$ is a complex diagonal matrix (see Bradaric et al. [27] for proof). In other words, the identified MIMO channel filtering matrix $\hat{\mathbf{H}}(e^{j\omega})$ is related to the channel filter matrix $\mathbf{H}(e^{j\omega})$ up to a frequency-dependent permutation ambiguity $\mathbf{P}(e^{j\omega})$ and a frequency-dependent diagonal scaling ambiguity $\Lambda(e^{j\omega})$.

Given assumption 4, it is clear that the ambiguity effect of $\Lambda(e^{j\omega})$ can be eliminated. In this case, without the effect of the permutation ambiguity, we can obtain the channel filter matrix $\mathbf{H}(e^{j\omega})$ by the expression

$$\tilde{\mathbf{H}}(e^{j\omega}) \cong \frac{\hat{\mathbf{H}}(e^{j\omega})}{\text{diag}\left\{\hat{\mathbf{H}}(e^{j\omega})\right\}}, \tag{7.209}$$

where $\text{diag}\left\{\hat{\mathbf{H}}(e^{j\omega})\right\} = \Lambda(e^{j\omega})$. However, in fact, $\tilde{\mathbf{H}}(e^{j\omega}) \neq \mathbf{H}(e^{j\omega})$. This is because the permutation matrix $\mathbf{P}(e^{j\omega})$ is unknown.

To obtain the permutation matrix $\mathbf{P}(e^{j\omega})$, we define the so-called invariance functions for each pair of rows (k, m), $k, m = 0, 1, \cdots, N - 1$, as follows:

$$I_1^N(\omega; k, m) \cong \sum_{j=0}^{N-1} \frac{\tilde{H}_{kj}(e^{j\omega})}{\tilde{H}_{mj}(e^{j\omega})} \tag{7.210}$$

$$I_2^N(\omega; k, m) \cong \sum_{j_1=0}^{N-1} \sum_{j_2>j_1}^{N-1} \frac{\tilde{H}_{kj_1}(e^{j\omega}) \; \tilde{H}_{kj_2}(e^{j\omega})}{\tilde{H}_{mj_1}(e^{j\omega}) \; \tilde{H}_{mj_2}(e^{j\omega})} \quad (7.211)$$

$$I_3^N(\omega; k, m) \cong \sum_{j_1=0}^{N-1} \sum_{j_2>j_1}^{N-1} \sum_{j_3>j_2}^{N-1} \frac{\tilde{H}_{kj_1}(e^{j\omega}) \; \tilde{H}_{kj_2}(e^{j\omega}) \; \tilde{H}_{kj_3}(e^{j\omega})}{\tilde{H}_{mj_1}(e^{j\omega}) \; \tilde{H}_{mj_2}(e^{j\omega}) \; \tilde{H}_{mj_3}(e^{j\omega})} \quad (7.212)$$

$$\vdots$$

$$I_N^N(\omega; k, m) \cong \prod_{j=0}^{N-1} \frac{\tilde{H}_{kj}(e^{j\omega})}{\tilde{H}_{mj}(e^{j\omega})}. \quad (7.213)$$

Note that these quantities are the same based on either $\tilde{\mathbf{H}}(e^{j\omega})$ or $\mathbf{H}(e^{j\omega})$.

Consider the quantity as

$$\prod_{j=0}^{N-1} H_{kj}(e^{j\omega}) H_{mj}^*(e^{j\omega}). \quad (7.214)$$

The phase in (7.214) equals the phase of $I_N^N(\omega; k, m)$ in (7.213) with a linear phase. Thus, under assumption 4, (7.214) can be computed from a scalar constant defined as $[c(k,m)]^2$, since an FIR sequence, which does not have zero-phase convolution components, can be reconstructed within a scalar from its phase only. Then, we define $P^N(\omega, k, m)$ to be the function that can be computed from the phase of $I_N^N(\omega; k, m)$:

$$P^N(\omega, k, m) \cong [c(k,m)]^2 \prod_{j=0}^{N-1} H_{kj}(e^{j\omega}) H_{mj}^*(e^{j\omega}), \quad (7.215)$$

and we also define the quantity

$$\begin{aligned} M_k(\omega, k, r) &\cong \left| P^N(\omega, k, r) I_N^N(\omega; k, r) \right|^{1/2} \\ &= c(k, r) \prod_{j=0}^{N-1} \left| H_{kj}(e^{j\omega}) \right|, \; r = 0, \cdots, N-1. \end{aligned}$$
$$(7.216)$$

Consider a polynomial

$$Q(I_i^N x^N) = x^N + \sum_{i=1}^{N-1}(-1)^i I_i^N(\omega; k, p)x^{N-i}, \qquad (7.217)$$

where $p = 1, 2, \cdots, N$. Also let $X_i^p(e^{j\omega})$ denote the ith root of the polynomial $Q(I_i^N x^N)$ in (7.217). Then $X_i^p(e^{j\omega})$ will have one of the following results:

$$\frac{H_{k1}(e^{j\omega})}{H_{p1}(e^{j\omega})}, \frac{H_{k2}(e^{j\omega})}{H_{p2}(e^{j\omega})}, \cdots, \frac{H_{kN}(e^{j\omega})}{H_{pN}(e^{j\omega})}. \qquad (7.218)$$

For the previously selected r, we define the element of the following set:

$$K(\omega_m; k, r) = \{k_{j_1 j_2 \cdots j_N}(\omega_m; k, r)\}, \quad m = 0, 1, \cdots, N-1, \qquad (7.219)$$

where

$$k_{j_1 j_2 \cdots j_N}(\omega_m; k, r) = \frac{M_{k1}(\omega_m; k, r)}{|X_{j_1}^1(e^{j\omega})| \cdots |X_{j_p}^p(e^{j\omega})| \cdots |X_{j_N}^N(e^{j\omega})|}, \qquad (7.220)$$

for $j_1, j_2, \cdots, j_N = 1, 2, \cdots, N$.

According to (7.218) with the assumptions 3 and 4, one and only one element of the set $K(\omega_m; k, r)$ is independent of the following frequency:

$$|X_{j_1}^1(e^{j\omega})| = \frac{|H_{k1}(e^{j\omega})|}{|H_{11}(e^{j\omega})|} = |H_{k1}(e^{j\omega})|, \qquad (7.221)$$

$$|X_{j_2}^2(e^{j\omega})| = \frac{|H_{k2}(e^{j\omega})|}{|H_{22}(e^{j\omega})|} = |H_{k2}(e^{j\omega})|, \qquad (7.222)$$

$$\vdots$$

$$|X_{j_N}^N(e^{j\omega})| = \frac{|H_{kN}(e^{j\omega})|}{|H_{NN}(e^{j\omega})|} = |H_{kN}(e^{j\omega})|. \qquad (7.223)$$

Thus, given a large enough N, we obtain

$$K(\omega_0; k, r) \cap K(\omega_1; k, r) \cap \ldots \cap K(\omega_{N-1}; k, r) = \{c(k, r)\}. \tag{7.224}$$

Solving (7.224) yields the scalar constant $c(k, r)$. Assume that there is not ambiguity at the certain frequency. Hence, the cross-channel $H_{kq}(e^{j\omega})$ can be obtained by

$$k_{j_1 j_2 \ldots j_N}(\omega_m; k, r) = c(k, r), \tag{7.225}$$

$$H_{kq}(e^{j\omega_m}) = X_{jq}^{q}(e^{j\omega_m}). \tag{7.226}$$

Therefore, an arbitrary cross-channel $h_{kq}[k]$, $k, q = 1, 2, \cdots, N$, of the MIMO discrete-time channel can be recovered.

In summary, assume that an estimated MIMO channel filtering matrix $\hat{\mathbf{H}}(e^{j\omega})$ in (7.208) is available by using some kinds of blind MIMO algorithms. Then the MIMO channel filtering matrix $\mathbf{H}(e^{j\omega})$ can be obtained by the following procedure:

1. Determine $\tilde{\mathbf{H}}(e^{j\omega})$ by using (7.209).
2. Compute the invariance functions $I_k^N(\omega; k, p)$ for $p, k = 1, 2, \cdots, N$.
3. Obtain $P^N(\omega; k, r)$ in (7.215).
4. Compute $M_{k1}(\omega; k, r)$ using (7.216)
5. Determine the roots of the polynomial $Q(I_i^N x^N)$ in (7.217).
6. Establish the set $K(\omega; k, r)$ in (7.224) and (7.225) for $\omega_m = (2\pi/N)m$, $m = 0, 1, 2, \cdots, N-1$.
7. Determine the scalar constant $c(k, r)$ by finding the only common element of $K(\omega_m; k, r)$, $m = 0, 1, \cdots, N-1$, in (7.224).
8. Find $H_{kq}(e^{j\omega_m})$ by selecting the root in (7.225) and (7.226).
9. Finally compute $h_{kq}[k]$ based on the results of $H_{kq}(e^{j\omega_m})$.

This blind identification procedure of the discrete-time MIMO channel can achieve a perfect system reconstruction in the absence of noise. However, in the presence of noise, this blind identification method for the discrete-time MIMO channel is relatively sensitive.

This is because this blind identification recovers the discrete-time MIMO channel from phase, which can be very sensitive to error with the presence of noise in the discrete-time MIMO channel.

7.7 Summary

In this chapter, channel estimation and blind identification were introduced. We first presented the characteristics of discrete-time channel models and estimators, with emphases on the maximum likelihood, least squares, generalized least squares, and MMSE estimators. In the special case with the noise covariance equal to $\sigma^2 \mathbf{I}$, the least squares estimator has been shown as the best linear unbiased estimator for the channel estimation. However, in a general case, the best linear unbiased estimator is the generalized least squares estimator. This can be found by using the new transformed channel model given by (7.42). Second, we introduced the fundamental theory of adaptive channel estimations and algorithms. The LMS and RLS adaptive algorithms were presented along with the analysis of their convergence, EMSE, and misadjustment. The convergence rate of the RLS algorithm is much faster than the convergence rate of the LMS algorithm. However, the LMS algorithm has much fewer computation complexities compared to the RLS algorithm. Furthermore, the LMS algorithm does not have a stablity issue, while the RLS algorithm does. Thus, the LMS algorithm is still the most often used adaptive algorithm for the channel estimation. We also described the mathematical treatments of the discrete-time SISO, SIMO, and MIMO channel models and estimations based on the HOS methods. The blind identifications for discrete-time SISO, SIMO, and MIMO channels were then introduced based on the second-order cyclostationary statistics using observations of the received output signals.

Traditional adaptive channel estimation as widely used for the wireless and wired digital communications uses a training sequence or a known input signal source. This approach can provide a reliable estimation method of an unknown channel for a communication receiver system. However, in a time-varying channel, such as in fast mobile communication, the adaptive approach may not be appropriate since this approach has to waste a fraction of the transmission time and frequency bandwidth to use the training sequence. An alternative way of estimating unknown SISO, SIMO, and MIMO channels is to use blind identification. Blind identification of the discrete-time SISO, SIMO, and MIMO channels has been successfully reported based on either HOS or the second-order cyclostationary statistics under the statistics assumptions. However, for the MIMO channel, blind identification is still needed to be further refined if one wishes to turn this identifiablity approach into a method of robust channel estimation.

Designing a good channel equalizer usually requires a knowledge of the channel characteristics, which involves channel estimation and identification. In Chapter 8, we will introduce channel equalizers for communication receivers in detail.

References

[1] Proakis, J. G., *Digital Communications*, 2nd ed., McGraw-Hill Book Company, New York, 1989.

[2] Mood, A. M., F. A. Graybill, and D. D. Boes, *Introduction to the Theory of Statistics*, 3rd ed., McGraw-Hill, Inc., New York, 1974.

[3] Mardia, K. V., J. T. Kent, and J. M. Bibby, *Multivariate Analysis*, 3rd ed., Academic Press, London, U.K., 1982.

[4] Kailath, T., *Lectures on Wiener and Kalman Filtering*, 2nd ed., Springer-Verlag, Wein, New York, 1981.

[5] Haykin, S., *Adaptive Filter Theory*, 3rd ed., Prentice Hall, Upper Saddle River, New Jersey, 1996.

[6] Miao, G. J., and M. A. Clements, *Digital Signal Processing and Statistical Classification*, Artech House, Norwood, Massachusetts, 2002.

[7] Hayes, M. H., *Statistical Digital Signal Processing and Modeling*, John Wiley & Sons, New York, 1996.

[8] Haykin, S., *Neural Networks: A Comprehensive Foundation*, Macmillan College Publishing Company, New York, 1994.

[9] Zhao, K., et al., "Slide Window Order-Recursive Least-Squares Algorithms," *IEEE Trans. on Signal Processing*, Vol. 42, No. 8, pp. 1961–1972, August 1994.

[10] Sayed, A. H., and T. Kailath, "A State-Space Approach to Adaptive RLS Filtering," *IEEE Signal Processing Magazine*, Vol. 11, pp. 18–60, 1994.

[11] Paulraj, A. J., and C. B. Papadias, "Space-Time Processing for Wireless Communications," *IEEE Signal Processing Magazine*, pp. 49–83, November 1997.

[12] Moulines, E., P. Duhamel, J. F. Cardoso, and S. Mayrargue, "Subspace Methods for Blind Identification of Multichannel FIR Filters," *IEEE Trans. on Signal Processing*, Vol. 43, pp. 516–525, 1995.

[13] Gesbert, D., et al., "From Theory to Practice: An Overview of MIMO Space-Time Coded Wireless Systems," *IEEE Journal on Selected Areas in Communications*, Vol. 21, No. 3, pp. 281–302, April 2003.

[14] Tugnait, J. K., "Identification and Deconvolution of Multichannel Linear Non-Gaussian Processes Using Higher Order Statistics and Inverse Filter Criteria," *IEEE Trans. on Signal Processing*, Vol. 45, No. 3, pp. 658–672, March 1997.

[15] Papadias, C. B., and A. J. Paulraj, "A Constant Modulus Algorithm for Multiuser Signal Separation in Presence of Delay Spread Using Antenna Arrays," *IEEE Signal Processing Letter*, Vol. 4, pp. 178–181, June 1997.

[16] Li Y., and K. J. R. Liu, "Adaptive Blind Source Separation and Equalization for Multiple-Input/ Multiple-Output Systems," *IEEE Trans. on Information Theory*, Vol. 44, pp. 2864–2876, November 1998.

[17] Chi, C. Y., and C. H. Chen, "Cumulant Based Inverse Filter Criteria for MIMO Blind Deconvolution: Properties, Algorithm, and Application to DS/CDMA Systems in Multipath," *IEEE Trans. on Signal Processing*, Vol. 49, pp. 1282–1299, July 2001.

[18] Chi, C. Y., et al., "Batch Processing Algorithms for Blind Equalization Using Higher-Order Statistics," *IEEE Signal Processing Magazine*, Vol. 20, No. 1, pp. 25–49, January 2003.

[19] Tong, L., G. Xu, and T. Kailath, "Blind Identification and Equalization Based on Second-Order Statistics: A Time Domain Approach," *IEEE Trans. on Information Theory*, Vol. 40, No. 2, pp. 340–349, March 1994.

[20] Tong, L., et al., "Blind Channel Identification Based on Second-Order Statistics: A Frequency-Domain Approach," *IEEE Trans. on Information Theory*, Vol. 41, No. 1, pp. 329–334, January 1995.

[21] Gorokhov, A., and P. Loubaton, "Subspace-Based Technique for Blind Separation of Convolutive Mixtures with Temporally Correlated Sources," *IEEE Trans. on Circuits and Systems I*, Vol. 44, pp. 813–820, September 1997.

[22] Abed-Meraim, K., and Y. Hua, "Blind Identification of Multi-Input Multi-Output System Using Minimum Noise Subspace," *IEEE Trans. on Signal Processing*, Vol. 45, pp. 254–258, January 1997.

[23] Hua, Y., and J. K. Tugnait, "Blind Identifiability of FIR MIMO Systems with Colored Input Using Second Order Statistics," *IEEE Signal Processing Letter*, Vol. 7, pp. 348–350, December 2000.

[24] Mendel, J. M., "Tutorial on Higher-Order Statistics (Spectra) in Signal Processing and System Theory: Theoretical Results and Some Applications," *Proceedings of the IEEE*, Vol. 79, No. 3, pp. 278–305, March 1991.

[25] Shamsunder, S., and G. B. Giannakis, "Multichannel Blind Signal Separation and Reconstruction," *IEEE Trans. on Speech and Audio Processing*, Vol. 5, pp. 515–527, November 1997.

[26] Diamantaras, K. I., A. P. Petropulu, and B. Chen, "Blind Two-Input-Two-Output FIR Channel Identification Based on Frequency Domain Second-Order Statistics," *IEEE Trans. on Signal Processing*, Vol. 48, No. 2, pp. 534–542, February 2000.

[27] Bradaric, I., A. P. Petropulu, and K. I. Diamantaras, "On Blind Identifiability of FIR-MIMO Systems with Cyclostationary Inputs Using Second Order Statistics," *IEEE Trans. on Signal Processing*, Vol. 51, No. 2, pp. 434–441, February 2003.

8

Adaptive Equalizers in Communication Receivers

8.1 Introduction

Adaptive equalization techniques have been developed over the last three decades for high-speed, single-carrier and multicarrier transmission over wireline and wireless channels, such as a twisted pair of copper wires, coaxial cable, optical fiber, microwave line of sight (LOS), and nonline of sight (NLOS). *Coherence bandwidth* is a statistical measure of the range of frequencies over which the channel passes all spectral components with approximately equal gain and linear phase [1]. If the modulation bandwidth exceeds the coherence bandwidth of the wireline or radio channel, intersymbol interference (ISI) occurs in such a way that modulation pulses are spread in time. For such channels, adaptive equalization can provide the means for combating ISI arising from the bandwidth or time-dispersive characteristic of the channel and allow us to use the available channel bandwidth more efficiently.

In a broad sense, the term adaptive equalization can be used to describe a signal processing operation for minimizing ISI. However, not every signal processing operation that minimizes ISI is adaptive equalization. For instance, having prematched filters is not adaptive equalization. In the wireless channel with random and time-varying fading, a variety of adaptive equalization techniques can be used to cancel interference while providing *diversity*. The diversity is used to compensate for fading channel impairments, and is usually implemented by using multiple antennas that are strategically spaced and connected to a common receiving system. In wireline channel, adaptive equalization techniques can be used for combating ISI arising from the bandwidth channel. In any case, equalizations must track the characteristics of the channel and thus are referred to as *adaptive channel equalizations*.

Modern communication systems require many signal processing techniques that improve the link performance in a variety of environments. One key part of signal processing techniques for the digital communication receiver commonly uses an adaptive equalization because it reduces ISI due to the limited channel bandwidth. Therefore, the design of such adaptive equalization is important since it controls the maximum quality attainable from the view of a communication receiver.

Equalization techniques for reducing ISI on limited bandwidth channels can be subdivided into two general categories: *linear* and *nonlinear* equalization. The basic linear equalization includes transversal and lattice equalizer structure. The nonlinear equalization methods may be subdivided into four types: *decision feedback equalization* (DFE), *maximum likelihood symbol detection* (MLSD), *maximum likelihood sequence estimation equalizer* (MLSEE), and *neural networks–based equalizer* (NNE). The DFE usually has transversal and lattice equalizer structures, while the MLSD and MLSEE have a transversal equalizer structure along with a channel estimator. The NNE has multiplayer feedforward networks, recurrent networks, and a lattice equalizer structure.

Linear equalizations have their numerical stability and faster convergence. However, linear equalizations usually do not perform well on bandlimited channels when the frequency passband has a deep spectral null. In order to compensate for such distortion, linear equalizations attempt to place too much gain in the vicinity of the spectral null, thereby enhancing the noise present in those frequency bands. On the other hand, nonlinear equalizations are used in applications where the channel distortion is too severe for a linear equalization to handle. However, nonlinear equalizations, such as DFE, may have their numerical instability due to error propagation. This leads to closed eye diagram in the digital communication receiver.

A more robust equalizer is the least mean square (LMS) algorithm-based equalizer where the criterion used is the minimization of the mean square error (MSE) between the desired equalizer output and the actual equalizer output. Another type of equalizer is the recursive least squares (RLS) algorithm-based equalizer where the least square error used is the time average. There are a number of variations of the LMS and RLS algorithms that are used for adapting an equalizer.

A *blind equalizer* performs channel equalization without a training sequence. In other words, the equalizer performs equalization on the data without a reference signal. Instead, the blind equalizer depends on the knowledge of the signal's structure and its statistics to perform the channel equalization. Hence, bandwidth is not wasted by its transmission. The drawback of the blind equalizer is that the equalizer will typically take longer to converge compared to a trained equalizer. Although various blind equalization algorithms exist, the one that has gained the most popularity is the constant modulus algorithm (CMA). The CMA minimizes the constant modulus cost function and then adjusts the taps of the equalizer in an attempt to minimize the difference between the squared magnitude of the samples and the so-called Godard dispersion constant. The CMA-based blind equalizer is rapidly gaining popularity in the wired and wireless communication receivers.

The organization of this chapter is as follows. In this section we briefly introduced the background, overview, and types of equalizations along with the corresponding adaptive algorithms.

In Section 8.2, we describe basic theory of linear equalizers, including channel equalizers and the mean square error criterion. In Section 8.3 we present adaptive linear equalizers and adaptive algorithms to adjust the tap coefficient for equalizers. In addition, we discuss the training methodology and tap length selection for adaptive equalizers. Subsequently, the fundamental fractionally spaced equalizer (FSE) is given in Section 8.4. In this section, we introduce the multirate communication system model and multichannel model-based equalizers. The emphases are given to the FSE minimum mean square error (MMSE), FSE-CMA, and FSE constant modulus noise functions as well as FSE constant modulus performances. Section 8.5 provides the theory of the decision feedback equalizer (DFE). We also describe the MMSE for DFE, predictive DFE, FSE-based DFE, and error propagation in the DFE. The space-time equalizers, including the time-only, space-only, and space-time MMSE equalizers, are given in Section 8.6. The diversity equalizer is described in Section 8.7. We introduce the basic theory of Rake receivers and equalized Rake receivers. Finally, a brief summary is given in Section 8.8.

8.2 Linear Equalizer

The most common type of channel equalizer used in practice to combat ISI arising from the bandlimited channel is a linear equalizer. The linear equalizer can be implemented as a finite impulse response (FIR) filter (also known as a transversal filter), with adjustable filter tap coefficients. This type of linear equalizer is the simplest type available.

A basic structure of a linear transversal equalizer that is most often used for equalization is shown in Figure 8.1. The input is the sequence $x[n]$ and the output is the estimated sequence $\hat{d}[n]$. The estimated sequence $\hat{d}[n]$ of the nth symbol is given by

$$\hat{d}[n] = \sum_{k=0}^{p} b_n[k]x[n-k], \qquad (8.1)$$

where $b_n[k]$ are the p filter coefficients of the nth symbol. Then

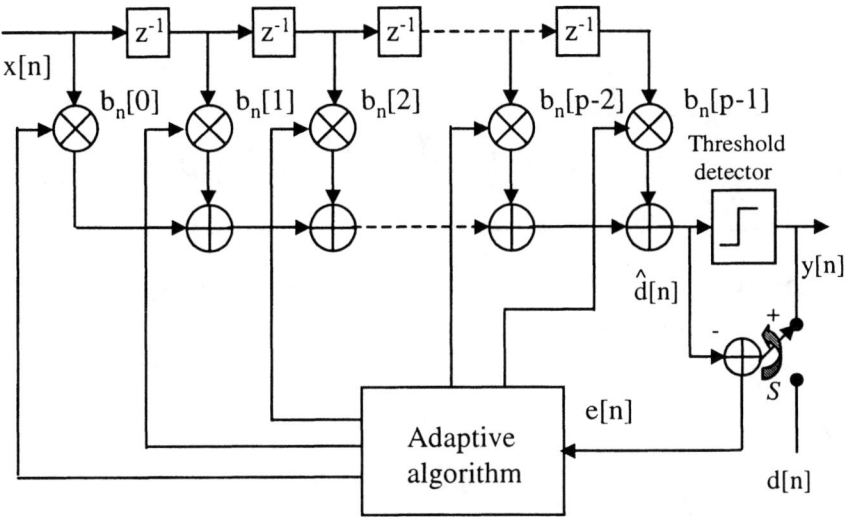

Figure 8.1 A linear equalizer with a transversal filter structure.

the estimated sequence $\hat{d}[n]$ is quantized to the nearest information symbol to form a decision sequence $y[n]$ by using a *threshold detector*.

The threshold detector that may be a two-level or a multilevel quantizer in the linear transversal equalizer is determined according to the value of the input sequence $x[n]$. For example, if the input signal $x[n]$ is a $+1$ or -1 information sequence, then the two-level quantizer can be used and is defined by

$$y[n] = \begin{cases} 1 & \hat{d}[n] \geq 0 \\ -1 & \hat{d}[n] < 0. \end{cases} \qquad (8.2)$$

The threshold detector in (8.2) is also referred to as a *slice* in some literatures.

A switch as shown in Figure 8.1 is used to connect with the decision sequence $y[n]$ or a training sequence $d[n]$. When the switch connects to the decision sequence $y[n]$, the linear transversal

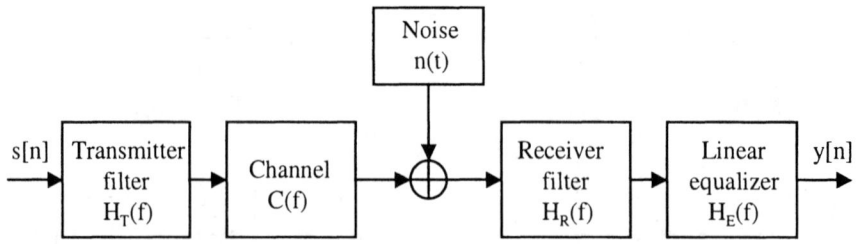

Figure 8.2 A block diagram of a communication system with a linear equalizer.

equalizer is referred to as the blind equalizer. In this case, no training sequence presents during the training mission. The error sequence $e[n]$ is formed by using the estimated sequence $\hat{d}[n]$ and the decision sequence $y[n]$. When the switch connects to the training sequence $d[n]$, the linear transversal equalizer is called a no-blind equalizer. In this case, the error sequence $e[n]$ is formed by using the estimated sequence $\hat{d}[n]$ and the training sequence $d[n]$.

If the decision sequence $y[n]$ is not equal to the estimated sequence $\hat{d}[n]$, the error sequence $e[n]$ is produced as follows:

$$e[n] = y[n] - \hat{d}[n]. \tag{8.3}$$

The error sequence $e[n]$ is used as an input for a block diagram of adaptive algorithm, which controls the update of p tap-weight coefficients of the FIR filter during the implementation.

8.2.1 Channel Equalizer

A linear equalizer can be used as a channel equalizer. Figure 8.2 shows a basic block diagram of a communication system employing a linear equalizer as the channel equalizer. The characteristics of this linear equalizer can be illustrated from a point of view in the frequency domain.

The transmission section contains a transmitting filter with a frequency response $H_T(f)$ in cascade with a channel frequency response $C(f)$ plus a noise $n(t)$. The receiver section consists of a receiver filter with frequency response $H_R(f)$ in cascade with a channel equalizing filter, which has a frequency response $H_E(f)$. Assume that the frequency response of the receiver filter $H_R(f)$ is matched to the transmitter filter response such that $H_R(f) = H_T^*(f)$. In addition, the product of frequency responses $H_R(f)H_T^*(f)$ is designed in such a way that there is no ISI at the sampling instant, $t = kT$, $k = 0, 1, ...$, and T the signal interval ($\frac{1}{T}$ is called the symbol rate) when $H_R(f)H_T^*(f) = H_{RC}(f)$, where $H_{RC}(f)$ is a raised cosine pulse spectral characteristic.

Note that the frequency response of channel is not ideal as shown in Figure 8.2. In order to eliminate the ISI, the desired condition for zero ISI is expressed as

$$H_T(f)C(f)H_R(f)H_E(f) = H_{RC}(f). \tag{8.4}$$

Given $H_R(f)H_T^*(f) = H_{RC}(f)$, the frequency response of a linear equalizer that compensates for the channel distortion is given by

$$\begin{aligned} H_E(f) &= \frac{1}{C(f)} \\ &= \frac{1}{|C(f)|}e^{-j\theta_c(f)}. \end{aligned} \tag{8.5}$$

Then the magnitude of the linear equalizer frequency response $H_E(f)$ is as follows:

$$H_E(f) = \frac{1}{|C(f)|}, \tag{8.6}$$

and its phase response is obtained by

$$\theta_E(f) = -\theta_c(f). \tag{8.7}$$

Thus, in this case, the linear equalizer is referred to as the *inverse channel filter* for the channel frequency response. The inverse

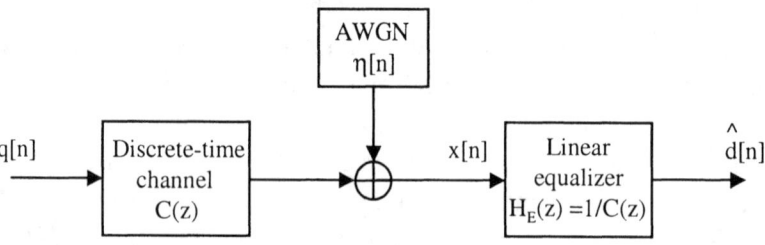

Figure 8.3 A block diagram of a discrete-time channel with a zero-forcing linear equalizer.

channel filter can completely eliminate ISI caused by the channel distortion because it forces the ISI to be zero at the sampling instant. As a result, such a linear equalizer is also called a *zero-forcing equalizer*.

Figure 8.3 shows a block diagram of a discrete-time channel with a zero-forcing equalizer. Assume that the discrete-time channel is a form of the FIR filter. Consequently, the transmission sequence $q[n]$ through the discrete-time channel results in an output sequence $x[n]$ that can be expressed as

$$x[n] = \sum_{k=0}^{N} c[k]q[n-k] + \eta[n], \qquad (8.8)$$

where $c[n]$ denotes a set of FIR filter coefficients of a discrete-time channel having a transfer function $C(z)$ and $\eta[n]$ is an additive white Gaussian noise (AWGN). Therefore, the estimate sequence $\hat{d}[n]$ to the threshold detector has a simple form

$$\hat{d}[n] = q[n] + \eta[n], \quad n = 0, 1, ..., \qquad (8.9)$$

where $q[n]$ is the desired transmission sequence.

Note that the FIR type of a zero-forcing equalizer usually does not completely eliminate ISI because it has a finite tap weight of the filter coefficients. However, the residual ISI can be further reduced when the length of filter coefficients of the zero-forcing equalizer is increased. In theory, the ISI trends to zero in the limit as the filter length $p \to \infty$.

Example 8.1

Consider a channel distorted pulse response denoted by $x(t)$ as the input to the equalizer as shown in Figure 8.2. The pulse response prior to the equalizer is given by the expression

$$x(t) = \frac{1}{1 + (5t/T)^2}, \qquad (8.10)$$

where $1/T$ is the symbol rate. The channel distorted pulse is sampled at the rate of $2/T$ and equalized by using a zero-forcing equalizer. We need to determine the three coefficients of a zero-forcing equalizer.

Since $X(f) = H_T(f)C(f)H_R(f)$ and $x(t)$ is the signal pulse corresponding to $X(f)$, then the equalized output signal pulse response is given by

$$y(t) = \sum_{n=-N}^{N} c[n] x(t - n\tau), \qquad (8.11)$$

where τ is the time delay between adjacent taps and is selected as $\tau = T/2$. The zero-forcing condition can now be applied to the samples of $y(t)$ taken at times $t = mT$. These samples are

$$y(mT) = \sum_{n=-N}^{N} c[n] x\left[mT - n\left(\frac{T}{2}\right)\right], \qquad m = 0, \pm 1, \pm 2, \cdots, \pm N. \qquad (8.12)$$

Notice that only $2N + 1$ sampled values of $y(t)$ can be controlled since there are $2N + 1$ equalizer coefficients. Therefore, the zero-forcing equalizer must satisfy the conditions

$$y(mT) = \sum_{n=-N}^{N} c[n] x\left[mT - n\left(\frac{T}{2}\right)\right]$$

$$= \begin{cases} 1, & m = 0 \\ 0, & m = \pm 1, \pm 2, \cdots, \pm N. \end{cases} \quad (8.13)$$

Equation (8.13) may be expressed in a matrix form as

$$\mathbf{y} = \mathbf{Xc}, \quad (8.14)$$

where $\mathbf{X}$ is the $(2N+1) \times (2N+1)$ matrix, $\mathbf{c}$ is the $(2N+1)$ tap-coefficient vector, and $\mathbf{y}$ is the $(2N+1)$ column vector. Thus, a set of $2N+1$ linear equations for the tap coefficient of the zero-forcing equalizer can be obtained.

Specifically, the matrix $\mathbf{X}$ with elements of $x(mT - nT/2)$, where m and n denote the columns and rows, respectively, is given by

$$\mathbf{X} = \begin{bmatrix} \frac{4}{29} & \frac{1}{26} & \frac{4}{229} \\ \frac{4}{29} & 1 & \frac{4}{29} \\ \frac{4}{229} & \frac{1}{26} & \frac{4}{29} \end{bmatrix}. \quad (8.15)$$

The output signal vector of the zero-forcing equalizer is given as

$$\mathbf{y} = [0 \ 1 \ 0]^T. \quad (8.16)$$

Thus, by inverting the matrix of $\mathbf{X}$, the linear equations $\mathbf{y} = \mathbf{Xc}$ can be solved to obtain the optimal solution for the zero-forcing equalizer as

$$\begin{aligned} \mathbf{c}_{opt} &= \mathbf{X}^{-1}\mathbf{y} \\ &= [-0.266 \ 1.073 \ -0.266]^T. \end{aligned} \quad (8.17)$$

8.2.2 Mean-Square-Error Criterion

The mean-square-error (MSE) criterion has seen widespread use for optimizing the filter tap coefficients of the linear equalizer. In the MSE criterion, the filter tap coefficients of the linear equalizer are adjusted to minimize the mean of the square error. The cost function ξ of the MSE criterion, which is formed based on Figure 8.4, is defined as

$$\begin{aligned} \xi &= E\{e[n]e^*[n]\} \\ &= E\{|e[n]|^2\} \\ &= E\{|q[n] - \hat{d}[n]|^2\}, \end{aligned} \quad (8.18)$$

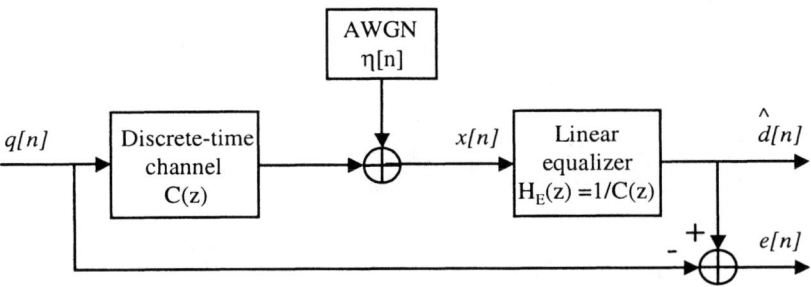

Figure 8.4 An error function of a discrete-time channel and a linear equalizer.

where the "E" notation represents the expected value and the cost function ξ is a quadratic function of the filter tap coefficients of the linear equalizer.

Assume that the linear equalizer has an infinite number of filter tap coefficients. In this case, the estimate sequence $\hat{d}[n]$ in (8.1) can be rewritten as

$$\hat{d}[n] = \sum_{k=-\infty}^{\infty} b_n[k] x[n-k]. \tag{8.19}$$

The filter tap coefficients $b_n[k]$ of the linear equalizer can be selected so that the error sequence $e[n]$ is orthogonal to the input sequence $x[n-k]$, $-\infty < k < \infty$, by invoking the theory of the orthogonality principle in a mean square estimation. Hence, we obtain the equation

$$E\{e[n]x^*[n-k]\} = 0, \quad -\infty < k < \infty. \tag{8.20}$$

Substituting the error sequence $e[n]$ into (8.20) yields

$$E\left[\left(q[n] - \sum_{k=-\infty}^{\infty} b_n[k]x[n-k]\right) x^*[n-k]\right] = 0. \tag{8.21}$$

This (8.21) is equivalent to

$$\sum_{k=-\infty}^{\infty} b_n[k] E\{x[n-k]x^*[n-k]\} = E\{q[n]x^*[n-k]\}, \quad -\infty < k < \infty. \tag{8.22}$$

Using the expression of $x[n]$ given in (8.8), we can rewrite (8.8) in terms of the expression of $x[n-k]$ as

$$x[n-k] = \sum_{l=0}^{N} c[l]q[n-k-l] + \eta[n-k]. \tag{8.23}$$

Now, we use (8.23) to develop the moments in (8.22) and obtain

$$E(x[n-k]x^*[n-k]) = \sum_{k=0}^{N} c[k] \sum_{j=0}^{N} c^*[-j] E\{|q[n-k-l]|^2\} + N_0 \delta_{kj}, \tag{8.24}$$

and

$$E(q[n]x^*[n-k]) = \sum_{k=0}^{N} c^*[-k] E\{q[n]q^*[n-k-l]\}. \tag{8.25}$$

For mathematical convenience, assume that the received signal power is unity. This implies that the expected values of $|q[n-k-l]|^2$ and $q[n]q^*[n-k-l]$ are also unity.

Substituting (8.24) and (8.25) into (8.21) yields

$$\sum_{i=-\infty}^{\infty} b_n[i] \left(\sum_{k=0}^{N} c[k] \sum_{j=0}^{N} c^*[-j] + N_0 \delta_{kj} \right) = \sum_{k=0}^{N} c^*[-k]. \tag{8.26}$$

Taking the z-transform of both sides in (8.26), with a time reversal of the z-transform properties (see Appendix A), we obtain

$$B(z)[C(z)C^*(1/z) + N_0] = C^*(1/z), \tag{8.27}$$

where $C(z)$ is the transfer function of the channel and N_0 is the noise spectral density. Therefore, the transfer function $B(z)$ of the linear equalizer based on the MSE solution is obtained by

$$B(z) = \frac{C^*(1/z)}{C(z)C^*(1/z) + N_0}. \tag{8.28}$$

Given the transfer function $B(z)$ of the linear equalizer in (8.28), the minimum value of the cost function ξ can be used for a measure of the residual ISI and additive noise. In this case, the minimum cost function ξ of the MSE criterion in (8.18) can be rewritten as

$$\begin{aligned} \xi_{min} &= E\{|e[n]|^2\} \\ &= E\{e[n]q^*[n]\} - E\{e[n]\hat{d}^*[n]\} \\ &= E\{e[n]q^*[n]\}, \end{aligned} \quad (8.29)$$

since $E\{e[n]\hat{d}^*[n]\} = 0$ by leveraging the theory of orgothonality conditions given in (8.20). It follows that the minimum cost function ξ is expressed as

$$\begin{aligned} \xi_{min} &= E\{e[n]q^*[n]\} \\ &= E\{|q[n]|^2\} - \sum_{k=-\infty}^{\infty} b_n[k] E\{x[n-k]q^*[n]\} \\ &= 1 - \sum_{k=-\infty}^{\infty} b_n[k] \sum_{j=-\infty}^{\infty} c[j]. \end{aligned} \quad (8.30)$$

Note that the summation in (8.30) is the convolution of filter tap coefficients $b_n[k]$ of the equalizer with the channel coefficients $c[n]$. If we denote $F(z)$ as the convolution of these two sequences in the z-domain, then

$$\begin{aligned} F(z) &= B(z)C(z) \\ &= \frac{C(z)C^*(1/z)}{C(z)C^*(1/z) + N_0}. \end{aligned} \quad (8.31)$$

Now let the term $f_0 = \sum_{k=-\infty}^{\infty} b_n[k] \sum_{j=-\infty}^{\infty} c[j]$. Thus, by the change in variable $z = e^{j\omega T}$ in (8.31), the term f_0 is obtained by

$$f_0 = \frac{T}{2\pi} \int_{-\pi/T}^{\pi/T} \frac{|C(e^{j\omega T})|^2}{|C(e^{j\omega T})|^2 + N_0} d\omega. \quad (8.32)$$

Thus, the desired MMSE ξ_{MMSE} of the linear equalizer is obtained by substituting (8.32) into the summation in (8.30)

$$\xi_{MMSE} = 1 - \frac{T}{2\pi} \int_{-\pi/T}^{\pi/T} \frac{|C(e^{j\omega T})|^2}{|C(e^{j\omega T})|^2 + N_0} d\omega$$

$$= \frac{T}{2\pi} \int_{-\pi/T}^{\pi/T} \frac{N_0}{|C(e^{j\omega T})|^2 + N_0} d\omega, \qquad (8.33)$$

where $C(e^{j\omega T})$ is the frequency response of the discrete-time channel and N_0 is the noise spectral density.

We observe that if $C(e^{j\omega T}) = 1$ then the ISI is eliminated. In this case, the MMSE ξ_{MMSE} is

$$\xi_{MMSE} = \frac{N_0}{1 + N_0}. \qquad (8.34)$$

Equation (8.34) indicates that $0 \leq \xi_{MMSE} \leq 1$. As a result, the relationship between signal-to-noise (SNR) and MMSE ξ_{MMSE} is obtained by

$$\text{SNR} = \frac{1 - \xi_{MMSE}}{\xi_{MMSE}}. \qquad (8.35)$$

Substituting (8.34) into (8.35) obtains

$$\text{SNR} = \frac{1}{N_0}. \qquad (8.36)$$

The relationship in (8.35) also holds even if there is residual ISI in addition to the noise.

We have derived the relationship of MMSE ξ_{MMSE} and SNR to the case in which the linear equalizer is with infinite filter tap coefficients. As a result, these expresses in (8.33), (8.34), and (8.35) can be serviced as upper bound limitations of the MMSE and SNR in theory for the linear equalizer. In the next section, we turn our attention to the case in which the transversal type of an equalizer has a finite tap coefficient, along with adaptive algorithms.

Example 8.2

Consider an equivalent discrete-time FIR channel model including two components a and b, where a and b satisfy the condition $a^2 + b^2 = 1$. Determine the MMSE and the corresponding output SNR of the linear equalizer.

The transfer function of the discrete-time channel in the z-transfer domain is given by

$$C(z) = a + bz^{-1}, \qquad (8.37)$$

and then

$$\begin{aligned}|C(z)|^2 &= C(z)C^*(1/z) \\ &= 1 + ab^*z + a^*bz^{-1}.\end{aligned} \qquad (8.38)$$

The corresponding frequency response is obtained as

$$\begin{aligned}|C(e^{j\omega T})|^2 &= 1 + ab^* e^{j\omega T} + a^* b e^{-j\omega T} \\ &= 1 + 2|a||b|\cos(\omega T + \phi),\end{aligned} \qquad (8.39)$$

where ϕ is the angle of ab^*.

The MMSE is given by (8.33) if a linear equalizer has an infinite number of coefficients. Evaluating the integral in (8.33) for the $|C(e^{j\omega T})|^2$ in (8.39) yields

$$\begin{aligned}\xi_{MMSE} &= \frac{T}{2\pi} \int_{-\pi/T}^{\pi/T} \frac{N_0}{|C(e^{j\omega T})|^2 + N_0} d\omega \\ &= \frac{N_0}{[N_0^2 + 2N_0 + (|a|^2 - |b|^2)^2]^{1/2}}.\end{aligned} \qquad (8.40)$$

In the special case, if $a = b = \frac{1}{\sqrt{2}}$, then the MMSE is

$$\xi_{MMSE} = \frac{N_0}{\sqrt{N_0^2 + 2N_0}}, \qquad (8.41)$$

and the corresponding output SNR is obtained by (8.35)

$$\begin{aligned}\text{SNR} &= \frac{1 - \xi_{MMSE}}{\xi_{MMSE}} \\ &= \left(1 + \frac{2}{N_0}\right)^{1/2} - 1 \\ &\approx \sqrt{\frac{2}{N_0}}, \quad N_0 \ll 1.\end{aligned} \qquad (8.42)$$

Notice that this result in (8.42) is compared with the output SNR of $\frac{1}{N_0}$ given in the ideal case of a channel without ISI. A significant loss in SNR occurs from the discrete-time channel model. This is because the characteristic of the discrete-time channel model has a spectral null at $\omega = \pi/T$ when using $a = b = 1/\sqrt{2}$.

8.3 Adaptive Linear Equalizer

An adaptive linear equalizer can be considered a time-varying filter with the tap weights of the filter coefficients that must constantly be adjusted at a time instant. For a time-varying channel (wireless channel) or a channel with unknown characteristics (wireline channel), an adaptive linear equalizer is designed to track the channel variations in such a way that (8.5) is approximately satisfied.

The general structure of an adaptive linear equalizer is again shown in Figure 8.1, where the n is used to denote a discrete-time index. Note that there is only one input $x[n]$ at any time instant. The value of the input $x[n]$ depends on the instantaneous state of the wireless channel or the wireline channel and the particular value of the noise. Hence, the input signal $x[n]$ is a random process. Practically, the adaptive linear equalizer as shown in Figure 8.1 has $p - 1$ delay elements, $p - 1$ complex additions, p taps, and p tunable complex multipliers, which are sometimes referred to as *tap weights*.

The tap weights of the transversal filter in the adaptive linear equalizer, which are described by their physical location in the delay line structure, have a subscript n to explicitly show that they vary with the discrete-time index. These tap weights are updated continuously by using an adaptive algorithm, either on a sample-by-sample basis or on a frame-by-frame basis. The latter case is referred to as the *block adaptive filter* [2].

The operation of adaptive algorithm is controlled by the error sequence $e[n]$. The error sequence $e[n]$ is produced by comparing the estimate output sequence of the linear equalizer $\hat{d}[n]$ with desired signal sequence $d[n]$ that is an exact scaled replica of the transmitted signal sequence $q[n]$. Then adaptive algorithm uses the error sequence $e[n]$ to minimize a cost function, which usually is an MSE.

To evaluate the adaptive linear equalizer in Figure 8.1, we define the input sequence $x[n]$ to the adaptive linear equalizer as a vector $\mathbf{x}[n]$ where

$$\mathbf{x}[n] = \{x[n], x[n-1], x[n-2], \cdots, x[n-p+1]\}, \quad (8.43)$$

and the tap weights of the filter coefficients in a vector form as

$$\mathbf{b}_n = \{b_n[0], b_n[1], b_n[2], \cdots, b_n[p-1]\}. \quad (8.44)$$

The estimate output of the adaptive linear equalizer is obtained in a vector notation by

$$\hat{\mathbf{d}}[n] = \mathbf{b}_n^T \mathbf{x}[n]. \quad (8.45)$$

Thus, the error signal $e[n]$ is written

$$\begin{aligned} e[n] &= d[n] - \hat{d}[n] \\ &= d[n] - \mathbf{b}_n^T \mathbf{x}[n]. \end{aligned} \quad (8.46)$$

Taking the expected value of the squared error signal $|e[n]|^2$ yields

$$\begin{aligned} \xi &= E\{|e[n]|^2\} \\ &= E\{|d[n] - \mathbf{b}_n^T \mathbf{x}[n]|^2\} \\ &= E\{|d[n]|^2\} - 2E\{\mathbf{x}[n]d^*[n]\}\mathbf{b}_n \\ &\quad + \mathbf{b}_n^T E\{\mathbf{x}[n]\mathbf{x}^T[n]\}\mathbf{b}_n, \end{aligned} \quad (8.47)$$

where it is assumed that the tap weights $\mathbf{b}_n$ of the filter coefficients are not time-varying and have converged to the optimum in a sense. If the input signal $\mathbf{x}[n]$ and desired signal $d[n]$ are stationary, then (8.47) can be rewritten as

$$\xi = E\{|d[n]|^2\} - 2\mathbf{r}_{dx}[n]\mathbf{b}_n + \mathbf{b}_n^T \mathbf{R}_x[n]\mathbf{b}_n, \quad (8.48)$$

where $\mathbf{r}_{dx}[n] = E\{\mathbf{x}[n]d^*[n]\}$ is the product of the cross-correlation function of the input signal $\mathbf{x}[n]$ and the desired transmitted signal

$d[n]$, and $\mathbf{R}_x[n] = E\{\mathbf{x}[n]\mathbf{x}^T[n]\}$ is the autocorrelation function of the input signal $\mathbf{x}[n]$.

Note that the MSE ξ in (8.48) is a multidimensional function. If the equalizer uses two tap weights, then the function of the MSE ξ is a bowl-shaped paraboloid with plotting the MSE ξ on the vertical axis and tap weights $b_n[0]$ and $b_n[1]$ on the horizontal axes. If the equalizer uses more than two tap weights, then the MSE function is a hyperparaboloid. In all cases, Widrow and Stearn [3] showed that the MSE function is concave upward so that a minimum value may be determined.

To minimize the MSE function ξ in (8.48), the derivate of ξ with respect to the weight vector $\mathbf{b}_n$ that is equal to zero yields

$$\mathbf{R}_x[n]\mathbf{b}_n = \mathbf{r}_{dx}[n]. \tag{8.49}$$

In this case, (8.49) is the same as the Wiener-Hopf solution that was shown in Chapter 7. As long as the matrix $\mathbf{R}_x[n]$ is invertible, then the optimum weight vector $\hat{\mathbf{b}}_n$ for the MMSE of a linear equalizer is obtained by

$$\hat{\mathbf{b}}_n = \mathbf{R}_x^{-1}[n]\mathbf{r}_{dx}[n]. \tag{8.50}$$

Thus, by minimizing the MSE function ξ in (8.48) in terms of the weight vector $\mathbf{b}_n$ of the filter coefficients, it becomes possible to adaptively tune the tap weights of the linear equalizer to provide a flat spectral response with minimum ISI for a communication receiver. This is due to the fact that minimizing the MSE function ξ leads to an optimal solution for the tap-weight vector $\mathbf{b}_n$.

8.3.1 Adaptive Algorithms for an Equalizer

An adaptive algorithm for a linear channel equalizer is required for channels whose characteristics are unknown and change with time. In this case, the ISI varies with time. It requires a specific adaptive algorithm to update the tap weights of the channel equalizer to reduce the ISI and track such time variations in the channel response. If the filtering characteristics of the channel are known and time-invariant, then the tap weights of the equalizer need be updated only once. To this extent, the adaptive algorithm of the equalizer is still needed to update the tap weights of the equalizer to compensate the channel distortion at the beginning of the transmission session. Therefore, the equalizers use adaptive algorithms to acquire the filter coefficients of the equalizer for a new signal, and then track it as needed.

In Chapter 7, we introduced methods with adaptive algorithms with great detail for adapting the tap weights of the filter coefficients for adaptive filters. These methods and algorithms include the Wiener-Hopf solution and the LMS and RLS algorithms, which have been extensively used for estimating optimal filter coefficients and/or adjusting the filter coefficients of the equalizer. Other algorithms also exist to adapt the filter coefficients such as Kalman filtering. For further algorithm developments, see [2–6].

This section outlines some practical issues of the adaptive algorithms with respect to equalizer algorithm performance. There are four factors that usually determine the performance of an adaptive algorithm:

- **Convergence rate**
  This factor is referred to as the number of iterations required for the adaptive algorithm of the equalizer in response to stationary inputs to converge close enough to the optimal solution or the Wiener-Hopf solution. A fast rate of convergence allows the algorithm to rapidly adapt the tap weights of the equalizer in a stationary environment of unknown statistics. In addition, it enables the algorithm to track channel variations while operating in a time-varying environment.

- **Misadjustment**
  This parameter is used to measure the amount that the final value of the MSE, averaged over an ensemble of adaptive equalizers, deviates from the MMSE in an optimal sense. The MSE is larger than the MMSE by an amount known as the excess mean square error (EMSE). Hence, the misadjustment is the ratio of the steady-state value of the EMSE to the MMSE.

- **Stability**
  In fixed-point implementation, some of the adaptive algorithms of the equalizers are considerably sensitive to round-off noise and quantization errors. These kinds of errors influence the stability of the algorithm for the equalizers during the real-time operations.
- **Computational complexity**
  This means the number of operations needed to make one iteration of the adaptive algorithm for the equalizer. A small number of operations result in less cost for the product.

In practice, the trade-off analysis is usually measured in terms of hardware complexity, chip area, processing consumption, and cost, in addition to convergence rate, misadjustment, stability, and computational complexity along with the equalizer structure and its adaptive algorithm. These factors are directly related to the filter length of the equalizer and which implementation structure is used.

In wireless communications, the characteristics of the radio channel and intended use of the subscriber equipment is also a key element. The speed of the mobile unit determines the channel fading rate and the Doppler spread that is directly related to the coherence time of the channel [7]. Hence, the choice of an adaptive algorithm along with its corresponding rate of convergence for a channel equalizer depends on the channel data rate and the coherence time [6, 7].

In wireline communications, the characteristics of the wireline channel, such as copper-wired telephone channel, are a predominant factor since the channel distortion is severe. The topology of a copperwire has complexities that usually contain multibridge taps and different types of copperwires. In addition, crosstalk between pairs in a multipair copperwire is the dominant impairment in any digital subscriber loop (DSL) system. The cause of this crosstalk is capacitive and inductive coupling between the wires [8]. Therefore, the choice of an adaptive equalizer structure and its algorithm, along with its corresponding convergence rate and numerical stability, is important for a channel equalizer in a DSL system.

8.3.2 Training Methodology

An adaptive algorithm requires knowledge of the "desired" response to form the error signal needed for the adaptive equalizer. In theory, the transmitted sequence is the desired response for an adaptive equalizer. In practice, however, the adaptive equalizer is physically separated from the transmitter of its ideal desired response since the adaptive equalizer is located in the communication receiver. For the adaptive channel equalizer, there are two methods in which a replica of the "desired" response may be generated locally to be used with the error signal in the communication receiver.

1. **Training method**
   This method is used during an initial training phase that takes place when the communication transmitter and receiver first establish a connection. A replica of the desired response is used from the signal sequence $d[n]$, as shown in Figure 8.1, when the transmitter sends a sequence that is known to the receiver. In fact, the generator of such a reference signal must be synchronized with the known transmitted sequence. In addition, Haykin [2] suggested using a test signal containing a pseudonoise sequence with a broad and even power spectrum because the pseudonoise sequence has noiselike properties and a periodically deterministic waveform.

2. **Decision-directed method**
   This method is to use the output sequence $y[n]$ of the threshold device in the receiver in Figure 8.1, as the transmitter sequence during the normal operation. Accordingly, if the output $y[n]$ is the correct transmitted sequence, then it may be used as the desired response to form the error sequence for the purpose of the adaptive equalizer. This method is said to be *decision-directed* because it is based on the decisions made by the communication receiver. This approach is also referred to as *blind training* since the communication receiver does not need a known transmitted sequence.

Note that, in the training method, the adaptive equalizer algorithm used to adjust the tap-weight settings of the equalizer corresponds mathematically to searching the unique minimum of a quadratic error-performance surface. The unimodal nature of this surface assures convergence of the adaptive algorithm. On the other hand, the decision-directed method modifies the error performance by using the estimated and unreliable output sequence into a multimodal one that may result in complex behavior. Haykin [9] and Miao [10] indicated that there are two types of error performance within local minima: (1) desired local minima, whose positions correspond to tap-weight settings, yielding the same performance as we obtained by using a known transmitted sequence; and (2) undesired local minima, whose positions correspond to tap-weight settings, yielding inferior equalizer performance.

A poor choice of the initial tap-weight settings may cause the adaptive equalizer to converge to an undesirable local minimum and stay there during an operation of the decision-directed method. Generally, for high performance, it is recommended that a linear adaptive equalizer be trained before it is switched to the decision-directed method. However, in a nonstationary and time-varying environment, it is impossible to consider the use of a transmitted training sequence. In such a case, the linear adaptive equalizer has to compensate the communication channel in an unsupervised mode. As a result, in this case, the operation of a linear adaptive equalizer is referred to as *blind equalization.* It is clear that the design of a blind equalizer is more challenging than a conventional adaptive equalizer because the transmitted sequence does not exist by some practical means.

8.3.3 Tap Length of Equalizer Coefficients

The computational complexities of a linear adaptive equalizer mainly depend on the tap length of the equalizer coefficients, which impacts the equalizer on the entire demodulator in the communication receiver. This brings into high profile the question of how long the equalizer's filter length must be to satisfactorily compensate for the dispersion of the channel. In theory, the symbol-based (sometimes called T-spaced) linear equalizer needs an infinite-length tap-weight setting to achieve the optimal solution for the equalizer. However, in fact, there are no clear answers available because the equalizer length depends on the type of channel to be equalized and the sample rate with respect to the transmitted signal bandwidth.

In practice, there are two approaches to selecting an equalizer length for compensating the dispersion of the channel [11]. The first method is to build a prototype and test the characteristics of the actual channel. For example, in a DSL system, the equalizer length can be determined in this way. The second method is to select the equalizer length with a presumption that the pulse response of the equalizer's convergence will approximately be the inverse of the channel bandwidth. In this case, if the channel can be well modeled with a finite length of a pulse response, then the equalizer length will often need to be three to five times longer than the delay spread of the channel.

Theoretically, however, this is not always true for an equalizer with fractional spacing based on a multichannel view of an oversampled signal. The fractional spacing changes the view of the equalization problem. Specifically, the length of the fractionally spaced equalizer can be made shorter and the associated computation may be substantially reduced. In the next section, we will introduce the theory of the fractionally spaced equalizer.

8.4 Fractionally Spaced Equalizer

In the previous sections, we have discussed the symbol-based (or T-spaced) linear equalizer structures, in which the spaces between adjacent equalizer taps are selected at the symbol rate. However, it has been shown that the symbol-based equalizer is very sensitive to the choice of receiver sampling time and phase [12–14]. This is because the received signal spectrum of the symbol-based equalizer is dependent on the choice of the sampling delay. In general, the symbol-based equalizer cannot compensate for the channel distortion inherent in the received signal spectrum associated with the sampling delay. This may result in a significant degradation to an equalizer performance.

In contrast to the symbol-based equalizer, it has been well recognized that a *fractionally spaced equalizer* (FSE) usually outperforms a symbol-based equalizer [15, 16] because of its insensitivity to receiver sampling time and phase. In fact, an FSE is based on sampling the incoming signal at least as fast as the Nyquist sampling rate.

Assume that the transmitted signal contains a pulse having a raised cosine spectrum with a rolloff factor β. Thus, its spectrum is given by

$$F_{max} = \frac{1+\beta}{2T}. \tag{8.51}$$

In this case, the incoming signal can be sampled at the communication receiver with a Nyquist sampling rate

$$\begin{aligned} F_S &= 2F_{max} \\ &= \frac{1+\beta}{T}, \end{aligned} \tag{8.52}$$

and then passed through a channel equalizer with an adjacent tap spacing at the reciprocal of the Nyquist sampling rate $T/(1+\beta)$. If the rolloff factor β is equal to 1, then we would refer the equalizer to a $T/2$-spaced equalizer. If the rolloff factor β is equal to 0.5, then we would have a $2T/3$-spaced equalizer, and so on. Generally, if an equalizer tap spacing has MT/N, where M and N are integers and $N > M$, then the equalizer is referred to as MT/N-spaced fractionally spaced equalizer. However, in practice, a $T/2$-spaced FSE is widely used in many applications.

8.4.1 Multirate Communication System Model

Consider the single-channel communication model with additive noise as shown in Figure 8.5 for a pulse amplitude modulation (PAM), phase-shift keying (PSK), or quadrature amplitude modulation (QAM) signal along with a T/L fractionally spaced equalizer,

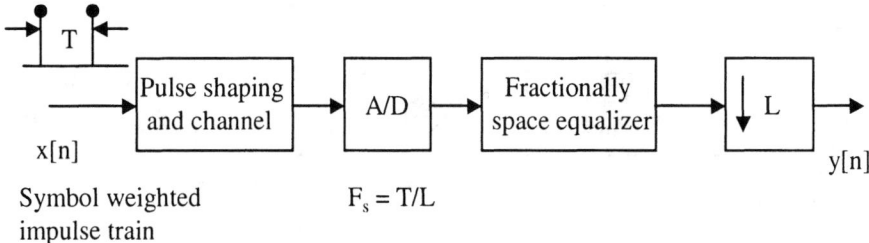

Figure 8.5 A block diagram of a communication system model with additive noise in cascade with a fractionally spaced equalizer and a downsampling by L.

where L is an integer. This model assumes that a single continuous-time filter is used for modeling the dispersive effects of the transmitter's pulse shaping and output filtering, channel propagation, and the receiver input filtering. A symbol sequence $x[n]$, which is spaced T seconds apart and weighted by the amplitude and phase of the symbols, is transmitted through a pulse shaping filter, modulated onto a propagation channel, and demodulated at the receiver. We also assume that all processing between the transmitter and receiver is linear time-invariant (LTI) and can thus be described by the continuous-time impulse response $c(t)$. The received signal $r(t)$ is also corrupted by additive channel noise denoted by $v(t)$. The output of this impulse-driven communication channel is then sampled at a rate of L/T Hz by using an A/D converter and applied to a fractional spaced equalizer with complex-valued pulse response. The output of the fractionally spaced equalizer is then decimated by a factor of L to produce the T-spaced output sequence $y[n]$. Decimation processing is accomplished by disregarding alternate samples, thus producing the baud-spaced "soft decisions" $y[n]$ every T seconds.

The partially continuous model of the pulse shaping and channel block as shown in Figure 8.5 can be accurately replaced by using the discrete-time multirate system model $c[n]$ as shown in Figure 8.6.

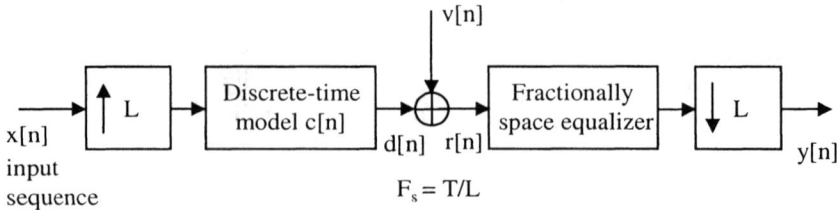

Figure 8.6 A discrete-time multirate communication system model, with additive noise in cascade with a fractionally spaced equalizer and a downsampling by L.

The PSK, PAM, or QAM symbol value, $x[n]$, is impressed on unit pulses in which each pair is upsampled by L. Then this zero-filled sequence is applied to a discrete-time system model $c[n]$ that models the channel propagation and filtering effects represented by the continuous filter. The received signal $r[n]$ is corrupted by additive channel noise denoted by the discrete-time random process $v[n] = v[n(T/L)]$. Assume that the discrete-time system model $c[n]$ in Figure 8.6 is an FIR filter with filter order of $L(L_c - 1)$, where L_c is the channel length. The output of the discrete-time communication system model is a multirate signal that is used as an input for the fractionally spaced equalizer. In addition, we assume that the fractionally spaced equalizer has the tap-weight order of $L(L_e - 1)$, where L_e is the equalizer length. Then the full rate output of the fractionally spaced equalizer is downsampled by a factor of L to create the T-spaced output $y[n]$ every T seconds.

The symbols $y[n]$ are the estimated value from the $r[n]$ by subsampling the output of $y[n]$. Mathematically, the relations of the input $x[n]$ and the output $r[n]$ are given by

$$r[n] = \sum_{k=0}^{N} c[n - kL]x[k] + v[n]$$
$$= d[n] + v[n], \qquad (8.53)$$

and the estimated symbols of the fractionally spaced equalizer are then obtained by

$$y[n] = \sum_{k=0}^{N} b[k]r[nL-k]. \qquad (8.54)$$

Note that the observations $r[n]$ are cyclostationary [17]. In other words, the correlation $C_r(n,m) = E\{r[n]r^*[n+m]\}$ is periodically time-varying with a period of L. If we assume that the random process of the input sequence $x[n]$ is zero mean and variance $C_x(0)$, with independent identical distribution (*i.i.d.*), and is also independent of the noise $v(n)$, then it follows from (8.53),

$$C_r(n,m) = C_x(0) \sum_{k}^{N} c[n-kL]c^*[n+m-kL] + C_v(n,m). \qquad (8.55)$$

Thus, the impulse response $c[n]$ of the discrete-time channel can be identified from the $C_r(n,m)$ by using (8.55). However, Tong et al. [18] and Ding et al. [19], almost at the same time, showed that the transfer function of the channel $C(z)$ should satisfy the *identifiability condition* that there are no zeros of the channel transfer function $C(z)$ equispaced on a circle with angle $\frac{2\pi}{L}$ separating one zero from the next.

This approach in (8.55) is sometimes referred to as *blind channel identification*. Note that fractionally spaced equalization can be performed by adopting an appropriate performance criterion as long as the impulse response $c[n]$ of the channel is estimated. In the next section, we introduce the requirement of a fractionally spaced equalizer to meet the zero-forcing criterion so that the entire transfer function can be identified based on the multichannel model, given the estimated impulse response $c[n]$ of the discrete-time channel.

8.4.2 Multichannel Model-Based Equalizer

The discrete-time multirate communication system model in Figure 8.6 uses the discrete-time channel model with the channel coefficients $c[n] = c[n(T/L)]$, the discrete-time fractionally spaced channel equalizer with the tap coefficients $b[n] = b[n(T/L)]$, and the discrete-time random process of additive noise $n[n] = n[n(T/L)]$. Since there is zero padding of the input sequence $x[n]$ and decimation at the output sequence $y[n]$, this allows us to redraw the multirate communication system model in Figure 8.6 as the multichannel model with single-rate of $1/T$ structure as shown in Figure 8.7. The discrete-time multirate channel can be broken into L subchannels, each of which consists of a finite impulse response (FIR) filter with the filter order of L_c. The coefficients of each FIR filter are a decimated version of the pulse response of the discrete-time multirate channel. The coefficients $c_0[n]$ of the first subchannel are equal to the discrete-time channel $c[nL]$, where $n = 0, 1, 2, \cdots, N/L$ and N/L is an integer. The coefficients $c_1[n]$ of the second subchannel are equal to $c[nL + 1]$. The coefficients $c_{L-1}[n]$ of the last subchannel are equal to $c[nL + L - 1]$. That is, $c_i[n] = c[nL+i]$, where $i = 0, 1, 2, \cdots, L-1$. Similarly, the discrete-time fractionally spaced equalizer is also decomposed into the L subspaced equalizers, $b_i[n] = b[nL+i]$, where $i = 0, 1, 2, \cdots, L-1$, and each one is built based on the decimation versions of the pulse response of the fractional spaced equalizer.

Now the question is: Under what conditions can all the subspaced equalizers, $b_i[nL + i]$, be chosen so that the propagation channel and other linear filtering embodied in $c_i[nL + i]$ can be perfectly equalized? The performance criterion is to require the subspaced equalizers to achieve the zero-forcing condition. In other words, this requires the transfer function from the input signal $x[n]$ to the output signal $y[n]$ to be the identity system as shown in Figure 8.6. If the zero-forcing condition is met, then the ISI can be completely removed. In this case, for the multichannel system in Figure 8.7, the solution of the zero-forcing condition in the discrete-time domain is given by

$$\sum_{i=0}^{L-1} \left\{ \sum_{k=0}^{N-1} c_i[k] b_i[n-k] \right\} = \delta[n-d], \tag{8.56}$$

where d is a positive integer that denotes the achieved delay. In

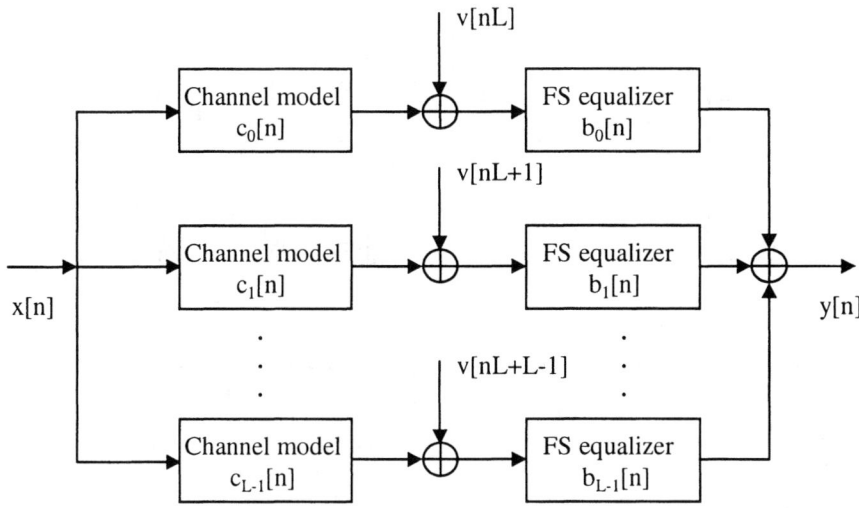

Figure 8.7 A discrete-time multichannel model-based fractionally spaced equalizer.

the frequency domain, the zero-forcing condition in (8.56) can be expressed as

$$\sum_{k=0}^{L-1} \left[B_k(e^{j\omega}) C_k(e^{j\omega}) \right] = e^{-jd\omega}. \tag{8.57}$$

If the channel $c[n]$ satisfies the identifiability condition, then the matrix of linear equations in (8.56) is a full rank [20]. Hence, the solutions for the subspaced equalizers $b[m]$ exist.

In practice, a common choice of L is equal to two for adaptive equalizers. In this case, the A/D converter in a communication receiver runs at twice the symbol rate, such as $2F_B$, resulting in equalizer taps spaced $T/2$ seconds apart. From the previous discussion, this $T/2$-spaced equalizer can be drawn as two subchannels, as shown in Figure 8.8. Given the subchannel with even coefficients $c_0[n] = c[2n]$ and odd coefficients $c_1[n] = c[2n+1]$,

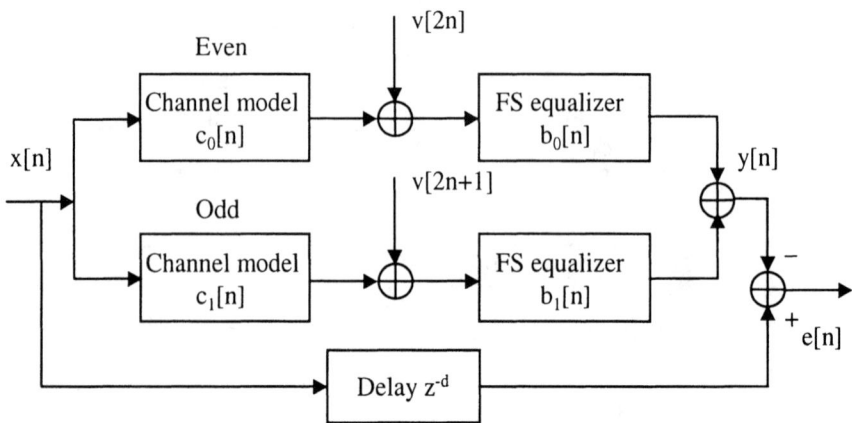

Figure 8.8 A two-channel system of a $T/2$ fractionally spaced equalizer.

where $n = 0, 1, \cdots, N$, the equalization objective is to choose the decomposed equalizer subchannel impulse responses $b_0[n] = b[2n]$ and $b_1[n] = b[2n+1]$ such that the output signal $y[n]$ approximates the unit delay $u[k-d]$ as closely as possible, where d is a positive integer for the achieved delay.

The desired result of the zero-forcing condition in the discrete-time domain is obtained by

$$\sum_{k=0}^{N-1} c_0[k]b_0[n-k] + \sum_{k=0}^{N-1} c_1[k]b_1[n-k] = \delta[n-d], \quad (8.58)$$

or equivalently in the frequency domain as

$$C_0(e^{j\omega})B_0(e^{j\omega}) + C_1(e^{j\omega})B_1(e^{j\omega}) = e^{-jd\omega}. \quad (8.59)$$

Either (8.58) or (8.59) is referred to as the zero-forcing $T/2$ fractionally spaced equalizer since the sum of the two convolutions is forced to zero for all delays except d. Thus, a channel can be perfectly equalized if two conditions can be met as follows:

(1) the two subchannels do not have common roots; and (2) the decimated equalizer order L_e must at least be equal to or greater than the decimated channel order $L_c - 1$. This indicates that if the delay spread of a channel can be measured or modeled to be less than some bound, then the fractionally spaced equalizer order can be shorted. This is contrast to the capabilities of a baud-spaced equalizer in which the zero-forcing cannot be achieved with a finite-length equalizer.

8.4.3 FSE-MMSE Function

Referring to Figure 8.6, in the presence of noise, the error signal can be formed as follows:

$$e[n] = y[n] - x[n - d], \qquad (8.60)$$

where d denotes a particular delay. We want to minimize the expected squared magnitude of the recovery error in (8.60) so that this criterion can be interpreted as the best compromise between ISI and noise amplification in an optimal sense of MMSE.

To formulate this error function more precisely, we define the P T-spaced samples of the source sequence into the vector

$$\mathbf{x}[n] = [x[n], x[n-1], x[n-2], \cdots, x[n-N]]^T, \qquad (8.61)$$

and the corresponding $2N$ fractionally sampled values of noise into vector

$$\begin{aligned}\mathbf{v}[n] = &[v[n-1], v[n-3], v[n-5], \cdots, v[n-2N+1],\\ &v[n], v[n-2], v[n-4], \cdots, v[n-2N+2]]^T,\end{aligned} \qquad (8.62)$$

where the odd noise samples are followed by the even noise samples.

We also define the $P \times N$ baud-spaced block Toeplitz convolution matrices $\mathbf{C}_{even}[n]$ and $\mathbf{C}_{odd}[n]$ for the even and odd channel coefficients, respectively, as shown in Figure 8.8, where

$P = M + N + 1$, as follows:

$$\mathbf{C}_{even}[n] = \begin{bmatrix} c[0] & 0 & \cdots & 0 \\ c[2] & c[0] & & \vdots \\ \vdots & c[2] & & 0 \\ c[2N-2] & \vdots & \ddots & 0 \\ 0 & c[2N-2] & & c[0] \\ \vdots & 0 & & c[2] \\ 0 & \vdots & & \vdots \\ 0 & 0 & \cdots & c[2N-2] \end{bmatrix}, \quad (8.63)$$

and

$$\mathbf{C}_{odd}[n] = \begin{bmatrix} c[1] & 0 & \cdots & 0 \\ c[3] & c[1] & & \vdots \\ \vdots & c[3] & & 0 \\ c[2N-1] & \vdots & \ddots & 0 \\ 0 & c[2N-1] & & c[1] \\ \vdots & 0 & & c[3] \\ 0 & \vdots & & \vdots \\ 0 & 0 & \cdots & c[2N-1] \end{bmatrix}. \quad (8.64)$$

Given a fractionally spaced equalizer with a finite length of $2N$, we form the even and odd taps of equalizer into the vectors, $\mathbf{b}_{even}[n]$ and $\mathbf{b}_{odd}[n]$, respectively,

$$\mathbf{b}_{even}[n] = [b[0], b[2], b[4], \cdots, b[2N-2]]^T, \quad (8.65)$$

and

$$\mathbf{b}_{odd}[n] = [b[1], b[3], b[5], \cdots, b[2N-1]]^T. \quad (8.66)$$

Defining the compound matrix in (8.63) and (8.64) and vector quantities in (8.65) and (8.66) is

$$\mathbf{C}[n] = [\mathbf{C}_{even}[n] \ \mathbf{C}_{odd}[n]], \quad (8.67)$$

and
$$\mathbf{b}[n] = [\mathbf{b}_{even}[n] \ \mathbf{b}_{odd}[n]]^T. \tag{8.68}$$

With these quantities from (8.61) to (8.68), the output sequence of the fractionally spaced equalizer can be written as

$$y[n] = \mathbf{x}^T[n]\mathbf{C}[n]\mathbf{b}[n] + \mathbf{v}^T[n]\mathbf{b}[n]. \tag{8.69}$$

Hence, substituting (8.69) into (8.60) yields

$$\begin{aligned} e[n] &= \mathbf{x}^T[n]\mathbf{C}[n]\mathbf{b}[n] + \mathbf{v}^T[n]\mathbf{b}[n] - x[n-d] \\ &= \mathbf{x}^T[n](\mathbf{C}[n]\mathbf{b}[n] - \mathbf{h}_d[n]) + \mathbf{v}^T[n]\mathbf{b}[n], \end{aligned} \tag{8.70}$$

where $\mathbf{h}_d[n] = [0, \cdots, 0, 1, 0, \cdots, 0]^T$ is the so called "zero-forcing" system impulse response under the *perfect source recovery* [21].

Assume that the signal processes and noise are $i.i.d.$ and jointly uncorrelated with respect to variances σ_x^2 and σ_v^2, respectively. Then the expected value of the magnitude-squared of the recovery error is obtained by

$$\begin{aligned} E\{|e[n]|^2\} &= \sigma_x^2 (\mathbf{C}[n]\mathbf{b}[n] - \mathbf{h}_d[n])^H (\mathbf{C}[n]\mathbf{b}[n] - \mathbf{h}_d[n]) \\ &\quad + \sigma_v^2 \mathbf{b}^H[n]\mathbf{b}[n], \end{aligned} \tag{8.71}$$

where $(\cdot)^H$ in the matrix denotes the Hermitian transpose. Therefore, the normalized mean square error (NMSE) cost function of the fractionally spaced equalizer based on (8.71) is

$$\begin{aligned} \xi_{NMSE} &= \frac{E\{|e[n]|^2\}}{\sigma_x^2} \\ &= (\mathbf{C}[n]\mathbf{b}[n] - \mathbf{h}_d[n])^H (\mathbf{C}[n]\mathbf{b}[n] - \mathbf{h}_d[n]) \\ &\quad + \frac{\sigma_v^2}{\sigma_x^2} \mathbf{b}^H[n]\mathbf{b}[n]. \end{aligned} \tag{8.72}$$

The ξ_{NMSE} in (8.72) is the sum of a zero-forcing measure, including the distance between the global channel equalizer impulse response $\mathbf{C}[n]\mathbf{b}[n]$ and the optimal zero-forcing $\mathbf{h}_d[n]$, and of a noise enhancement measure $\frac{\sigma_v^2}{\sigma_x^2}\mathbf{b}^H[n]\mathbf{b}[n]$, which is the amount of the received noise by the equalizer.

Minimizing ξ_{NMSE} in (8.72) with respect to the equalizer coefficient vector $\mathbf{b}[n]$ yields

$$\frac{\partial \xi_{NMSE}}{\partial \mathbf{b}[n]} = 2\mathbf{C}^H[n]\mathbf{C}[n]\mathbf{b}[n] - 2\mathbf{C}^H[n]\mathbf{h}_d[n] + 2\frac{\sigma_v^2}{\sigma_x^2}\mathbf{b}[n]. \quad (8.73)$$

Equating (8.73) to zero, we obtain

$$\mathbf{b}[n] = \frac{\mathbf{C}^H[n]\mathbf{h}_d[n]}{\mathbf{C}^H[n]\mathbf{C}[n] + \frac{\sigma_v^2}{\sigma_x^2}\mathbf{I}}. \quad (8.74)$$

Substituting (8.74) into (8.72) yields the MMSE of the fractionally spaced equalizer as

$$\xi_{MMSE} = \mathbf{h}_d^H \left(\mathbf{I} - \frac{\mathbf{C}[n]\mathbf{C}^H[n]}{\mathbf{C}^H[n]\mathbf{C}[n] + \frac{\sigma_v^2}{\sigma_x^2}\mathbf{I}} \right) \mathbf{h}_d. \quad (8.75)$$

Note that (8.75) still contains a function of system delay d. In other words, the degree of the system delay d can affect the MMSE performance of a fractionally spaced equalizer. Johnson [22] showed that the optimum system delay d corresponds to the index of the minimum diagonal element in (8.75). Thus, the optimal system delay d is formally given by

$$d = \arg\min_d \left\{ \mathbf{I} - \frac{\mathbf{C}[n]\mathbf{C}^H[n]}{\mathbf{C}^H[n]\mathbf{C}[n] + \frac{\sigma_v^2}{\sigma_x^2}\mathbf{I}} \right\}. \quad (8.76)$$

Therefore, we recommend that the preselection of the system delay d is important for a fractionally spaced equalizer in terms of the MMSE performance because it describes the typical adaptive equalization scenario when a training sequence is available.

Example 8.3

Assume that an FSE has an infinite-length tap coefficient and L is an integer of the oversampling factor (refer to Figures 8.6 and 8.7). The sampled output of the antialias filter can be decomposed into L interleaved sequences with the z-transforms $Q_0(z), Q_1(z), Q_2(z), \cdots, Q_{L-1}(z)$ at a sampling rate of $1/T$, where $Q_i(z)$ corresponds to the sample sequence $q[kT - iT/L]$. Also, assume that $V_i(z)$ is the transform of a symbol rate sampled white noise sequence with autocorrelation function $R_{nn}(z) = \frac{LV_0}{2}$ per dimension, and these noise sequences are independent of one another. Determine the MMSE of the infinite-length FSE.

The interleaved sequences $H_i(z)$ can be expressed as

$$Q_i(z) = C_i(z)X(z) + V_i(z), \qquad (8.77)$$

where $C_i(z)$ is the transform of the symbol-rate-spaced ith phase of the $c(t)$ channel sequence $c[kT + (i-1)T/L]$. A column vector of the transform is given by

$$\mathbf{Q}(z) = \mathbf{C}(z)X(z) + \mathbf{V}(z), \qquad (8.78)$$

where

$$\mathbf{C}(z) = [C_0(z), C_1(z), \cdots, C_{L-1}(z)]^T, \qquad (8.79)$$

$$\mathbf{V}(z) = [V_0(z), V_1(z), \cdots, V_{L-1}(z)]^T, \qquad (8.80)$$

and

$$\mathbf{Q}(z) = [Q_0(z), Q_1(z), \cdots, Q_{L-1}(z)]^T. \qquad (8.81)$$

If the FSE output is at a sampling rate of $1/T$, then the interleaved tap weights of the FSE can be written in a row vector as

$$\mathbf{B}(z) = [B_0(z), B_1(z), \cdots, B_{L-1}(z)]. \qquad (8.82)$$

Thus, the FSE output is obtained by

$$Y(z) = \mathbf{B}(z)\mathbf{Q}(z), \qquad (8.83)$$

and the error sequence $\mathbf{E}(z)$ can then be formed as

$$\mathbf{E}(z) = X(z) - Y(z). \qquad (8.84)$$

Notice that $\mathbf{E}(z)$ in (8.84) should be orthogonal to $\mathbf{Q}(z)$ by using the orthogonality condition. As a result, the expectation of the vectors $\mathbf{E}(z)$ and $\mathbf{Q}(z)$ are

$$\begin{aligned} E\{\mathbf{E}(z)\mathbf{Q}^*(1/z)\} &= \mathbf{R}_{xQ}(z) - \mathbf{B}(z)\mathbf{R}_{QQ}(z) \\ &= 0, \end{aligned} \quad (8.85)$$

where

$$\begin{aligned} \mathbf{R}_{xQ}(z) &= E\{X(z)\mathbf{Q}^*(1/z)\} \\ &= \mathcal{E}_{\mathbf{X}}\mathbf{C}^*(1/z), \end{aligned} \quad (8.86)$$

and

$$\begin{aligned} \mathbf{R}_{QQ}(z) &= E\{\mathbf{Q}(z)\mathbf{Q}^*(1/z)\} \\ &= \mathcal{E}_{\mathbf{X}}\mathbf{C}(z)\mathbf{C}^*(1/z) + \frac{LV_0}{2}\mathbf{I}. \end{aligned} \quad (8.87)$$

Therefore, by solving (8.85), the infinite-length FSE-MMSE tap-coefficient setting is obtained by

$$\begin{aligned} \mathbf{B}(z) &= \mathbf{R}_{xQ}(z)\mathbf{R}_{QQ}^{-1}(z) \\ &= \frac{\mathbf{C}^*(1/z)}{\mathbf{C}(z)\mathbf{C}^*(1/z) + L/SNR} \end{aligned} \quad (8.88)$$

The corresponding error sequence has autocorrelation function as

$$\begin{aligned} R_{ee}(z) &= \mathcal{E}_{\mathbf{X}} - \mathbf{R}_{xQ}(z)\mathbf{R}_{QQ}^{-1}(z)\mathbf{R}_{Qx}(z) \\ &= \frac{LV_0/2}{\mathbf{C}(z)\mathbf{C}^*(1/z) + L/\text{SNR}}. \end{aligned} \quad (8.89)$$

Therefore, the infinite-length FSE-MMSE is then obtained by

$$\xi_{FSE-MMSE} = \frac{T}{2\pi}\int_{-\pi/T}^{\pi/T}\left[\frac{LV_0/2}{|\mathbf{C}(e^{-j\omega T})|^2 + L/\text{SNR}}\right]d\omega. \quad (8.90)$$

Notice that (8.90) is exactly the same as the MMSE given in (8.33) for the MMSE linear equalizer as long as the sampling rate exceeds twice the highest frequency of $X(e^{j\omega T})$.

8.4.4 FSE Constant Modulus Algorithm

Most equalization techniques, such as those using the LMS algorithm, use a training sequence to allow the communication receiver to adjust the fractionally spaced equalizer coefficients. One obvious drawback of such an approach is that the training causes a reduction of the useful information rate with respect to the total information rate. In addition, there are many digital communications system applications, such as broadcast and wireless configurations, for which training data is not available. Therefore, blind fractionally spaced equalization without a training sequence, particularly the CMA, has received increasing interest since the last decade.

The CMA with application in QAM signals was originally proposed by Godard [23]. Sometimes the CMA is also referred to as the Godard algorithm. The CMA was later further developed with applications for FM signals by Treichler and Agee [24]. It has been shown that the fractionally spaced equalizer adapted using the CMA can converge globally to minimize ISI under conditions [25, 26]. Even for channels with deep spectral nulls, the CMA-based fractionally spaced equalizer does not require a large number of the equalizer tap coefficients and can then converge faster [27].

The multichannel vector representation of the blind adaptive fractionally spaced equalizer using the CMA is shown in Figure 8.9. The channel output of a communication system can be described using the baseband representation as

$$x(t) = \sum_{n=0}^{\infty} x[n]c(t - nT - t_0) + v(t), \quad (8.91)$$

where T is the symbol period and t_0 is a constant of the system delay. Assume that the input signal $x[n]$ is an independent and identically distributed sequence and the channel output may be corrupted by channel noise $v(t)$, which is independent of the input signal $x[n]$. The input sequence $x[n]$ is passed through an LTI channel with an impulse response $c(t)$ by the transmitter. The communication receiver attempts to recover the input sequence $x[n]$ from the measurable channel output.

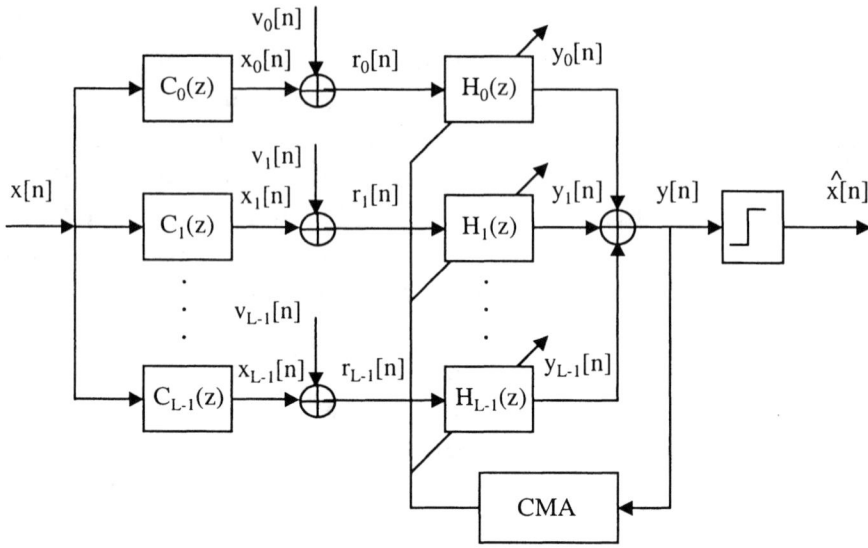

Figure 8.9 A multichannel vector representation of the blind adaptive fractionally spaced equalizer using the CMA.

Define L as an integer and the oversampling symbol interval as $\Delta = T/L$. It has been shown by Gardner [28] that the oversampling can provide channel diversity if the channel bandwidth is greater than the minimum symbol rate of $1/(2T)$. Thus, the output sequence of the sampled channel is given by

$$x[k\Delta] = \sum_{n=0}^{\infty} x[n]c[k\Delta - nL\Delta - t_0] + v[k\Delta]. \quad (8.92)$$

Let the subchannel frequency response in the z-domain be

$$C_i(z) = \sum_{n}^{L-1} c_i[n]z^{-n}, \quad (8.93)$$

where the subchannel impulse response in the discrete-time domain

is
$$c_i[n] \cong c[k\Delta + i\Delta - t_0], \quad i = 0, 1, 2, \cdots, L. \tag{8.94}$$

The oversampled channel output $x[k\Delta]$ in (8.92) may be divided into L linearly independent subsequences as

$$x_i[n] = x[n] * c_i[n] + v[nT + i\Delta], \quad i = 0, 1, 2, \cdots, L. \tag{8.95}$$

Hence, these L subsequences $x_i[n]$ can be viewed as stationary outputs of L discrete-time FIR channels $c_i[n]$ with a common input sequence $x[n]$. This approach is known as the single-input multiple-output (SIMO) channel.

The adjustable fractionally spaced equalizer $H_i(z)$ is provided for each subsequence $x_i[n]$ of the multichannel. In this case, each subequalizer $H_i(z)$ is an FIR filter as

$$H_i(z) = \sum_{n=0}^{N-1} h_i[n] z^{-n}, \quad i = 0, 1, 2, \cdots, L. \tag{8.96}$$

Thus, the L FIR filter outputs $y_i[n]$ are summed to create the stationary equalizer output

$$y[n] = \sum_{i=0}^{L-1} y_i[n]. \tag{8.97}$$

The constant modulus (CM) criterion for the fractionally spaced equalizer is expressed as

$$\xi_{CM} = E\{(|y[n]|^2 - \gamma)^2\}, \tag{8.98}$$

where $y[n]$ are the fractionally spaced equalizer output sequences that are coincident with symbol instances and γ is the CM dispersion constant defined as

$$\gamma = \frac{E\{|x[n]|^4\}}{E\{|x[n]|^2\}}. \tag{8.99}$$

Defining the fractionally spaced equalizer coefficients into the vector as

$$\mathbf{h}_0[n] = [h_0[1], h_0[2], \cdots, h_0[N-1]],$$
$$\mathbf{h}_1[n] = [h_1[1], h_1[2], \cdots, h_1[N-1]],$$
$$\vdots \qquad \vdots$$
$$\mathbf{h}_{L-1}[n] = [h_{L-1}[1], h_{L-1}[2], \cdots, h_{L-1}[N-1]]. \quad (8.100)$$

Then the compound matrix of the fractionally spaced equalizer is given by

$$\mathbf{b}[n] = [\mathbf{h}_0[n] \; \mathbf{h}_1[n] \; \cdots \; \mathbf{h}_{L-1}[n]]. \quad (8.101)$$

In order to adaptively adjust the fractionally spaced coefficients $\mathbf{b}[n]$ without a training sequence, the CMA, which can be implemented to jointly update the L FIR filters to minimize the CM cost function in (8.98), is the stochastic, gradient search rule that descends this cost surface by updating the fractionally spaced equalizer coefficients $\mathbf{b}[n]$ according to

$$\mathbf{h}_i[n+1] = \mathbf{h}_i[n] + \mu_i(\gamma - |y[n]|^2)y[n]\mathbf{r}_i^*[n], \; i = 0, 1, 2, \cdots, L, \quad (8.102)$$

where μ is a small, positive, tunable step size and $\mathbf{r}_i[n] = [r_i[n], r_i[n-1], \cdots, r_i[n-N+1]]$ is a vector of received samples.

Johnson et al. [29] showed that the mean of the CM cost function in (8.98) represents the average performance of CMA in (8.102).

The step size μ_i in (8.102) determines the distance covered and speed towards the negative gradient of the constant modulus cost surface on each iteration. Making a large step size μ_i will adjust the equalizer taps rapidly, but it will also increase the amount of EMSE as discussed in Chapter 7, which is the amount of error that is in excess of the optimal case. This is because the tap values take such large steps that they constantly overshoot the minimum of the constant modulus cost surface. As a result, instead of converging, the values of tap-weight equalizer will rattle around the bottom of one of the constant modulus cost surface bowls. The rattling phenomenon is referred to as *stochastic jitter* [30]. On the other hand, the small step size μ_i will make the tap values update very slowly, thereby increasing the time for convergence. The step size μ_i must be selected carefully. A balance is recommended so that the equalizer is able to track channel variations while keeping the EMSE relatively small.

8.4.5 FSE-CM Noisy Cost Function

In this section, we are interested in the FSE-CM criterion in terms of equalizability. In other words, we want to know a measure of the FSE-CM performance in equalizing a given channel in the presence of additive noise and then compare the FSE-CM MSE with an analytical expression of a lower equalizability bound, MMSE in (8.75) for a given SNR. This comparison should provide a meaningful measure of the FSE-CM criterion performance.

Note that, from averaging theory, the extreme of the FSE-CM cost function ξ_{CM} in (8.98) is the possible convergence setting of the CMA in (8.102). Understanding the FSE-CM cost function allows us to analyze the mean convergence points of the FSE CMA with additive noise conditions.

In order to have a better understanding of the equalizer setting with minimizing the FSE-CM criterion, we can establish an expression for the FSE-CM cost function ξ_{CM} in (8.98) with conditions of an additive noise channel. Assume that the channels are a set of causal FIRs and the noise $v_i[n]$ is independent from the input signal $x[n]$. Fijalkow et al. [31] showed that the FSE-CM cost function in (8.98) can be expressed as

$$\xi_{CM} = \Gamma_0(\mathbf{f}[n]) + \Phi(\mathbf{b}[n]), \qquad (8.103)$$

where

$$\Phi(\mathbf{b}[n]) = \frac{\sigma_v^2}{\sigma_x^2}\mathbf{b}^H[n]\mathbf{b}[n]\left[2\left(k_g\mathbf{f}^H[n]\mathbf{f}[n] - k_s\right) + k_g\frac{\sigma_v^2}{\sigma_x^2}\mathbf{b}^H[n]\mathbf{b}[n]\right], \qquad (8.104)$$

where k_g is a Gaussian signal kurtosis that is equal to 3 in the case of real-valued signals and is equal to 2 in the complex-valued signals,

and
$$\mathbf{f}[n] = \mathbf{C}[n]\mathbf{b}[n], \tag{8.105}$$

and
$$\Gamma_0(\mathbf{f}[n]) = E\{(k_s - |\mathbf{f}^T[n]x[n]|^2)^2\}, \tag{8.106}$$

which is the noise-free cost function, and $k_s = \frac{E\{|x[n]|^4\}}{(\sigma_x^2)^2}$ is the input signal kurtosis that is used to measure the source signal "compactness." Equation (8.103) is referred to as the *FSE-CM noisy cost function*.

The minimization of the FSE-CM cost function of (8.103) in the noisy context is equivalent to the minimization of noise-free cost function $\Gamma_0(\mathbf{f}[n])$ over the vector $\mathbf{f}[n]$ regularized by the second term of the additional deterministic factor, $\Phi(\mathbf{b}[n])$. Note that for most input sequences $x[n]$, for example, a PAM or QAM signal, $k_s < k_g - 1$. This indicates $\mathbf{f}^T[n]\mathbf{f}[n] > k_s/k_g$ if $\mathbf{f}[n]$ is not too close to the origin $\mathbf{f}[n] = 0$, which is the maximum of ξ_{CM}. Therefore, the deterministic factor $\Phi(\mathbf{b}[n])$ has the desirable property to be a positive convex quadratic function in terms of $\mathbf{b}[n]$. Then, the main consequence of the regularization is to forbid the FSE norm to be too high and to reduce the noise enhancement contribution in the FSE output signal $y[n]$. Comparing (8.103) with the FSE MSE criterion in (8.72), we note that the constraint on the term $\mathbf{b}^H[n]\mathbf{b}[n]$ is much stronger, thereby limiting all noise enhancement.

8.4.6 FSE-CM Performances

For a PAM or QAM signal, the relation between MSE cost function of the FSE-CM and the FSE NMSE in (8.72) is approximately given by Endres [32]

$$\xi_{CM-MSE} \approx 2\sigma_x^2(k_g - k_s)\xi_{NMSE}. \tag{8.107}$$

Minimizing ξ_{CM-MSE} in (8.107) with respect to $\mathbf{b}[n]$ obtains

$$\mathbf{b}[n] = \frac{\mathbf{C}^H[n]\mathbf{h}_d[n]}{\mathbf{C}^H[n]\mathbf{C}[n] + \frac{\sigma_v^2}{\sigma_x^2}\mathbf{I}}. \tag{8.108}$$

It is interesting to note that (8.108) is exactly equal to the optimal vector $\mathbf{b}[n]$ given in (8.74) because the term of $2\sigma_x^2(k_g - k_s)$ in (8.107) does not include a direct relationship with the FSE taps $\mathbf{b}[n]$. Therefore, substituting (8.108) into (8.107) yields the approximate MMSE of the FSE-CM as

$$\begin{aligned}\xi_{CM-MMSE} &\approx 2\sigma_x^2(k_g - k_s)\xi_{MMSE} \\ &= 2\sigma_x^2(k_g - k_s)\left[\mathbf{h}_d^H\left(\mathbf{I} - \frac{\mathbf{C}[n]\mathbf{C}^H[n]}{\mathbf{C}^H[n]\mathbf{C}[n] + \frac{\sigma_v^2}{\sigma_x^2}\mathbf{I}}\right)\mathbf{h}_d\right]\end{aligned}$$
(8.109)

Therefore, in this case, the MMSE of the FSE-CM receiver may be approximately upper bounded by $\frac{\xi_{CM-MMSE}}{2\sigma_x^2(k_g-k_s)}$.

It is well established that the system delay d in the combined channel equalizer can influence FSE-CM MSE performance with variations of several orders of magnitude being typical. It has been understood that the FSE-CM MSE performance in (8.109) also depends on the system delay d on the FSE-CM cost function in terms of describing the relative depth between FSE-CM local minima associated with different system delays. In general, there does not exist a closed-form expression for the FSE-CM local minima settings since the FSE-CM cost function depends on the fourth-order moment of the FSE vector. Endres et al. [32] found a closed-form estimate of the FSE-CM local minima for noiseless binary phase-shift keying (BPSK) signal by approximating the FSE-CM cost function with a second-order Taylor series expanded about the length-constrained Wiener settings. However, our results in (8.109) also suggest that there exist FSE-CM local minima in close proximity to those MSE local minima that correspond to better-performing system delays.

8.5 Decision Feedback Equalizer

A decision feedback equalizer employs previous decisions to eliminate the ISI caused by previously detected symbols on the

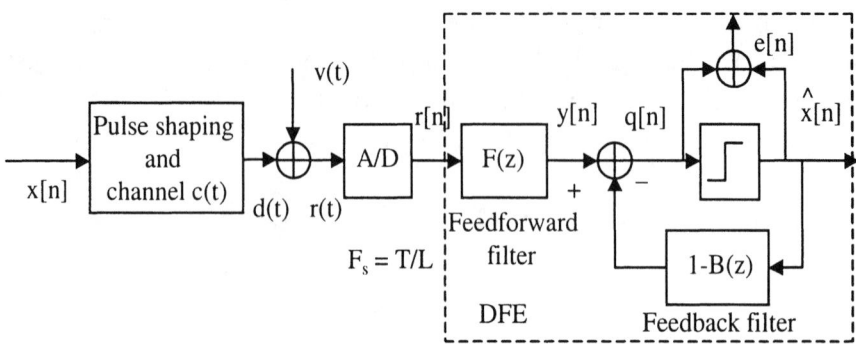

Figure 8.10 A generalized block diagram of a decision feedback equalizer.

current symbol to be detected. In other words, the basic idea behind the DFE is that the ISI, induced on future symbols, can be estimated and subtracted out before detection of subsequent symbols once an information symbol has been detected and decided upon. The DFE is inherently a nonlinear communication receiver. However, the DFE can be analyzed using linear techniques if we assume that all previous decisions are correct. Thus, in this section, we derive an MMSE solution and provide the performance analysis for the DFE.

A simple block diagram of a DFE-based communication receiver is shown in Figure 8.10, where the DFE section consists of two filters and one decision element. The first filter $F(z)$ is called a linear *feedforward filter* (FFF) and it is generally a fractionally spaced FIR filter with adjustable tap coefficients. Its input is the received filtered signal $r(t)$ sampled at some rate that is a multiple of the symbol rate, for example, at rate of $2/T$. The output of the FFF is denoted by $y[n]$, and the input to the decision element is denoted by $q[n]$. The second filter is a causal, *feedback filter* (FBF), with $1 - B(z)$, where $b_0 = 1$. The FBF is implemented as an FIR filter with symbol-spaced taps having adjustable tap coefficients. Its input is the set of previously detected symbols. The output of the FBF is subtracted from the output $y[n]$ of the FFF to form the input to the detector. The FFF will try to shape the channel output signal in such a way that it is a causal signal. The FBF will then subtract any trailing ISI. Note that any bias removal in Figure 8.10 is absorbed by the decision element.

The DFE has a total of N (from $-N_1$ to N_2) taps in the FFF and N_3 taps in the FBF, and its output in the discrete-time domain can be expressed as

$$q[k] = \sum_{n=-N_1}^{N_2} f^*[n]y[k-n] - \sum_{i=1}^{N_3} b[i]\hat{x}[k-i], \qquad (8.110)$$

where $f^*[n]$ and $b[n]$ are the adjustable coefficients of the FFF and FBF, respectively. Hence, the $\hat{x}[k]$ is decided from $q[k]$ once it is obtained by using (8.110). Therefore, $\hat{x}[k]$ along with previous decision $\hat{x}[k-1]$, $\hat{x}[k-2]$, $\cdots$, are fed back into the equalizer, and $\hat{x}[k+1]$ is then obtained based on (8.110).

Miao and Clements [6] provided a numeric example of a conventional DFE with a T-spaced in the FFF and FBF using the LMS algorithm to adjust the tap coefficient of the equalizer.

8.5.1 MMSE for DFE

In this section, we derive the MMSE for DFE in the z-transfer domain under the assumption that previous decisions are correct. However, in practice, this assumption may not be true, and can be a significant weakness of DFE. However, the analysis becomes intractable if the errors are included in the decision feedback section.

The error signal $e[n]$ in the DFE section can be expressed as

$$e[n] = x[n] - q[n]. \qquad (8.111)$$

Then, the MSE of the error signal $e[n]$ is a function of the tap coefficients $f[n]$ and $b[n]$ chosen and is written

$$\begin{aligned} \xi_{DFE} &= E\{|e[n]|^2\} \\ &= E\{|x[n] - q[n]|^2\}. \end{aligned} \qquad (8.112)$$

The MMSE for DFE jointly optimizes the tap settings of both the FFF, $f[n]$, and FBF, $b[n]$, to minimize the MSE in (8.112). Thus, the MMSE for the DFE is obtained by

$$\xi_{DFE-MMSE} \equiv \min_{f[n],b[n]} \left\{ E\{|x[n] - q[n]|^2\} \right\}. \quad (8.113)$$

Note that the error sequence $e[n]$ in (8.111) can be written in the z-transfer domain as

$$\begin{aligned} E(z) &= X(z) - Q(z) \\ &= X(z) - \{Y(z) - [1 - B(z)]X(z)\} \\ &= X(z) - \{R(z)F(z) - [1 - B(z)]X(z)\} \\ &= B(z)X(z) - F(z)R(z). \end{aligned} \quad (8.114)$$

Given any fixed function of FBF $B(z)$, minimizing the MSE of the error signal in (8.114) leads to

$$E\{E(z)R^*(z^{-1})\} = 0, \quad (8.115)$$

because the error signal $e[n]$ at any time k is uncorrelated with any equalizer input signal $y[n]$ by using the orthogonality principle. This allows us to have the relation as

$$B(z)P_{xr}(z) - F(z)P_{rr}(z) = 0, \quad (8.116)$$

where

$$P_{xr}(z) = E\{X(z)R^*(z^{-1})\}, \quad (8.117)$$

is the power spectrum of the cross-correlation function $R_{xy}[k]$ and

$$P_{rr}(z) = E\{R(z)R^*(z^{-1})\}, \quad (8.118)$$

is the power spectrum of the autocorrelation function $R_x[k]$. Thus, for any function of FBF, $B(z)$ with $b_0 = 1$, the function of $F(z)$ in (8.116) is obtained by

$$F(z) = \frac{B(z)P_{xr}(z)}{P_{rr}(z)}. \quad (8.119)$$

Using (8.114), we obtain the power spectrum of the autocorrelation function for the error sequence

$$\begin{aligned} P_{ee}(z) &= E\{E(z)E^*(z^{-1})\} \\ &= E\{|B(z)X(z) - F(z)R(z)| \\ &\quad \cdot |B^*(z^{-1})X^*(z^{-1}) - F^*(z^{-1})R^*(z^{-1})|\} \\ &= B(z)P_{xx}(z)B^*(z^{-1}) - 2B(z)P_{xr}(z)F^*(z^{-1}) \\ &\quad + F(z)P_{rr}(z)F^*(z^{-1}). \end{aligned} \qquad (8.120)$$

substituting (8.119) into (8.120) obtains the power spectrum of the autocorrelation function for the error sequence

$$P_{ee}(z) = B(z)\left[P_{xx}(z) - \frac{P_{xr}(z)P^*_{xr}(z^{-1})}{P^*_{rr}(z^{-1})}\right]B^*(z^{-1}). \qquad (8.121)$$

Thus, it can be shown that the DFE power spectrum of the autocorrelation function for the error sequence with arbitrary monic function $B(z)$ is given by

$$P_{ee}(z) = B(z)P_{MMSE-LE}(z)B^*(z^{-1}), \qquad (8.122)$$

where

$$\begin{aligned} P_{MMSE-LE}(z) &= \left[P_{xx}(z) - \frac{P_{xr}(z)P^*_{xr}(z^{-1})}{P^*_{rr}(z^{-1})}\right] \\ &= \frac{N_0}{|C(z)|^2 + N_0}. \end{aligned} \qquad (8.123)$$

Equation (8.123) is the power spectrum for the error sequence of an MMSE linear equalizer. Note that the inverse power spectrum of the autocorrelation in (8.123) can be expressed into a spectral factorization as

$$\frac{1}{N_0}|C(z)|^2 + 1 = \gamma_0 G(z)G^*(z^{-1}), \qquad (8.124)$$

where γ_0 is a positive real number and $G(z)$ is a filter response. The function $G(z)$ is called a *canonical filter response* if it is causal

($g[k] = 0$ for $k < 0$), monic ($g[0] = 1$), and minimum-phase. In this case, the function $G^*(z^{-1})$ is referred to as *anticanonical filter response*. Using the spectral factorization in (8.124), (8.122) can be expressed as

$$\begin{aligned} P_{ee}(z) &= \frac{B(z)B^*(z^{-1})N_0}{|C(z)|^2 + N_0} \\ &= \left[\frac{B(z)B^*(z^{-1})}{G(z)G^*(z^-)}\right]\left(\frac{1}{\gamma_0}\right) \\ &= \left|\frac{B(z)}{G(z)}\right|^2 \left(\frac{1}{\gamma_0}\right). \end{aligned} \qquad (8.125)$$

The factional polynomial inside the squared norm in (8.125) is necessarily monic and causal, and the squared norm has a minimum value of 1 if and only if $B(z) = G(z)$. This leads to

$$P_{ee}(z) \geq \frac{1}{\gamma_0}. \qquad (8.126)$$

This result in (8.126) states that the error sequence for the MMSE-DFE is "white" sequence when minimized since $P_{ee}(z)$ is a constant and has MMSE-DFE of $\frac{1}{\gamma_0}$. In addition, taking $\frac{T}{2\pi}\int_{-\pi/T}^{\pi/T} \ln$ on both sides of (8.124), we can obtain

$$\frac{T}{2\pi}\int_{-\pi/T}^{\pi/T} \ln\left[\frac{1}{N_0}|C(e^{j\omega T})|^2 + 1\right] d\omega = $$
$$\ln(\gamma_0) + \frac{T}{2\pi}\int_{-\pi/T}^{\pi/T} \ln\left[G(e^{j\omega T})G^*(e^{-j\omega T})\right] d\omega. \qquad (8.127)$$

Since

$$\frac{T}{2\pi}\int_{-\pi/T}^{\pi/T} \ln\left[G(e^{j\omega T})G^*(e^{-j\omega T})\right] d\omega \geq 0, \qquad (8.128)$$

then (8.127) can be expressed as

$$\ln(\gamma_0) \leq \frac{T}{2\pi}\int_{-\pi/T}^{\pi/T} \ln\left[\frac{1}{N_0}|C(e^{j\omega T})|^2 + 1\right] d\omega. \qquad (8.129)$$

The equation in (8.129) leads to a famous expression for the MMSE of DFE that was first derived by Salz [33] in 1973,

$$\xi_{MMSE-DFE} = \exp\left\{\frac{T}{2\pi}\int_{-\pi/T}^{\pi/T} \ln\left[\frac{N_0}{|C(e^{j\omega T})|^2 + N_0}\right] d\omega\right\}. \tag{8.130}$$

The corresponding output SNR of DFE is obtained by

$$\begin{aligned}\text{SNR}_{DFE} &= \frac{1-\xi_{MMSE-DFE}}{\xi_{MMSE-DFE}} \\ &= -1 + \exp\left\{-\frac{T}{2\pi}\int_{-\pi/T}^{\pi/T} \ln\left[\frac{N_0}{|C(e^{j\omega T})|^2 + N_0}\right] d\omega\right\}. \end{aligned} \tag{8.131}$$

Note that if $|C(e^{j\omega T})|^2 = 1$, this indicates the absence of ISI. Hence, in this case, we obtain

$$\xi_{MMSE-DFE} = \frac{N_0}{1+N_0}. \tag{8.132}$$

Thus, the corresponding output SNR of DFE is

$$\text{SNR}_{DFE} = \frac{1}{N_0}. \tag{8.133}$$

The receiver SNR for the MMSE-DFE can be defined by

$$\text{SNR}_{R-DFE} \cong \frac{\mathcal{E}_X}{E\{|e[n]|^2\}}, \tag{8.134}$$

where $\mathcal{E}_X$ is called the average energy of the complex signal constellation and the denominator in (8.134) is the MMSE for a DFE. Therefore, the receiver SNR for the MMSE can now be easily obtained by

$$\begin{aligned}\text{SNR}_{R-DFE} &= \frac{\mathcal{E}_X}{\xi_{MMSE-DFE}} \\ &= \mathcal{E}_X \cdot \exp\left\{-\frac{T}{2\pi}\int_{-\pi/T}^{\pi/T} \ln\left[\frac{N_0}{|C(e^{j\omega T})|^2 + N_0}\right] d\omega\right\}. \end{aligned} \tag{8.135}$$

In a similar way, if $|C(e^{j\omega T})|^2 = 1$ in the absence of ISI, the receiver SNR for a MMSE-DFE is

$$\text{SNR}_{R-DFE} = \mathcal{E}_X \left(1 + \frac{1}{N_0}\right). \quad (8.136)$$

It has been shown that the MMSE for a DFE in (8.130) is smaller than the MMSE for a linear equalizer in (8.33) unless the squared-norm term of $|C(e^{j\omega T})|^2$ is a constant where adaptive equalization is not needed in this case [12]. Moreover, a DFE has significantly smaller MMSE than a linear equalizer if there are nulls in the squared-norm term of $|C(e^{j\omega T})|^2$. To that end, a linear equalizer is well behaved when the channel spectrum is comparatively flat. However, if the channel is severely distorted or exhibits nulls in the spectrum, the performance of a linear equalizer deteriorates and the MMSE of a DFE is much better than a linear equalizer. In addition, a linear equalizer has difficulty equalizing a nonminimum phase channel, where the strongest energy arrives after the first arriving signal component. Therefore, a DFE yields a significant improvement in performance relative to the linear equalizer having the same number of tap coefficients. Thus, a DFE is more appropriate for severely distorted wired and wireless communication channels.

Example 8.4

In this example, we reconsider Example 8.2 for considering the discrete-time FIR channel. It is interesting to compare the MMSE DFE with the value of MMSE given by the MMSE linear equalizer.

Using (8.130), the MMSE DFE for this channel is obtained by

$$\begin{aligned}\xi_{MMSE-DFE} &= \exp\left\{\frac{T}{2\pi}\int_{-\pi/T}^{\pi/T} \ln\left[\frac{N_0}{\Psi(\omega)}\right] d\omega\right\} \\ &= N_0 \exp\left\{-\frac{T}{2\pi}\int_{-\pi/T}^{\pi/T} \ln\left[\Psi(\omega)\right] d\omega\right\}\end{aligned}$$
$$(8.137)$$

where $\Psi(\omega) = 1 + 2|a||b|\cos(\omega T + \phi) + N_0$. After the integral in (8.137), the MMSE DFE is

$$\xi_{MMSE-DFE} = \frac{2N_0}{1 + N_0 + [(1+N_0)^2 - 4|ab|^2]^{1/2}}. \quad (8.138)$$

Notice that $\xi_{MMSE-DFE}$ is minimized when $|a| = |b| = 1/\sqrt{2}$. Thus,

$$\begin{aligned} \xi_{MMSE-DFE} &= \frac{2N_0}{1 + N_0 + [(1+N_0)^2 - 1]^{1/2}} \\ &\approx 2N_0, \quad N_0 \ll 1. \end{aligned} \quad (8.139)$$

The corresponding output SNR is

$$\begin{aligned} \text{SNR}_{DFE} &= \frac{1 - 2N_0}{2N_0} \\ &\approx \frac{1}{2N_0}, \quad N_0 \ll 1. \end{aligned} \quad (8.140)$$

This indicates that there is a 3-dB degradation in DFE output SNR because of the presence of ISI. However, in comparison, the performance loss of the MMSE linear equalizer is more severe than the MMSE DFE. Notice that the output SNR of the MMSE linear equalizer as given by (8.42) is about $\Upsilon_{SNR} \approx \sqrt{2/N_0}$ for $N_0 \ll 1$.

8.5.2 Predictive DFE

Another form of DFE contributed by Belfiore and Park [34] is known as a *predictive DFE* in Figure 8.11. This system also consists of an FFF as in the conventional DFE while the FBF is driven by an input sequence generated by using the difference of the output of the decision detector and the output of the FFF directly. Thus, the FBF is referred to as a *noise predictor* because it predicts the noise and the residual ISI contained in the signal at the FFF output and subtracts it from the decision detector output after some feedback delay.

Given the infinite length of the FFF in the predictive DFE, Proakis [12] showed that the power density spectrum of the noise

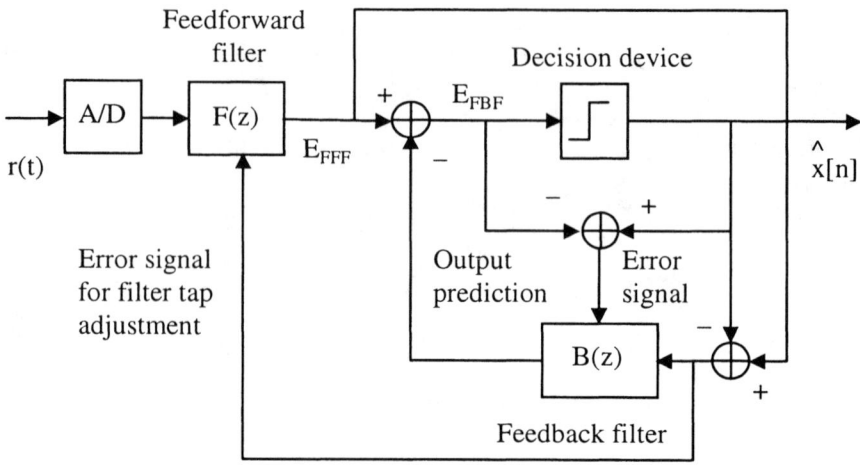

Figure 8.11 A block diagram of a predictive decision feedback equalizer.

output of the FFF is given by

$$P_{noise}(e^{j\omega T}) = \frac{N_0|C(e^{j\omega T})|^2}{[N_0+|C(e^{j\omega T})|^2]^2}, \quad |\omega| \leq \tfrac{\pi}{T}. \quad (8.141)$$

The residual ISI of the FFF output has the power density spectrum as

$$P_{ISI}(e^{j\omega T}) = \left|1 - \frac{|C(e^{j\omega T})|^2}{N_0+|C(e^{j\omega T})|^2}\right|^2$$

$$= \frac{N_0^2}{[N_0+|C(e^{j\omega T})|^2]^2}, \quad |\omega| \leq \tfrac{\pi}{T}. \quad (8.142)$$

Note that the sum of these two spectra (8.141) and (8.142) represents the power density spectrum of the total noise and ISI at the output of the FFF section. Hence, adding (8.141) and (8.142) together yields

$$E_{FFF}(e^{j\omega T}) = \frac{N_0}{N_0+|C(e^{j\omega T})|^2}, \quad |\omega| \leq \tfrac{\pi}{T}. \quad (8.143)$$

Based on observation, if $|C(e^{j\omega T})|^2 = 1$, then the channel is ideal. Thus, the MSE can not be reduced further. However, if the channel has distortion, then the power of the error sequence at the output of the FFF can be reduced by using linear prediction based on past values of the error sequence.

Assume that $B(e^{j\omega T})$ is the frequency response of the infinite length FBF (linear predictor). Then, the error at output of the predictor is obtained by

$$\begin{aligned} E_{FBF}(e^{j\omega T}) &= E_{FFF}(e^{j\omega T}) - E_{FFF}(e^{j\omega T})B(e^{j\omega T}) \\ &= E_{FFF}(e^{j\omega T})\left[1 - B(e^{j\omega T})\right]. \end{aligned} \quad (8.144)$$

The MSE of this error function in (8.144) is

$$\xi = \frac{1}{2\pi} \int_{-\pi/T}^{\pi/T} |1 - B(e^{j\omega T})|^2 |E(e^{j\omega T})|^2 d\omega. \quad (8.145)$$

Proakis [12] showed that minimizing the MSE of this error function in (8.145) over the predictor coefficients $b[n]$ yields an optimum predictor in the form as

$$B(e^{j\omega T}) = 1 - \frac{G(e^{j\omega T})}{g[0]}, \quad (8.146)$$

where $G(e^{j\omega T})$ is known as the solution of the spectral factorization using

$$\frac{1}{|E(e^{j\omega T})|^2} = G(e^{j\omega T})G^*(e^{j\omega T}), \quad (8.147)$$

and

$$G(e^{j\omega T}) = \sum_{n=0}^{\infty} g[n]e^{-j\omega nT}. \quad (8.148)$$

Thus, the output of infinite length FBF linear predictor is a white noise sequence with a power spectral density of $\frac{1}{g^2[0]}$. The corresponding MMSE of the predictive DFE is obtained by

$$\xi_{MMSE-PDFE} = \exp\left\{\frac{T}{2\pi} \int_{-\pi/T}^{\pi/T} \ln\left[\frac{N_0}{|C(e^{j\omega T})|^2 + N_0}\right] d\omega\right\}. \quad (8.149)$$

Note that (8.149) is equal to (8.130). Therefore, this indicates that the MSE performance of the predictive DFE is identical to the conventional DFE if their lengths are infinite.

The performance of the predictive DFE is suboptimum if the lengths of the two FFF and FBF filters are finite. This is because the optimizations of the FFF and FBF predictors in the predictive DFE are done separately, while the optimizations of the conventional DFE tap coefficients in the FFF and FBF are done jointly. Therefore, the conventional DFE yields the minimum MSE while the MSE of predictive DFE is at least as large as that of the conventional DFE. Although the predictive DFE has this suboptimality, it has been suggested by Proakis [12] that it is suitable as an equalizer for trellis-coded signals where the conventional DFE is not as suitable.

The FBF in the predictive DFE can be implemented by using a lattice structure given by Zhou et al. [35]. In this case, the RLS algorithm as discussed in the previous chapter can be used to produce fast convergence for the predictive DFE.

8.5.3 FSE-DFE

The structures of the conventional and predictive DFEs, which were analyzed earlier, employ a T-spaced FFF filter for the feedforward section. Such a structure optimality is based on the assumption that the receiver analog filter preceding the DFE is matched to the channel-corrupted pulse response and its output is sampled at the optimum time instant. However, in practice, the channel response is not known a priori so that it is not possible to design an ideal matched filter. Therefore, practical application often uses the fractionally spaced FFF filter while the FBF filter tap spacing still remains T-spaced for the feedback section.

Figure 8.12 shows a fractionally spaced FFF filter based on DFE. The signal is sampled at the communication receiver at a rate of $2/T$ and then is passed through a $T/2$-spaced FFF equalizer. It consists of a serial-to-parallel (S/P) converter and two FFF filters, $F_0(z)$ and $F_1(z)$, in the feedforward filter section. The S/P converter is to convert one sequence $r[n]$ (for $n = 0, 1, 2, \cdots, 2N$) into

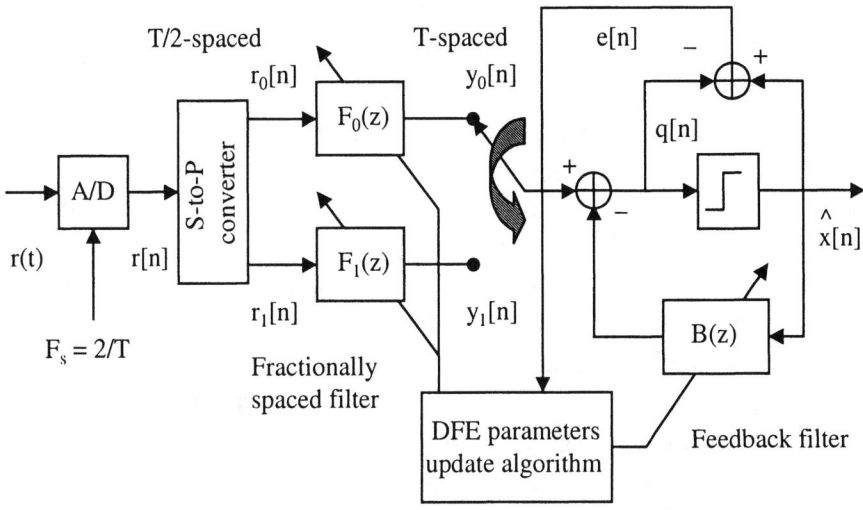

Figure 8.12 A block diagram of a fractionally spaced filter-based decision feedback equalizer.

two sequences, $r_0[n]$ and $r_1[n]$, where $r_0[n] = r[2n]$ and $r_1[n] = r[2n+1]$ for $n = 0, 1, 2, \cdots, N$. Two FFF filters, $F_0(z)$ and $F_1(z)$, are operated in parallel, where $F_0(z)$ and $F_1(z)$ contain all the even and odd tap coefficients, respectively. The filter outputs $y_0[n]$ and $y_1[n]$ are at T-spaced. Then the FBF operates T-spaced when the switch connects to $y_0[n]$, and produces an error sequence by using the difference between $q[n]$ and $\hat{x}[n]$ after a decision detector. Hence, the tap coefficients of the filter $F_0(z)$ in the feedforward section and $B(z)$ in the feedback section are updated by using an adaptive DFE algorithm. The next step is to connect the switch to $y_1[n]$. In a similar way, the FBF is operated T-spaced and creates the error sequence. Then, the tap coefficients of the filter $F_1(z)$ in the feedforward section and $B(z)$ in the feedback section are updated based on the adaptive DFE algorithm by using the error sequence. Therefore, this FSE-DFE system can achieve the $T/2$-spaced FFF equalizer in the feedforward section while the FBF filter still operates T-spaced in the feedback section. This system is sometimes referred to as a fractionally $T/2$-spaced DFE. This approach is also called the polyphase-based filter bank DFE.

In general, if an FFF equalizer tap spacing has MT/N, where M and N are integers and $N > M$, then the equalizer is referred to as fractionally MT/N-spaced DFE. Miao and Qi [36] developed the fractionally $2T/3$-spaced DFE using the LMS algorithm for updating the filter tap coefficients with applications in DSL modems.

The fractionally spaced based DFE has often been implemented as an adaptive blind equalizer by using CMA technique. Utilizing the CMA with the FSE-DFE is rapidly gaining popularity. This is because the FSE-DFE has been proven to suppress timing phase sensitivity and noise enhancement. In addition, the CMA based FSE-DFE may be able to overcome some undesirable local convergence and can be applied to channels with zeros on the unit circle without noise amplification. Even when some subchannels have deep spectral nulls, the CMA-based FSE-DFE is capable of compensating spectral nulls as long as these zeros are not common to all subchannels while it does not require a large number of tap coefficients, thereby converging fast. Moreover, the CMA-based FSE-DFE can equalize nonconstant modulus signals and those with a constant modulus. This makes it a prime candidate for use with the varity of signal formats found in the high-speed broadband communication modems.

8.5.4 Error Propagation

So far, we have not discussed the effect of decision errors in the feedback section of the DFE. The result of an incorrect decision in the feedback section of the DFE, which produces additional errors that would not have occurred if the first decision had been correct, is referred to as *error propagation*. Error propagation in the DFE can be a major concern in practical application for a communication receiver structure, especially if constellation-expanding codes, or convolutional codes, are used in concatenation with the DFE.

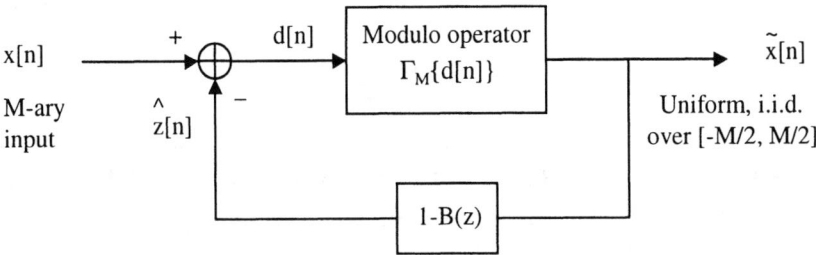

Figure 8.13 A block diagram of the Tomlinson precoder.

Therefore, in this section, we introduce the so-called *precoding* solution to the error-propagation problem of the DFE.

The precoding essentially moves the feedback section of the DFE to the transmitter with a minimal transmit-power-increase penalty without decreasing the DFE SNR. In other words, the basic idea of the precoding is to move the DFE feedback section to the transmitter so that the decision errors are impossible to happen. However, the precoding usually requires a feedback channel. The precoding is also known as the *Tomlinson precoder* or the *Tomlinson-Harashima precoder*.

Figure 8.13 illustrates the block diagram of the Tomlinson precoder for a real-signal case. The Tomlinson precoder is used in the transmitter as a preprocessor to the modulator, and maps the data symbol $x[n]$ into another data symbol $\tilde{x}[n]$ that is in turn applied to the modulator. The Tomlinson precoder is eventually a device that is used to eliminate error propagation. However, straightforward moving of the feedback filter $1/B(z)$ to the transmitter could result in increasing transmit power significantly. In order to prevent most of this power increase, a modulo operator $\Gamma_M\{x\}$ is employed to bound the value of $\tilde{x}[n]$ before the modulator.

The modulo operator $\Gamma_M\{x\}$ is a nonlinear function that defines an M-ary input constellation (PAM and QAM square) with uniform spacing d, and is given by

$$\Gamma_M\{x\} = x - Md \left\lfloor \frac{x + Md/2}{Md} \right\rfloor, \tag{8.150}$$

where $\lfloor y \rfloor$ denotes the largest integer that is less than or equal to the value of y, and $\Gamma_M\{x\}$ can be an arbitrary value. The operation of the Tomlinson precoder produces an internal signal as shown in Figure 8.13,

$$d[n] = x[n] - \sum_{k=1}^{\infty} b[k]\tilde{x}[n-k], \tag{8.151}$$

where

$$\begin{aligned} \tilde{x}[n] &= \Gamma_M\{d[n]\} \\ &= \Gamma_M\{x[n] - \sum_{k=1}^{\infty} b[k]\tilde{x}[n-k]\}. \end{aligned} \tag{8.152}$$

The Tomlinson precoder output $\tilde{x}[n]$ is an approximately *i.i.d.* if the M-ary input is an *i.i.d.* In addition, the output sequence of the Tomlinson precoder is an approximately uniform distribution over the interval $[-Md/2, Md/2)$.

Example 8.5

Assume that a precoder of modulo operator is for M-ary input PAM signals with $M = 5$ and uniform spacing $d = 2$, and $B(z) = 0.5z^{-1}$ as illustrated in Figure 8.13. We want to determine the precoder output of the modulo operator when an input $x[n]$ is a sequence $\{5, 4.5, 2.1, 1, -1.5, -6, 2, 9.1\}$.

Using (8.150), (8.151), and (8.152) with an initial value $\tilde{x}[-1] = 0$, the result of precoder of the modulo operator can be expressed as

$$\Gamma_M\{x\} = x - 10 \left\lfloor \frac{x+5}{10} \right\rfloor. \tag{8.153}$$

Since $B(z) = 0.5z^{-1}$, then $b[1] = 0.5$. Thus,

$$d[n] = x[n] - \sum_{k=1}^{\infty} b[k]\tilde{x}[n-k]$$

$$\begin{aligned} &= x[n] - b[1]\tilde{x}[n-1] \\ &= x[n] - 0.5\tilde{x}[n-1]. \end{aligned} \quad (8.154)$$

The Tomlinson precoder output is then given by

$$\begin{aligned} \tilde{x}[n] &= \Gamma_M\{d[n]\} \\ &= x[n] - 0.5\tilde{x}[n-1] \\ &\quad -10\left\lfloor \frac{x[n] - 0.5\tilde{x}[n-1] + 5}{10} \right\rfloor. \end{aligned} \quad (8.155)$$

Therefore, substituting the input sequence $x[n]$ into (8.155), the output of the Tomlinson precoder is obtained as
$\tilde{x}[n] = \{-5, -3, 3.6, 0, -1.5, -5.25, 4.625, -3.2125\}$.

8.6 Space-Time Equalizer

Space-time processing (or smart antennas) provides an effective way against cochannel interference (CCI), which usually arises from cellular frequency reuse. The space-time processing that combats CCI can have a major impact on overall wireless communications performance. This is because space-time processing with multiple antennas in the receiver and transmitter is a promising way of mitigating CCI by exploiting the spatial-temporal dimension, thereby significantly improving average signal power, mitigating fading, and reducing CCI and ISI. As a result, this can greatly improve the capacity, coverage, and quality of wireless communications.

A space-time equalizer operates simultaneously on all the antennas, processing signal samples both in space and time domains. This technique allows the receiver to exploit the difference between the desired signal and CCI to reduce CCI that arises from the cellular frequency reuse, thereby increasing the quality and capacity of wireless networks. Time-only signal processing corresponds to equalizers that use a weighted sum of signal samples, while space-only processing corresponds to simple beamforming that uses a weighted sum of antenna outputs. In this section, we first examine

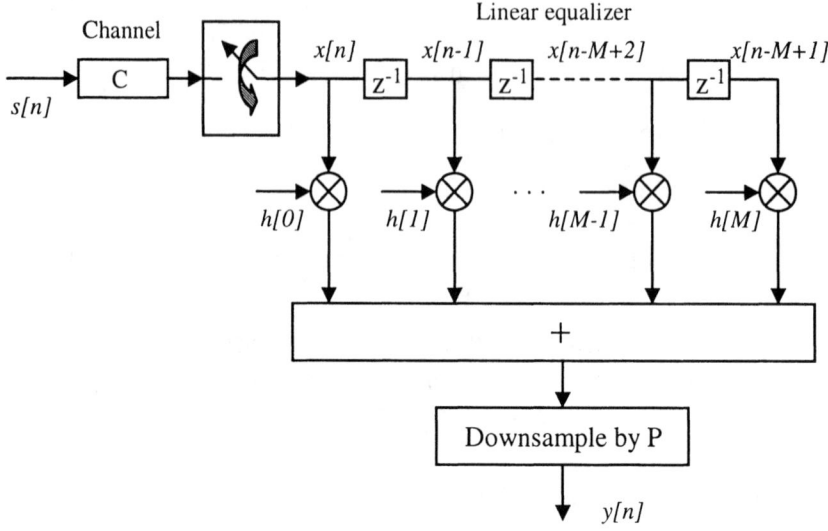

Figure 8.14 A block diagram of a time-only equalizer.

the time-only and space-only equalizers, respectively, and then present the space-time equalizers at the receiver.

8.6.1 Time-Only Equalizer

A time-only equalizer based on the baseband temporal structure is depicted in Figure 8.14. A continuous-time received signal $x(t)$ is sampled at baud rate or a higher rate to produce a discrete-time signal output $x[n]$. Then the discrete-time signal $x[n]$ is filtered through a linear equalizer to produce the discrete-time output $y[n]$. Note that in single-user environments, the equalizer is used to minimize the channel ISI. However, in the wireless channel, the equalizer also has to consider minimizing CCI.

Assuming that there are Q users and no additive noise in the single-input single-output (SISO) environments, the discrete-time channel with a single-sensor case, and no oversampling is given by

$$x[n] = \sum_{k=1}^{Q} \mathbf{c}_k \mathbf{s}_k[n], \tag{8.156}$$

where a row vector $(1 \times N)$ $\mathbf{c}_k$ is the FIR channel and a column vector $(N \times 1)$ $\mathbf{s}_k[n]$ is

$$\mathbf{s}_k[n] = [s_k[n], s_k[n-1], \cdots, s_k[n-N+1]]^T. \tag{8.157}$$

As shown in Figure 8.14, the linear equalizer $\mathbf{h}$ is an $(M \times 1)$ column vector, and its output can then be expressed as

$$y[n] = \mathbf{h}^H \mathbf{x}[n], \tag{8.158}$$

where the $(M \times 1)$ received signal vector is given by

$$\mathbf{x}[n] = [x[n], x[n-1], \cdots, x[n-M+1]]^T. \tag{8.159}$$

Furthermore, we can write $\mathbf{x}[n]$ as follows:

$$\mathbf{x}[n] = \sum_{k=1}^{Q} \mathbf{C}_k \tilde{\mathbf{s}}_k[n], \tag{8.160}$$

where $\mathbf{C}_k$ is the $M \times (M+N-1)$ Toeplitz matrix,

$$\mathbf{C}_k = \begin{bmatrix} \mathbf{c}_k & 0 & \cdots & 0 \\ 0 & \mathbf{c}_k & 0 & \vdots \\ \vdots & 0 & \ddots & 0 \\ 0 & \cdots & 0 & \mathbf{c}_k \end{bmatrix}, \tag{8.161}$$

and

$$\tilde{\mathbf{s}}_k[n] = [s_k[n], ..., s_k[n-N-M+2]]^T. \tag{8.162}$$

Combining (8.158) and (8.160) yields the equalizer output as

$$y[n] = \mathbf{h}^H \mathcal{C} \mathcal{S}[n], \tag{8.163}$$

where
$$\mathcal{C} = [\mathbf{C}_1, \mathbf{C}_2, ..., \mathbf{C}_Q] \tag{8.164}$$
and
$$\mathcal{S} = [\tilde{\mathbf{s}}_1^T[n], \tilde{\mathbf{s}}_2^T[n], ..., \tilde{\mathbf{s}}_Q^T[n]]^T. \tag{8.165}$$

In order to eliminate the ISI and CCI, (8.163) is needed to satisfy the zero-forcing condition:
$$\mathbf{h}^H \mathcal{C} = \delta[n - d], \tag{8.166}$$

where d is a positive integer of the achieved delay. However, (8.166) cannot be satisfied because $\mathcal{C}$ is the $M \times Q(N + M - 1)$ matrix that is not a full-column rank. This indicates that joint linear ISI and CCI can not be cancelled with a time-only equalizer when the received signal is sampled at the symbol rate. However, with the oversampling of P, the channel matrix $\mathcal{C}$ will be the full-column rank of block-Toeplitz matrix with size of $MP \times Q(N + M - 1)$ if the equalizer length holds the condition given by Paulraj and Papadias [37]:
$$M(P - 1) \le Q(N - 1), \tag{8.167}$$
and if the polynomials corresponding to the rows of C have no common roots (see Moulines et al. [38]).

In practice, the temporal channel for signal and CCI cannot be well separated if the symbol waveforms have a small excess bandwidth. As a result, the channel matrix $\mathcal{C}$ is ill conditioned, and CCI cancellation will lead to excessive noise enhancement. Therefore, time-only equalizers provide only a small degree of CCI reduction that is a function of baud synchronization offset between the signal and the CCI, the excess bandwidth of the symbol waveform, and the multipath channel response for the signal and the CCI.

8.6.2 Space-Only Equalizer

A space-only equalizer corresponds to simple beamforming that uses a weighted sum of antenna outputs. Figure 8.15 shows a

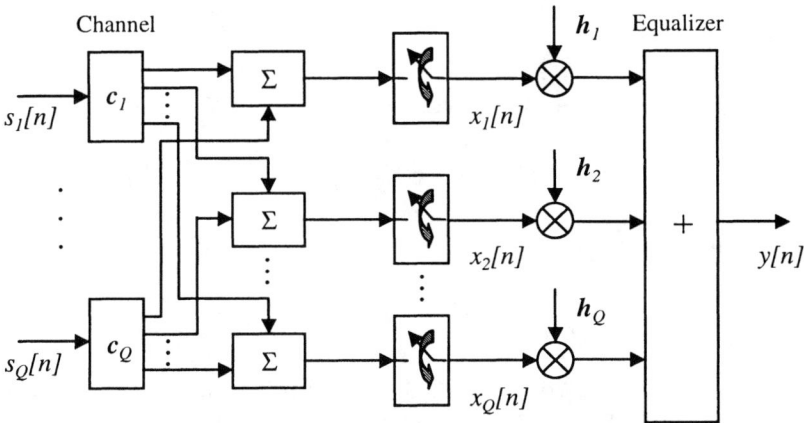

Figure 8.15 A block diagram of a space-only equalizer.

typical array beamforming structure in which the antenna outputs are weighted and combined to generate the beamformer output.

Assume that there are Q users and no multipath. The received signal is given by

$$\mathbf{x}[n] = \sum_{k=1}^{Q} \mathbf{c}_k s_k[n], \qquad (8.168)$$

where the $M \times N$ vector $\mathbf{c}_k$ denotes the space-only channel that is N tap-coefficient length for each user, in the absence of delay spread. Thus, $s_k[n]$ is a scalar.

The space-only equalizer of the beamformer output is obtained by

$$y[n] = \mathbf{h}^H \mathbf{x}[n], \qquad (8.169)$$

where $\mathbf{h}$ is a $M \times 1$ weight vector. Integrating (8.168) and (8.169) yields the space-only equalizer of the beamformer output as

$$y[n] = \sum_{k=1}^{Q} \mathbf{h}^H \mathbf{c}_k s_k[n], \qquad (8.170)$$

or equivalently in the vector form as

$$y[n] = \mathbf{h}^H \mathbf{C}\mathbf{s}[n], \tag{8.171}$$

where the channel matrix is given by

$$\mathbf{C} = [\mathbf{c}_1, \mathbf{c}_2, \cdots, \mathbf{c}_Q], \tag{8.172}$$

and the input signal matrix is

$$\mathbf{s}[n] = [s_1[n], s_2[n], \cdots, s_Q[n]]^T. \tag{8.173}$$

If we assume that $s_i[n]$ is the desired signal, then the zero-forcing condition cancelling all CCI and ISI is obtained by

$$\mathbf{h}^H \mathbf{C} = \delta[n - i], \tag{8.174}$$

where ith represents the ith user. However, this requires that $M \geq NQ$ and the channel matrix $\mathbf{C}$ must be of full-column rank. As a result, purely spatial combining is able to perfectly cancel both ISI and CCI.

In practice, this is impossible because real channels having multipath require too many antenna elements. In addition, if either the multipath angle spread is small or the desired signal and CCI are not well separated, then the column of $\mathbf{C}$ becomes ill-conditioned, resulting in excessive noise enhancement. The space-only equalizer can effectively be used to eliminate CCI, but its effectiveness against ISI depends on the angle spread of multipath. Therefore, in general, space-time equalizers should be used to improve the performance over both time-only and space-only equalizers.

8.6.3 Space-Time MMSE Equalizer

In this section, we introduce the space-time equalizer that combines time-only and space-only equalizers as discussed earlier. We present the single-user case in which we are interested in demodulating the signal of interest. Thus, we treat interference from other users as unknown additive noise. This approach can also be referred to as

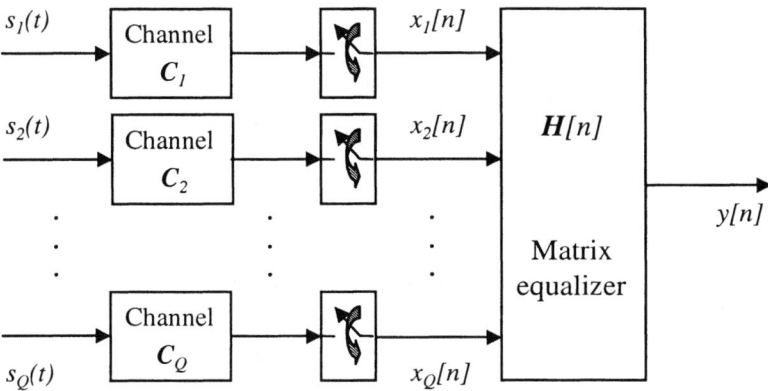

Figure 8.16 A block diagram of a space-time MMSE equalizer.

interference suppression. Figure 8.16 shows a block diagram of a space-time MMSE equalizer.

In the presence of CCI, the signal components are likely to be both spatially and temporally correlated because of delay spread. The temporal correlation complicates the implementation structure at the receiver. In this case, a space-time equalizer based on an MMSE solution is attractive; this is a solution in which the space-time equalizer combines the input in space and time to generate an output that minimizes the squared error between itself and the desired signal.

In the space-time equalizer, the tap coefficient $\mathbf{H}[k]$ of the equalizer beamformer (or matrix equalizer) has the following expression,

$$\mathbf{H}[k] = \begin{bmatrix} h_{11}[k] & h_{12}[k] & \cdots & h_{1P}[k] \\ h_{21}[k] & h_{22}[k] & \cdots & h_{2P}[k] \\ \vdots & \vdots & \ddots & \vdots \\ h_{M1} & h_{M2}[k] & \cdots & h_{MP}[k] \end{bmatrix}, \qquad (8.175)$$

where M represents the number of taps in each of the channels and P is the length of the equalizer. The received signal matrix $\mathbf{X}[k]$ is given by

$$\mathbf{X}[k] = \begin{bmatrix} x_{11}[k] & x_{12}[k] & \cdots & x_{1P}[k] \\ x_{21}[k] & x_{22}[k] & \cdots & x_{2P}[k] \\ \vdots & \vdots & \ddots & \vdots \\ x_{M1} & x_{M2}[k] & \cdots & x_{MP}[k] \end{bmatrix}. \quad (8.176)$$

Then the scalar of the equalizer beamformer, $y[n]$, is obtained by

$$y[k] = \text{tr}\{\mathbf{H}^H[k]\mathbf{X}[k]\}, \quad (8.177)$$

where "tr" means the trace of a matrix.

The space-time MMSE equalizer chooses the space-time equalizer tap coefficient so that the MMSE $\mathcal{E}_{STE}$ can be achieved as

$$\begin{aligned} \mathcal{E}_{STE} &= \arg\{\min_{\mathbf{H}} E||y[k] - s^*[k-d]||^2\} \\ &= \arg\{\min_{\mathbf{H}} E||\text{tr}\{\mathbf{H}^H[k]\mathbf{X}[k]\} - s^*[k-d]||^2\}, \end{aligned} \quad (8.178)$$

where the superscript "*" denotes a complex conjugate and d is a delay factor to be selected to the center of the space-time equalizer that will strongly affect performance. Equation (8.178) can be solved by using the least-squares method based on projection theory,

$$E\{\mathbf{X}[k](\text{tr}\{\mathbf{X}^H[k]\mathbf{H}[k]\} - s^*[k-d])\} = 0. \quad (8.179)$$

Solving (8.179) yields

$$\mathbf{H}[k] = \left(E\{\mathbf{X}[k]\mathbf{X}^H[k]\}\right)^{-1} E\{\mathbf{X}[k]s^*[k-d]\}. \quad (8.180)$$

Note that if the interference and noise are independent of the signal, then the transmitted signal is a white sequence. Thus, we have

$$\begin{aligned} E\{\mathbf{X}[k]s^*[k-d]\} &= [0,...,0,\tilde{\mathbf{H}}[k],0,...,0]^T \\ &= \bar{\mathbf{H}}[k], \end{aligned} \quad (8.181)$$

where $\tilde{\mathbf{H}}[k] = [h_1[k], h_2[k]..., h_M[k]]$ and the number of zeros preceding and succeeding $\bar{\mathbf{H}}[k]$ depends upon the selection of d. Therefore, (8.180) can be rewritten as

$$\mathbf{H}[k] = \mathbf{R}_{XX}^{-1}[k]\bar{\mathbf{H}}[k], \qquad (8.182)$$

where $\mathbf{R}_{XX}^{-1}[k] = \left(E\{\mathbf{X}[k]\mathbf{X}[k]^H\}\right)^{-1}$.

In practice, we compute the finite sample estimate of $\mathbf{R}_{XX}[k]$ and $\bar{\mathbf{H}}[k]$ using the received training samples. In case of the channels with a slow time-varying over a slot period, an optimal matrix equalizer during the training period is obtained by using the entire time slot. If the channels have a fast time-varying, then the optimal matrix equalizer during the training period should be calculated by using, for example, a decision-directed adaptive algorithm [39, 40]. Equation (8.182) can be adaptively solved by using popular algorithms including LMS, RLS, and numerous variants. In Chapter 7, we discussed the trade-off analysis of adaptive algorithms and their computational complexities.

The space-time MMSE equalizer combines the strengths of time-only and space-only equalizers to exchange CCI and ISS reduction against noise enhancement. This technique can primarily cancel CCI in the spatial domain and ISI either in the space domain or in the time domain depending on where it can be done more efficiently. However, the performance of the space-time MMSE equalizer scheme is influenced by the dominance of the CCI and ISI as well as the nature of the channel structure.

8.7 Diversity Equalizer

Diversity in the transmission is to use the multiple channels from a single information source to several communication receivers. Usually, diversity methods lead to a lower probability of error for the same transmitted message. This is mainly because a greater channel-output minimum distance between possible (noiseless) output data symbols can be achieved with a larger number of channel output dimensions. However, ISI between successive transmissions

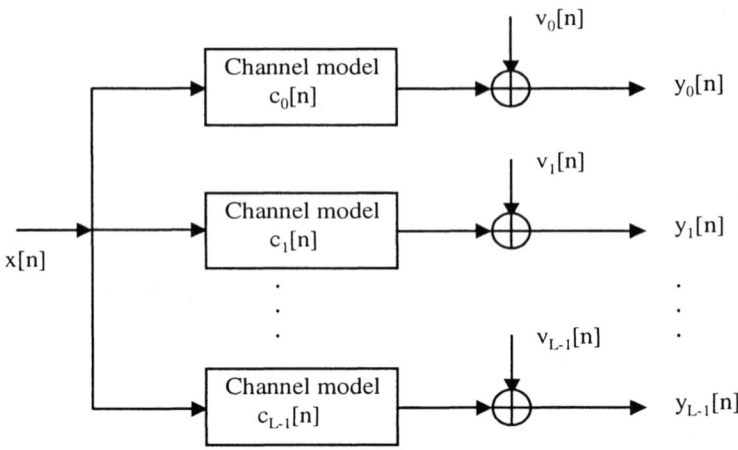

Figure 8.17 A block diagram of a discrete-time basic diversity channel model.

along with interference between the diversity multidimensions may cause a potentially complex optimum receiver and detector. Thus, equalization can again allow productive use of suboptimal detectors along with other signal processing methods, such as a Rake receiver. The equalization in the case of diversity is referred to as a *diversity equalizer*.

8.7.1 Fundamentals of a Rake Receiver

The basic discrete-time diversity channel is shown in Figure 8.17. The channel input is labeled by $x[n]$ while the channel output caused by the same channel input has labels $y_i[n]$, $i = 0, 1, 2, \cdots, L-1$. The additive noises in the channels are denoted by $v_i[n]$, $i = 0, 1, 2, \cdots, L-1$. These channel outputs can be created intentionally by retransmission of the same input sequence $x[n]$ at different times and/or center frequencies. On the other hand, propagation delay spread or multipath in the radio channel merely provides multiple versions of the transmitted signal at the communication receiver.

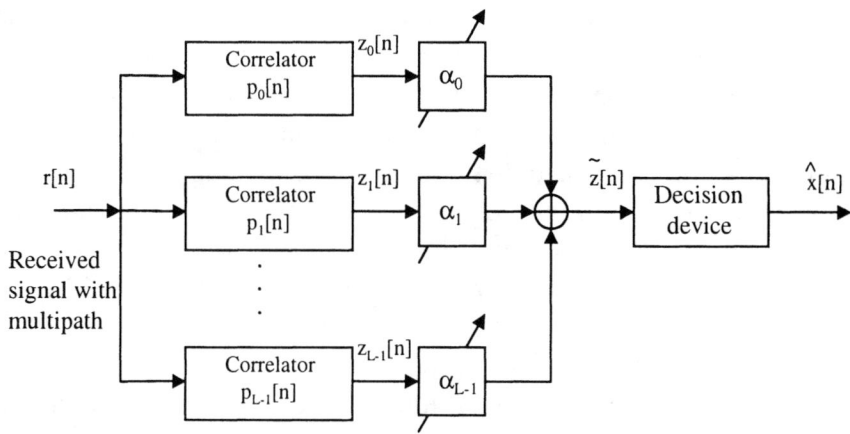

Figure 8.18 A block diagram of a discrete-time L-branch Rake receiver.

Thus, spatial diversity often occurs in wireless transmission where L spatially separated antennas may all receive the same transmitted signal with different filtering and noises that are at least partially independent. The spatially separated antennas are also referred to as the technology of *smart antennas*.

A Rake receiver is a set of parallel matched filters or correlators in which each operating on one of the diversity channels in a diversity transmission system is followed by a summing and decision device as shown in Figure 8.18. The basic idea of such a Rake receiver was first proposed by Price and Green [41] based on the analogy of the various matched filters being the "*fingers*" of a garden rake and the sum corresponding to the collection of the fingers at the rake's pole handle. The original Rake concept was conceived in connection with a spread-spectrum CDMA system for wireless communications that achieved diversity essentially in a code-division dimension discussed by Viterbi [42] and Glisic and Vucetic [43]. Recently, a variety type of the Rake receivers has been used in ultra wideband (UWB) communications, dual-mode UWB and wireless local area network communications, and MIMO wireless sensor networks communications [44–46]. The Rake receiver is sometimes referred to as a *diversity combiner* to address other lower-performance suboptimal combining methods, which do not maximize the overall SNR strength through the matched filter. In addition, the Rake receiver is also called *maximal combining* if it applies a matched filter only to the strongest of the L diversity paths to save receiver complexity. In this case, the equivalent channel is thus the channel corresponding to this maximum-strength path individually.

A Rake receiver utilizes multiple correlators to separately detect the L strongest multipath components in the spread-spectrum CDMA communication systems. In other words, each correlator detects a time-shifted version of the original CDMA transmission, and each finger of the Rake correlators to a portion of the signal that is delayed by at least one chip in time from the other fingers. The outputs of each correlator are then weighted by α_i, $i = 0, 1, 2, \cdots, L-1$, and summed to provide a better estimate of the transmitted signal that is provided by a signal message. The weighting coefficients are based on the power or the SNR from each correlator output. If the output of a special correlator has a large power or a large SNR, it will then be assigned a larger weighting factor. In the case of a maximal ration combining diversity scheme, the overall signal $\tilde{z}[n]$ after a summing device is obtained by

$$\tilde{z}[n] = \sum_{k=0}^{L-1} \alpha_k z_k[n]. \qquad (8.183)$$

The weighting coefficients, α_k, are usually normalized for the output signal power of the correlator in such a way that the sum of the weighting coefficients is unity as

$$\alpha_k = \frac{z_k^2[n]}{\sum_{i=0}^{L-1} z_i^2[n]}. \qquad (8.184)$$

However, in the case of using adaptive equalizers and diversity combining, there are other ways to generate the weighting coefficients for the Rake receiver. Therefore, choosing weighting coefficients by using the actual outputs of the correlators yields a better Rake performance.

Assume that we transmit only one direct-sequence spread spectrum signal with binary symbols 1 and 0 mapping into bipolar symbols $b[m]$ of $+1$ and -1, respectively, and that the demodulator provides perfect estimates of the channel. We also assume that the data symbol duration T is a multiple of the chip duration T_c. The processing (or spreading) gain, defined here as the number of pseudorandom chips per data symbol, is defined as

$$G = \frac{T}{T_c} = \frac{B_c}{B}, \qquad (8.185)$$

where B_c is the spread spectrum signal bandwidth and B is the baseband signal bandwidth. The overall signal $\tilde{z}[n]$ (also called a decision variable) at the input of the decision device is given by

$$\tilde{z}[n] = \sum_{l=1}^{L} \left[\sum_{k=1}^{G} \sqrt{E_c} b[k] \alpha_l^2 + \sum_{k=1}^{G} v[k] \alpha_l \right], \qquad (8.186)$$

where $v[k]$ is Gaussian noise and E_c is the transmitted energy per chip.

For a fixed set of the weighting coefficients α_l, $l = 1, 2, ..., L$, the decision variable $\tilde{z}[n]$ has the mean value

$$E\{\tilde{z}[n]\} = G\sqrt{E_c} \sum_{l=1}^{L} \alpha_l^2, \qquad (8.187)$$

and the variance

$$\text{Var}\{\tilde{z}[n]\} = \frac{GN_0}{2} \sum_{l=1}^{L} \alpha_l^2. \qquad (8.188)$$

The average probability of bit error P_b can be expressed as

$$P_b = \frac{1}{2} Pr\{\tilde{z}[n] > 0 | b = -1\} + \frac{1}{2} Pr\{\tilde{z}[n] < 0 | b = +1\}. \qquad (8.189)$$

If we assume that the probabilities of transmitting symbols -1 and $+1$ are equal, we obtain

$$Pr\{\tilde{z}[n] > 0 | b = -1\} = Pr\{\tilde{z}[n] < 0 | b = +1\}. \qquad (8.190)$$

The average probability of bit error can then be written as

$$P_b = Pr\{\tilde{z}[n] < 0 | b = +1\}. \qquad (8.191)$$

Assume that the number of chips per data symbol is sufficiently large. The decision variable $\tilde{z}[n]$ can be approximated according to the central limit theorem. Thus, the probability of error for the fixed weighting coefficients α_l is the probability of the decision variable $\tilde{z}[n]$ in (8.195) given by

$$P_b(\gamma_b) = Q\left(\sqrt{\frac{(E\{\tilde{z}[n]\})^2}{\text{Var}\{\tilde{z}[n]\}}}\right), \qquad (8.192)$$

where $Q(x)$ is the complementary error function expressed as

$$Q(x) = \frac{1}{\sqrt{2\pi}} \int_0^\infty e^{-t^2/2} dt. \qquad (8.193)$$

Substituting (8.187) and (8.188) into (8.192), we obtain

$$\begin{aligned} P_b(\gamma_b) &= Q\left(\sqrt{\frac{2GE_c}{N_0} \sum_{l=1}^L \alpha_l^2}\right) \\ &= Q\left(\sqrt{\frac{2E_b}{N_0} \sum_{l=1}^L \alpha_l^2}\right) \\ &= Q\left(\sqrt{2\gamma_b}\right). \end{aligned} \qquad (8.194)$$

where E_b is the energy per bit and γ_b is the *total bit energy-to-noise ratio* given by

$$\gamma_b = \frac{E_b}{N_0} \sum_{l=1}^L \alpha_l^2 = \sum_{l=1}^L \gamma_l, \qquad (8.195)$$

where
$$\gamma_l = \frac{\alpha_l^2 E_b}{N_0}. \tag{8.196}$$

If α_l is Rayleigh distributed, α_l^2 has a chi-square distribution with two degrees of freedom [6, 43, 47]. Thus, γ_l is also chi-square distributed. Accordingly, the density function is

$$p(\gamma_l) = \frac{1}{\bar{\gamma}_l} \exp\left\{-\frac{\gamma_l}{\bar{\gamma}_l}\right\}, \tag{8.197}$$

where $\bar{\gamma}_l$ is the average received bit energy-to-noise ratio defined as

$$\bar{\gamma}_l = \frac{E_b}{N_0} E\{\alpha_l^2\}, \tag{8.198}$$

where $E\{\alpha_l^2\}$ is the average value of α_l^2. To compute the density of γ_b, we note that the characteristic function of γ_l is

$$\psi_{\gamma_l}(jv) = \frac{1}{1 - jv\bar{\gamma}_l}. \tag{8.199}$$

Since the fading of the L channels is statistically independent, γ_l, $l = 1, 2, ..., L$, are statistically independent. Thus, the characteristic function of γ_b is

$$\psi_{\gamma_b}(jv) = \prod_{l=1}^{L} \frac{1}{1 - jv\bar{\gamma}_l}. \tag{8.200}$$

Glisic and Vucetic [43] obtained the probability density function of γ_b by taking the inverse Fourier transform of (8.200)

$$p(\gamma_b) = \frac{1}{(L-1)!\bar{\gamma}_l^L} \gamma_b^{(L-1)} \exp\left\{-\frac{\gamma_b}{\bar{\gamma}_l}\right\}. \tag{8.201}$$

Thus, with Rayleigh fading and the distinct mean square values of $\bar{\gamma}_l$, the average probability of bit error is

$$\begin{aligned} P_b &= \int_0^\infty P_b(\gamma_b) p(\gamma_b) d\gamma_b \\ &\approx \binom{2L-1}{L} \prod_{l=1}^{L} \left(\frac{1}{2\bar{\gamma}_l}\right). \end{aligned} \tag{8.202}$$

Let us assume that the mean square value $E\{\alpha_l^2\}$ is exactly the same for all tap values. Using (8.195) and (8.198), we obtain the relationship between $\bar{\gamma}_b$ and $\bar{\gamma}_l$

$$\bar{\gamma}_b = \bar{\gamma}_l. \tag{8.203}$$

Thus, the average probability of bit error in (8.202) can be rewritten in terms of the average bit energy-to-noise ratio $\bar{\gamma}_b$ as

$$P_b \approx \binom{2L-1}{L} \prod_{l=1}^{L} \left(\frac{1}{2\bar{\gamma}_b}\right). \tag{8.204}$$

Figure 8.19 shows an average probability of a bit error for a direct-sequence spread spectrum system with a BPSK modulation over a Rayleigh fading channel for a Rake receiver: (a) $L = 5$ in fading, (b) $L = 4$ in fading, (c) $L = 3$ in fading, (d) $L = 2$ in fading, and (e) $L = 1$ in fading. Note that the average probability of bit error of (8.204) decreases when the L is increased. When the channel becomes more dispersive, a greater diversity gain is achieved. The number of taps actually used in the Rake receiver can be less than the channel length L. However, such a Rake receiver will not capture all the received signal energy and may suffer from some loss in performance. Therefore, increasing the number of taps for the Rake receiver will enhance its performance.

8.7.2 Adaptive Rake Receiver

The performance of the conventional Rake receiver of Figure 8.18 uses a maximal ratio combining, which is a traditional approach to determine the weighting coefficients for the conventional Rake receiver. In this section, we introduce an adaptive Rake receiver followed by an adaptive linear equalizer as shown in Figure 8.20. The adaptive Rake receiver employs an MMSE combining method to improve the performance of the Rake receiver in the presence of interference, such as ISI and interpath interference (IPI). In this approach, the weighting coefficients α_l, $l = 0, 2, \cdots, L-1$, of the adaptive Rake receiver are updated in an optimal sense such that the

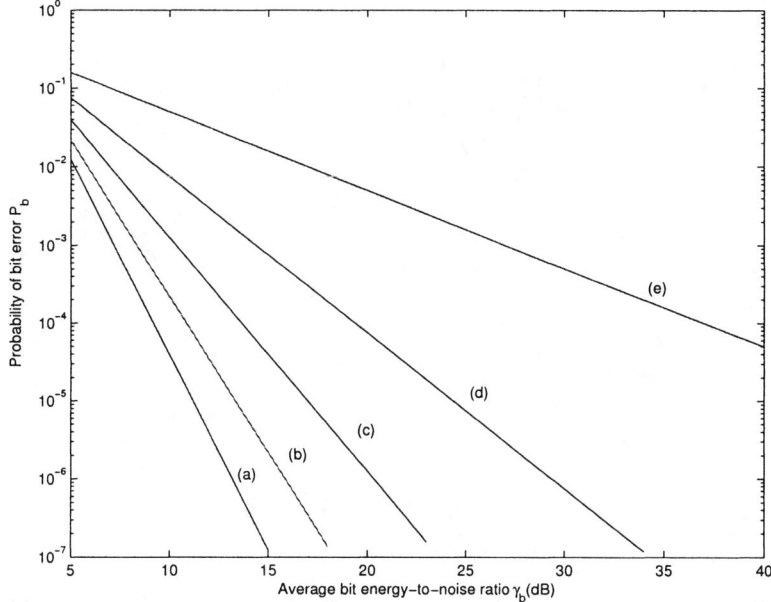

Figure 8.19 A probability of bit error for a DS spread spectrum system with a BPSK modulation over a Rayleigh fading channel for a Rake receiver.

cost function ξ of the MSE criterion is minimized,

$$\xi = E\{|\tilde{z}[n] - b[n]|^2\}, \qquad (8.205)$$

where

$$\tilde{z}[n] = \sum_{l=0}^{L-1} \alpha_l z_l[n], \qquad (8.206)$$

and $b[n]$, $n = 1, 2, \cdots, M$, is a training sequence to be transmitted.

To find optimal estimators of the weighting coefficients α_l, $l = 0, 2, \cdots, L - 1$, we shall employ a method of least squares. For each sample observation $(\tilde{z}[n], b[n])$, the method of least squares requires that we consider the sum of the M squared deviations.

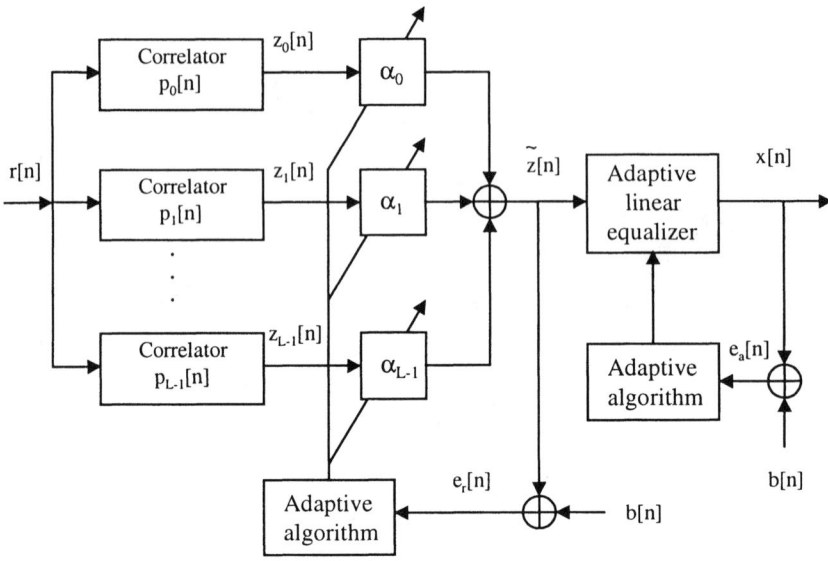

Figure 8.20 An adaptive Rake receiver followed by an adaptive linear equalizer.

This criterion is also referred to as the cost function ξ of the MSE in (8.205) given by

$$\xi = \frac{1}{M} \sum_{n=0}^{M-1} |\tilde{z}[n] - b[n]|^2$$

$$= \frac{1}{M} \sum_{n=0}^{M-1} \left| \sum_{l=0}^{L-1} \alpha_l z_l[n] - b[n] \right|^2. \qquad (8.207)$$

In matrix notation, (8.207) can be rewritten as

$$\xi = (\mathbf{Z}\boldsymbol{\alpha} - \mathbf{B})^H (\mathbf{Z}\boldsymbol{\alpha} - \mathbf{B}). \qquad (8.208)$$

Expanding this, we obtain

$$\xi = \mathbf{B}^H \mathbf{B} - \boldsymbol{\alpha}^H \mathbf{Z}^H \mathbf{B} - \mathbf{B}^H \mathbf{Z}\boldsymbol{\alpha} + \boldsymbol{\alpha}^H \mathbf{Z}^H \mathbf{Z}\boldsymbol{\alpha}. \qquad (8.209)$$

Note that $\mathbf{B}^H \mathbf{Z}\boldsymbol{\alpha}$ is 1×1 matrix. Hence, it is equal to its *Hermitian transpose* that is $\boldsymbol{\alpha}^H \mathbf{Z}^H \mathbf{B}$ according to the theorem of matrices in Appendix B. Thus, we find

$$\xi = \mathbf{B}^H \mathbf{B} - 2\boldsymbol{\alpha}^H \mathbf{Z}^H \mathbf{B} + \boldsymbol{\alpha}^H \mathbf{Z}^H \mathbf{Z}\boldsymbol{\alpha}. \tag{8.210}$$

In order to determine the value of $\boldsymbol{\alpha}$ that minimizes ξ, we differentiate with respect to $\alpha_l, l = 0, 1, 2, ..., L - 1$. Let

$$\frac{\partial}{\partial \boldsymbol{\alpha}} = \begin{bmatrix} \frac{\partial}{\partial \alpha_0} \\ \frac{\partial}{\partial \alpha_1} \\ \vdots \\ \frac{\partial}{\partial \alpha_{L-1}} \end{bmatrix}. \tag{8.211}$$

Then it follows that

$$\frac{\partial}{\partial \boldsymbol{\alpha}} = -2\mathbf{Z}^H \mathbf{B} + 2\mathbf{Z}^H \mathbf{Z}\boldsymbol{\alpha}. \tag{8.212}$$

Equating to zero gives the matrix form of the least squares normal equations,

$$\mathbf{Z}^H \mathbf{Z}\boldsymbol{\alpha} = \mathbf{Z}^H \mathbf{B}. \tag{8.213}$$

Thus, by using the inverse of the matrix $\mathbf{Z}^H \mathbf{Z}$, we obtain the optimized weighting coefficients for the Rake receiver

$$\boldsymbol{\alpha} = (\mathbf{Z}^H \mathbf{Z})^{-1} \mathbf{Z}^H \mathbf{B}. \tag{8.214}$$

We refer to the solution of (8.214) as the *MMSE* Rake receiver.

Neter et al. [48] noted that whenever the columns of $\mathbf{Z}^H \mathbf{Z}$ are linearly dependent, the normal equations of (8.213) will be linearly dependent. In that case, no unique solutions can be obtained for α_l, $l = 0, 1, 2, \cdots, L - 1$. Fortunately, in most cases, the columns of $\mathbf{Z}^H \mathbf{Z}$ are linearly independent, thereby leading to unique solutions for $\alpha_l, l = 0, 1, 2, \cdots, L - 1$.

8.7.3 Equalized Rake Receiver

Equalization has been used previously to reduce any loss due to ISI and interference between symbols when the channel bandwidth is limited. Clearly, the spread-spectrum CDMA signal will often experience ISI interference between chips and between symbols in actual transmission because the channel is bandlimited to cause band limitations of multipath fading or other frequency-dispersive transmission effects. Usually, the Rake receiver ignores this ISI by choosing instead to treat the ISI as another user, thereby increasing the number of overall users. In addition, the effect of IPI that exists for Rake receiver in the direct-sequence spread spectrum system is another important factor in system performance. This is because the spreading ratio in WCDMA systems may be as low as four in order to increase the high-speed data transmission, such as 2 Mbps, within a fixed bandwidth. However, using an equalized Rake structure can reduce those interferences. Simulation results [49–52] in previous years showed that the performance of the WCDMA system by using the equalized Rake receiver can be significantly improved over the conventional Rake receiver.

As shown in Figure 8.20, this is a joint multipath MMSE Rake receiver combined with an MMSE linear equalizer, which is used to mitigate residual interference for WCDMA systems. Let the coefficients of the linear equalizer be $c_n[k]$, $k = -N, -(N-1), ..., -1, 0, 1, ..., N-1, N$. Then the output of the linear equalizer is given by

$$x[n] = \sum_{k=-N}^{N} c_n[k]\tilde{z}[n-k]. \qquad (8.215)$$

In order to minimize the MSE function, we choose the equalizer coefficients as an unbiased estimator so that the cost function ξ of the MSE criterion is minimized,

$$\xi = E\{|x[n] - b[n]|^2\}, \qquad (8.216)$$

where $b[n]$, $n = 1, 2, ..., M$, is a training sequence to be transmitted.

Then, (8.216) can be rewritten as

$$\xi = \frac{1}{M} \sum_{n=0}^{M-1} |x[n] - b[n]|^2$$

$$= \frac{1}{M} \sum_{n=0}^{M-1} \left| \sum_{k=-N}^{N} c_n[k]\tilde{z}[n-k] - b[n] \right|^2. \quad (8.217)$$

We now express (8.217) in the matrix form

$$\xi = (\tilde{\mathbf{Z}}\mathbf{C} - \mathbf{B})^H(\tilde{\mathbf{Z}}\mathbf{C} - \mathbf{B}). \quad (8.218)$$

By using the method of least squares in (8.214), we obtain the optimal coefficients for the MMSE linear equalizer

$$\mathbf{C} = (\tilde{\mathbf{Z}}^H\tilde{\mathbf{Z}})^{-1}\tilde{\mathbf{Z}}^H\mathbf{B}. \quad (8.219)$$

Combining solutions of the MMSE Rake receiver in (8.214) and the MMSE linear equalizer in (8.219) produces the joint multipath diversity receiver for the WCDMA systems. The performance of the joint multipath diversity receiver should be much better than the performance of the conventional Rake receiver in theory. However, the updated weighting coefficients of the MMSE Rake receiver and coefficients of the MMSE linear equalizer are implemented in a separate way. Thus, it is possible to further improve the performance by jointly updating the weighting coefficients of the MMSE Rake receiver and coefficients of the MMSE linear equalizer at the same time.

A generalized block diagram of equalized Rake receiver is shown in Figure 8.21. In that case, the weighting coefficients α_l, $l = 1, 2, .., L - 1$, simple summing device, and linear equalizer of Figure 8.20 are replaced by using a matrix equalizer with L input and one output. Let the matrix equalizer $\mathbf{W}$ have its $1 \times L$ coefficients corresponding to $\mathbf{w}_0$, $\mathbf{w}_1$, ..., $\mathbf{w}_{L-1}$. Each of the L coefficients

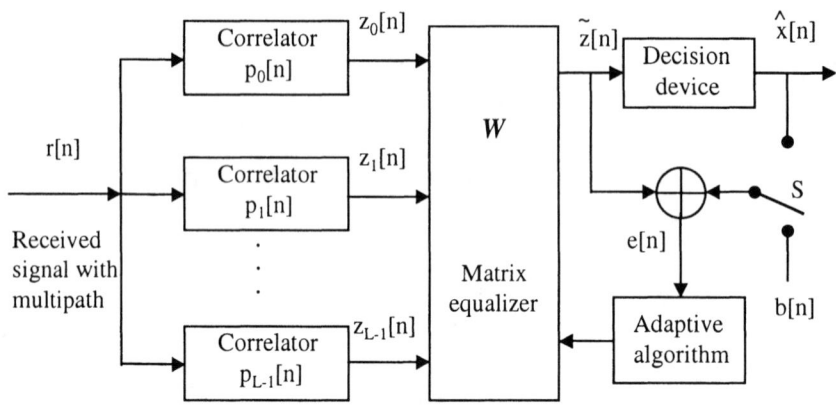

Figure 8.21 A block diagram of a general equalized Rake receiver: (a) a training-based operation if a switch connects to $b[n]$ and (b) a blind-based operation if the switch connects to $\hat{x}[n]$.

contains N values for each of the L equalizers, that is,

$$\mathbf{W} = \begin{bmatrix} \mathbf{w}_0[n] \\ \mathbf{w}_1[n] \\ \vdots \\ \mathbf{w}_{L-1}[n] \end{bmatrix}. \tag{8.220}$$

Thus, we obtain the output value of $\tilde{z}[n]$ as

$$\tilde{z}[n] = \sum_{i=0}^{L-1} \left[\sum_{k=0}^{N-1} z_i[n] w_i[n-k] \right]. \tag{8.221}$$

In order to find an optimal solution of the equalizer coefficients $\mathbf{w}_i$, $i = 1, 2, ..., L-1$ for the equalized Rake receiver, we choose the equalizer coefficients so that the cost function ξ of the MSE criterion is minimized,

$$\xi = E\{|\tilde{z}[n] - b[n]|^2\}, \tag{8.222}$$

where $b[n]$, $n = 1, 2, ..., M$, is a training sequence to be transmitted. Then, (8.222) can be rewritten as

$$\xi = \frac{1}{M} \sum_{n=0}^{M-1} |\tilde{z}[n] - b[n]|^2$$

$$= \frac{1}{M} \sum_{n=0}^{M-1} \left| \sum_{i=0}^{L-1} \left[\sum_{k=0}^{N-1} z_i[n] w_i[n-k] \right] - b[n] \right|^2. \quad (8.223)$$

Equation (8.223) can be expressed in the matrix form

$$\xi = \left(\sum_{i=0}^{L-1} \mathbf{z}_i \mathbf{w}_i - \mathbf{B} \right)^H \left(\sum_{i=0}^{L-1} \mathbf{z}_i \mathbf{w}_i - \mathbf{B} \right). \quad (8.224)$$

Expanding (8.224), we obtain

$$\xi = \mathbf{B}^H \mathbf{B} - \left(\sum_{i=0}^{L-1} \mathbf{z}_i \mathbf{w}_i \right)^H \mathbf{B} - \mathbf{B}^H \left(\sum_{i=0}^{L-1} \mathbf{z}_i \mathbf{w}_i \right)$$

$$+ \left(\sum_{i=0}^{L-1} \mathbf{z}_i \mathbf{w}_i \right)^H \left(\sum_{i=0}^{L-1} \mathbf{z}_i \mathbf{w}_i \right). \quad (8.225)$$

Since $\left(\sum_{i=0}^{L-1} \mathbf{z}_i \mathbf{w}_i \right)^H \mathbf{B} = \mathbf{B}^H \left(\sum_{i=0}^{L-1} \mathbf{z}_i \mathbf{w}_i \right)$, it is equal to

$$\xi = \mathbf{B}^H \mathbf{B} - 2 \left(\sum_{i=0}^{L-1} \mathbf{z}_i \mathbf{w}_i \right)^H \mathbf{B} + \left(\sum_{i=0}^{L-1} \mathbf{z}_i \mathbf{w}_i \right)^H \left(\sum_{i=0}^{L-1} \mathbf{z}_i \mathbf{w}_i \right). \quad (8.226)$$

To find the values of $\mathbf{w}_i$ that minimize ξ in (8.226), we differentiate with respect to w_{ij}, $i = 0, 1, 2, ..., L-1$ and $j = 1, 2, ..., N$. Let,

$$\frac{\partial}{\partial \mathbf{w}_i} = \begin{bmatrix} \frac{\partial}{\partial w_{0j}} \\ \frac{\partial}{\partial w_{1j}} \\ \vdots \\ \frac{\partial}{\partial w_{(L-1)j}} \end{bmatrix}, \quad j = 1, 2, ..., N. \quad (8.227)$$

Then it follows that

$$\frac{\partial}{\partial \mathbf{w}_i} = -2 \left[\sum_{i=0}^{L-1} \mathbf{z}_i^H \right] \mathbf{B} + 2 \left[\sum_{i=0}^{L-1} \mathbf{z}_i^H \right] \left[\sum_{i=0}^{L-1} \mathbf{z}_i \mathbf{w}_i \right]. \quad (8.228)$$

Equating (8.228) to zero, we obtain the matrix form of the least squares normal equations,

$$\left[\sum_{i=0}^{L-1} \mathbf{z}_i^H\right] \left[\sum_{i=0}^{L-1} \mathbf{z}_i \mathbf{w}_i\right] = \left[\sum_{i=0}^{L-1} \mathbf{z}_i^H\right] \mathbf{B}. \qquad (8.229)$$

Let $\mathbf{Z} = [\mathbf{z}_0, \mathbf{z}_1, ..., \mathbf{z}_{L-1}]$. Then (8.229) can be rewritten as

$$\left[\sum_{i=0}^{L-1} \mathbf{z}_i^H\right] \mathbf{ZW} = \left[\sum_{i=0}^{L-1} \mathbf{z}_i^H\right] \mathbf{B}. \qquad (8.230)$$

Thus, by using the inverse of the matrix $\left[\sum_{i=0}^{L-1} \mathbf{z}_i^H\right] \mathbf{Z}$, we obtain the optimal coefficients of the matrix equalizer for the equalized Rake receiver

$$\mathbf{W} = \left[\sum_{i=0}^{L-1} \mathbf{z}_i^H \mathbf{Z}\right]^{-1} \left[\sum_{i=0}^{L-1} \mathbf{z}_i^H\right] \mathbf{B}, \qquad (8.231)$$

which also can be called the *MMSE equalized Rake receiver*.

Such a general equalized Rake receiver must work at least as well as the performance of the Rake receiver, which can be considered a special case for spread-spectrum communication systems.

8.8 Summary

In this chapter, we introduced adaptive equalization in communication receivers. First, we presented a class of linear equalizers for channel equalization known as zero-forcing equalizers, and then adaptive linear equalizers. Second, we described the fractional spaced equalizer from a point of multichannel model view. Third, the decision feedback equalizer, which is a nonlinear equalizer, was introduced along with T-space and fractional space for the FFF filters. Then we introduced diversity equalizers, including the basic theory of Rake receivers, adaptive Rake receivers, and equalized Rake receivers that have been used in spread spectrum and UWB communication receivers.

A linear equalizer is the simplest type and the most common type of channel equalizer used in practice to combat ISI arising from the bandlimited channel. In the linear equalizer, the emphasis in this chapter is given on the zero-forcing equalizer for channel equalizer. The performance of the infinite-length tap-coefficient zero-forcing equalizer that can completely eliminate ISI is expressed in terms of the MMSE and SNR at its output, which serves as a rule for comparison with other types of equalizers. In addition, the linear equalizer has been shown to have stability and a faster convergence. However, a zero-forcing equalizer usually does not completely eliminate ISI because it has a finite-length tap-weight setting. Furthermore, the linear equalizer does not perform well on bandlimited channels if the frequency passband has deep spectral null because it attempts to place too much gain in the vicinity of the spectral null, thereby increasing the noise present in the frequency bands.

In adaptive equalizers, there are a number of variation algorithms that exist for adapting an equalizer. The zero-forcing algorithm developed by Lucky [39] is one used for eliminating the ISI in wired communications. However, the zero-forcing algorithm may excessively amplify noise at frequencies where the folded channel spectrum has high attenuation. Thus, the zero-forcing algorithm-based equalizer neglects the effect of noise altogether, and is not often used for wireless communications. A more robust equalizer uses the LMS algorithm where the criterion used is to minimize the MMSE between the desired equalizer output and the actual equalizer output. The LMS algorithm has low computational complexity, but it tends to have a slow convergence and poor tracking in some cases. An RLS with its variation algorithms has a much better convergence and tracking ability than the LMS algorithm. But the RLS algorithm usually has high computational complexity and may be unstable in some cases.

In blind equalizers, a CMA algorithm has complexity about like the LMS level. The CMA-based blind equalizer minimizes the constant modulus cost function and adjusts the taps of the equalizer in an attempt to minimize the difference between the samples' squared magnitude and the Godard dispersion constant. The CMA has been well proven to be able to open an eye-diagram and achieve convergence for the blind equalizer in the communication receivers. However, the CMA-based equalizer may converge to less than an optimum setting if the selection of the initial tap values is not appropriate. This will greatly affect the CMA's performance because the constant modulus cost surface is multimodal.

An FSE samples the incoming signal at least as fast as the Nyquist rate. In some cases, the FSE has a tap spacing of MT/N, $N > M$, but the output of the equalizer is still computed at the symbol rate. The FSE can compensate for the channel distortion before aliasing effects occur because of the symbol rate sampling. In addition, the FSE is insensitive to timing delay during any arbitrary timing phase. Moreover, the FSE can be considered as a polyphase filter bank-based equalizer that can be used in SIMO and MIMO environments. Thus, shorter FSE tap lengths can be selected to perfectly equalize the channel. Hence, the FSE can achieve a performance superior to that of a symbol rate-based equalizer. Using a CMA adaptive algorithm, the FSE has been widely used in blind equalizers.

A DFE is a nonlinear equalizer that is used in wired and wireless applications where the channel distortion is too severe for a linear equalizer to handle. In addition, the DFE has a significantly smaller MMSE than a linear equalizer. However, the DFE may seriously suffer error propagation because the result of an incorrect decision in the feedback section produces additional errors. Therefore, in practice, so-called precoding is used to remove the error-propagation problem.

A space-time equalizer is a combination equalizer based on both time-only and space-only equalizers. The time-only equalizer, which is used to minimize the channel ISI, is developed based on a baseband temporal structure. In practice, the temporal channel structure for signals and CCI cannot be well separated if the symbol waveforms have a small excess bandwidth. Thus, CCI cancellation by using the time-only equalizer will lead to excessive noise enhancement, that is, the time-only equalizer can only provide a small degree of CCI reduction. On the other hand, the space-only equalizer corresponds to simple beamforming by using a weighted sum of antenna outputs. In theory, we have shown that a purely spatial combining of the space-only equalizer is able to perfectly cancel both ISI and CCI. However, in practice, this is impossible because real channels having a multipath will require too many antenna elements. Furthermore, this also requires that the multipath angle spread is not small or that the desired signal and CCI are well separated. In fact, the space-only equalizer can effectively be used to eliminate CCI while its effectiveness against ISI depends on the angle spread of a multipath. Therefore, a space-time equalizer, which operates simultaneously on all antennas and processes signal samples both in space and time domains, should be used to improve the performance. Especially, a space-time MMSE equalizer is able to exchange CCI and ISS reduction against noise enhancement. The space-time MMSE equalizer can primarily cancel CCI in the spatial domain and ISI either in the space or time domains depending on where it can be done more efficiently. In other words, performance of the space-time MMSE equalizer is influenced by the dominance of the CCI, the ISI, and the nature of the channel structure.

A diversity equalizer can be used to reduce ISI between successive transmissions and interference between diversity multi-dimensions in a communication receiver. A type of the diversity equalizer is an equalized Rake receiver, which has been shown to have been significantly improved to reduce interpath interference over a conventional Rake receiver in WCDMA systems in addition to enhancing SNR by combining received signals with multipath. However, a drawback of the equalized Rake equalizer is greater computational complexity than that of the conventional Rake receiver. Therefore, selecting the equalized Rake receiver over the conventional Rake receiver in a particular communication system depends on a trade-off of design and analysis.

References

[1] Gibson, J. D., (ed.), *The Mobile Communications Handbook*, 2nd ed., CRC Press LLC, Danvers, Massachusetts, 1999.

[2] Haykin, S., *Adaptive Filter Theory*, 3rd ed., Prentice Hall, Upper Saddle River, New Jersey, 1996.

[3] Widrow, B., and S. D. Stearn, *Adaptive Signal Processing*, Prentice Hall, Upper Saddle River, New Jersey, 1985.

[4] Proakis, J. G., "Adaptive Equalization Techniques for Acoustic Telemetry Channels," *IEEE Trans. on Oceanic Engineering*, Vol. 16, No. 1, pp. 21–31, January 1991.

[5] Proakis, J. G., "Adaptive Equalization for TDMA Digital Mobile Radio," *IEEE Trans. on Vehicular Technology*, Vol. 40, No. 2, pp. 333–341, May 1991.

[6] Miao, G. J., and M. A. Clements, *Digital Signal Processing and Statistical Classification*, Artech House, Norwood, Massachusetts, 2002.

[7] Rappaport, T. S., *Wireless Communications: Principles and Practice*, Prentice Hall, Upper Saddle River, New Jersey, 1996.

[8] Bingham, J. A., *ADSL, VDSL, and Multicarrier Modulation*, John Wiley & Sons, New York, 2000.

[9] Haykin, S., *Neural Networks: A Comprehensive Foundation*, Macmillan College Publishing Company, New York, 1994.

[10] Miao, J., *Component Feature-Based Digital Waveform Analysis and Classification*, UMI Dissertation Service, A Bell & Howell Company, Ann Arbor, Michigan, 1996.

[11] Treichler, J. R., I. Fijalkow, and C. R. Johnson, "Fractionally Spaced Equalizers: How Long Should They Really Be?" *IEEE Signal Processing Magazine*, pp. 65–81, May 1996.

[12] Proakis, J. G., *Digital Communications*, 2nd ed., McGraw-Hill Book Company, New York, 1989.

[13] Ling, F., "On Training Fractionally Spaced Equalizers Using Intersymbol Interpolation," *IEEE Trans. on Communications*, Vol. 37, No. 10, pp. 1096–1099, October 1989.

[14] Ling, F., and S. U. H. Qureshi, "Convergence and Steady-State Behavior of a Phase-Splitting Fractionally Spaced Equalizer," *IEEE Trans. on Communications*, Vol. 38, No. 4, pp. 418–425, April 1990.

[15] Ungerboeck, G., "Fractional Tap-Spacing Equalizer and Consequences for Clock Recovery in Data Modems," *IEEE Trans. on Communications*, Vol. COM-24, pp. 856–864, August 1976.

[16] Gitlin, R. D., and S. B. Weinstein, "Fractionally-Spaced Equalization: An Improved Digital Transversal Equalizer," *Bell Systems Technology Journal*, Vol. 60, pp. 275–296, February 1981.

[17] Giannakis, G. B., and S. D. Halford, "Blind Fractionally-Spaced Equalization of Noisy FIR Channels: Adaptive and Optimal Solutions," *Proc. IEEE Int. Conf. on on Acoustics, Speech and Signal Processing*, Vol. 3, pp. 1972–1975, May 1995.

[18] Tong, L., G. Xu, and T. Kailath, "Blind Identification and Equalization Based on

Second-Order Statistics: A Time Domain Approach," *IEEE Trans. on Information Theory*, Vol. 40, No. 2, pp. 340–349, March 1994.

[19] Ding, Z., and Y. Li, "On Channel Identification Based on Second-Order Cyclic Spectra," *IEEE Trans. on Signal Processing*, pp. 1260–1264, May 1994.

[20] Slock, D. T. M., "Blind Fractionally-Spaced Equalization, Perfect-Reconstruction Filter Banks and Multichannel Linear Prediction," *Proc. of IEEE Int. Conf. Acoustics, Speech and Signal Processing*, Vol. 4, pp. 585–588, April 1994.

[21] Johnson, C. R., et al., "Blind Equalization Using the Constant Modulus Criterion: A Review," *Proc. of IEEE*, Vol. 86, pp. 1927–1950, October 1998.

[22] Johnson, C. R., et al., "On Fractionally Spaced Equalizer Design for Digital Microwave Radio Channels," *Proc. of Asilomar Conf. Signals and Systems Comput.*, pp. 290–294, Pacific Grove, California, October 1995.

[23] Godard, D. N., "Self-Recovering Equalization and Carrier Tracking in Two-Dimensional Data Communication Systems," *IEEE Trans. on Communications*, Vol. COMM-28, pp. 1867–1875, November 1980.

[24] Treichler, J. R., and B. G. Agee, "A New Approach to Multipath Correction of Constant Modulus Signals," *IEEE Trans. on Acoustic, Speech, and Signal Processing*, Vol. ASSP-31, pp. 459–472, April 1983.

[25] Cusani, R., and A. Laurenti, "Convergence Analysis of the CMA Blind Equalizer," *IEEE Trans. on Communications*, Vol. 43, No. 2, pp. 1304–1307, April 1995.

[26] Li, Y., and Z. Ding, "Global Convergence of Fractionally Spaced Godard (CMA) Adaptive Equalizers," *IEEE Trans. on Signal Processing*, Vol. 44, No. 4, pp. 818–826, April 1996.

[27] Li, Y., and Z. Ding, "Global Convergence of Fractionally Spaced Godard Equalizers," *Proceedings of 28th Asilomar Conference on Signals, Systems and Computers*, Vol. 1, pp. 617–621, October 1994.

[28] Gardner, W. A., "A New Method of Channel Identification," *IEEE Trans. on Communications*, Vol. 39, pp. 813–817, August 1991.

[29] Johnson, C. R., S. Dasgupta, and W. A. Sethares, "Averaging Analysis of Local Stability of a Real Constant Modulus Algorithm Adaptive Filter," *IEEE Trans. on Acoustic, Speech, and Signal Processing*, Vol. 36, pp. 900–910, June 1988.

[30] Litwin, L. R., "Blind Channel Equalization," *IEEE Potential*, pp. 9–12, October 1999.

[31] Fijalkow, I, A. Touzni, and J. R. Treichler, "Fractionally Spaced Equalization Using CMA Robustness to Channel Noise and Lack of Disparity" *IEEE Trans. on Signal Processing*, Vol. 45, No. 1, pp. 56–66, January 1997.

[32] Endres, T. J., et al., "Robustness to Fractionally-Spaced Equalizer Length Using the Constant Modulus Criterion," *IEEE Trans. on Signal Processing*, Vol. 47, No. 2, pp. 544–548, February 1999.

[33] Salz, J., "Optimum, Mean-Square Decision Feedback Equalization," *Bell Systems Technology Journal*, Vol. 52, pp. 1341–1373, October 1973.

[34] Belfiori, C. A., and J. H. Park, "Decision Feedback Equalization," *Proceedings of IEEE*, Vol. 67, pp. 1143–1156, August 1979.

[35] Zhou, K., J. G. Proakis, and F. Ling, "Decision Feedback Equalization of Time

Dispersive Channels with Coded Modulation," *IEEE Trans. on Communications*, Vol. 38, pp. 18–24, January 1990.

[36] Miao, G. J., and X. Qi, "Decision-Feedback Equalizer," U.S. Patent Pending, Publication No. 2004/0120394, pp. 1–14, June 2004.

[37] Paulraj, A. J., and C. B. Papadias, "Space-Time Processing for Wireless Communications," *IEEE Signal Processing Magazine*, pp. 49–83, November 1997.

[38] Moulines, E., P. Duhamel, J. F. Cardoso, and S. Mayrargue, "Subspace Methods for the Blind Identification of Multichannel FIR Filters," *IEEE Trans. on Signal Processing*, Vol. SP-43, pp. 516–525, 1995.

[39] Lucky, R. W., "Automatic Equalization for Digital Communication," *Bell System Technical Journal*, Vol. 44, pp. 547–588, 1965.

[40] Qureshi, S. U. H., "Adaptive Equalization," *Proceedings of the IEEE*, Vol. 53, No. 12, pp. 1349–1387, September 1985.

[41] Price, R., and P. E. Green, "A Communication Technique for Multipath Channel," *Proceedings of the IRE*, pp. 555–570, March 1958.

[42] Viterbi, A. J., *CDMA: Principles of Spread Spectrum Communication*, Addison-Wesley Publishing Company, Reading, Massachusetts, 1995.

[43] Glisic, S., and B. Vucetic, *Spread Spectrum CDMA Systems for Wireless Communications*, Artech House, Norwood, Massachusetts, 1997.

[44] Miao, G. J., "Analog-to-Digital Converter Bank Based Ultra Wideband Communications," U.S. Patent No. 6,744,832, pp. 1–13, June 2004.

[45] Miao, G. J., "Dual-Mode Ultra Wideband and Wireless Local Area Network Communications," U.S. Patent No. 7,046,716, pp. 1–19, May 2006.

[46] Miao, G. J., "Multiple-Input Multiple-Output Wireless Sensor Networks Communications," U.S. Patent No. 7,091,854, pp. 1–13, August 2006.

[47] Stüber, G. L., *Principles of Mobile Communication*, 2nd ed., Kluwer Academic Publisher, Boston, Massachusetts, 2001.

[48] Neter, J., W. Wasserman, and M. H. Kutner, *Applied Linear Regression Models*, Richard D. Irwin, Inc., Homewood, Illinois, 1983.

[49] Ghauri, I., and D. T. M. Slock, "Linear Receivers for the DS-CDMA Downlink Exploiting Orthogonality of Spreading Sequences," *Proceedings of 32nd Asilomar Conf. Signals, System and Computers*, Vol. 1, pp. 650–654, Monterey, California, November 1998.

[50] Tantikovit, S., and A. U. H. Sheikh, "Joint Multipath Diversity Combining and MLSE Equalization (Rake-MLSE Receiver) for WCDMA Systems," *IEEE 51st Proceedings of Vehicular Technology Conference*, Vol. 1, pp. 435–439, Tokyo, Spring 2000.

[51] Krauss, T, M. Zoltowski, and G. Leus, "Simple MMSE Equalizers for CDMA Downlink to Restore Chip Sequence: Comparison to Zero-Forcing and RAKE," *Proceedings IEEE Int. Conf. on Acoustics, Speech and Signal Processing*, Vol. 5, pp. 2865–2868, Istanbul, Turkey, June 2000.

[52] Mudulodu, S., G. Leus, and A. Paulraj, "An Interference-Suppressing Rake Receiver for the CDMA Downlink," *IEEE Signal Processing Letters*, Vol. 11, No. 5, pp. 521–524, May 2004.

9

Multicarrier Modulation, DMT, and OFDM

9.1 Introduction

Reliably and efficiently transmitting signals with high-speed data over imperfectly wired and wireless channels is still one of the central problems of signal processing in digital communications. Recently, one successful approach to achieving high-speed data transmission either over a wired or a wireless channel uses a *channel partitioning* method to divide the transmission channel into a number of orthogonal subchannels or subcarriers. The channel partitioning method constructs a set of parallel subchannels, which are largely independent. The channel partitioning method is often referred to as *multicarrier modulation* (MCM) and is also called *multichannel modulation*.

In digital communications, MCM can be classified as discrete multitone (DMT) or orthogonal frequency division multiplexing (OFDM), depending on the applications in wired or wireless channels, respectively. Both DMT and OFDM have the same channel partitioning where their carrier spacing is selected in such a way that each of the subcarriers is orthogonal to the other subcarriers. The difference between DMT and OFDM is that DMT usually uses a dynamic loading algorithm to assign a proportionately large fraction of the digital information to subchannel signals with the largest signal-to-noise ratio (SNR) of the subchannel output, while OFDM carries a fixed number of bits on all subchannels.

Over the last decade, DMT has been used in wired digital communications, including ADSL and VDSL [1–3], over slowly time-varying wired channels, such as telephone subscriber lines. DMT is particularly useful for DSL to combat the impairments of the loop topologies of the telephone subscriber lines. These impairments are mainly due to signal attenuation, crosstalk, including near-end crosstalk (NEXT) and far-end crosstalk (FEXT), signal reflections, radio frequency interference (amateur and AM radio), and impulse noise. Recently, DMT has been suggested for use in wired power lines for broadband communications [4].

OFDM has been used in wireless time-varying channels, especially coded OFDM (COFDM), which allows recovery of lost subchannels caused by time-varying notches due to multipath fading and ISI. COFDM is a multicarrier modulation technology and particularly well suited to provide reliable reception of signals affected by strong distortions. COFDM has been used in digital audio broadcasting (DAB) [5], digital video broadcasting (DVB) [6, 7], and portable digital television (DTV) [8]. OFDM played a key role in multicarrier modulation technology with applications in wireless local area networks (LANs), such as IEEE 802.11a in 1999 and 802.11g in 2003 [9, 10], and in wireless local and metropolitan area networks (MANs), such as IEEE 802.16 in 2005 [11].

This chapter is organized as follows. In this section, we briefly provide an overview and the background of multicarrier modulation, DMT, and OFDM. Section 9.2 introduces the fundamentals of DMT modulation, including multitone transmission, geometric SNR, and the optimization of energy minimums and bit loading maximums. Section 9.3 presents fast Fourier transform (FFT)–based OFDM in which we describe OFDM systems, OFDM modulation and demodulation, and A/D converter resolution requirements. In addition, we introduce equalized OFDM, including time-domain and frequency-domain equalizers. In Section 9.4, filter bank–based OFDMs are introduced, with an emphasis on filter bank transmultiplexers, the DFT filter bank, the polyphase-based discrete Fourier transform (DFT) filter bank, the maximally decimated DFT transmitter filter bank, and a perfect reconstruction of the DFT filter bank. A brief summary of this chapter is given in Section 9.5.

9.2 Fundamentals of Discrete Multitone Modulation

The basic concept of transmitting data by dividing it into serveral interleaved bit streams and using these to modulate several carriers was published more than 50 years ago by Doelz et al. [12]. Since then, interest has continuously increased because digital modems based on the basic concept are used for high-speed data transmission over wired and wireless channels. The principle of the technique has been called by different names such as orthogonally multiplexed QAM [13], OFDM [14], MCM [15], vector coding [16], structured channel signaling (SCS) [17], DMT modulation [18], and discrete wavelet multitone (DWMT) modulation [19]. However, we will refer to it by generic names: MCM, DMT, or OFDM. Unless otherwise stated, the discussion in this section will concentrate on the special forms of DMT and OFDM modulations.

The fundamental goal of DMT modulation is to partition a transmission channel with ISI into a set of orthogonal and memoryless subchannels, each with its own subcarrier. A sequence of transmission data is then transmitted through each subchannel, which is independent of the other subchannels. It is expected that the channel response within each subchannel is ideally flat, as long as the channel is partitioned sufficiently well. Thus, DMT modulation is a special case of channel partitioning where each subchannel has a frequency index and all of subchannels are independent.

9.2.1 Multitone Transmission

A general basic structure of DMT transmission is shown in Figure 9.1. A block of serial data bits b_i ($i = 1, 2, \cdots, N$)

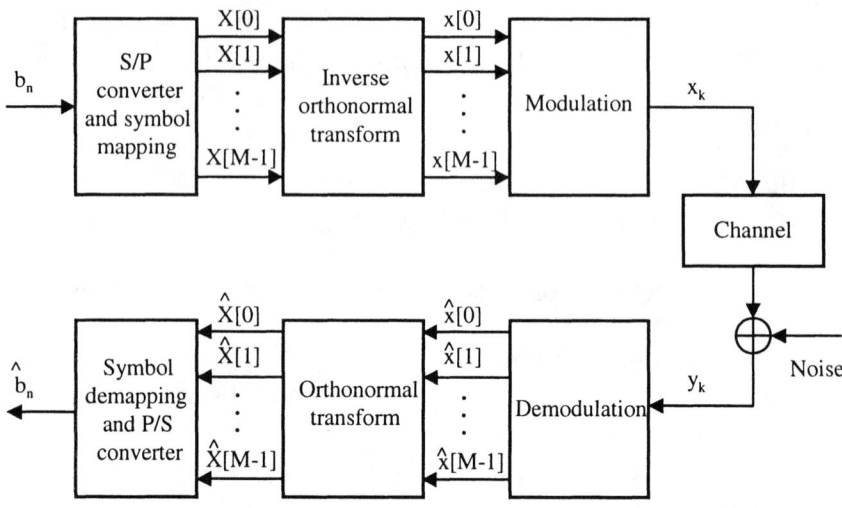

Figure 9.1 The basic structure of discrete multitone modulation.

is divided and mapped into a set of parallel symbol sequences $\{X[0], X[1], \cdots, X[M-1]\}$ in the frequency domain by using a serial-to-parallel (S/P) converter and symbol mapping. The output parallel sequences $\{X[0], X[1], \cdots, X[M-1]\}$ are passed through an inverse orthonormal transform to produce several parallel sequences $\{x[0], x[1], \cdots, x[M-1]\}$ in the time domain. Each of the parallel sequences $\{x[0], x[1], \cdots, x[M-1]\}$ is modulated onto a subcarrier. $\{x_k\}$, which is the modulated data sequence, passes through a channel with noise and $\{y_k\}$ is the received sequence in the time domain. $\{\hat{x}[0], \hat{x}[1], \cdots, \hat{x}[M-1]\}$, which are the demodulated parallel estimated sequences, are generated from the received sequence $\{y_k\}$ by using a demodulation, and are passed through an orthonormal transform to produce several parallel estimated symbols sequences $\{\hat{X}[0], \hat{X}[1], \cdots, \hat{X}[M-1]\}$ in the frequency domain. After the several parallel estimated sequences $\{\hat{X}[0], \hat{X}[1], \cdots, \hat{X}[M-1]\}$ are demapped into the estimated parallel bit sequences, they are formed into the block of

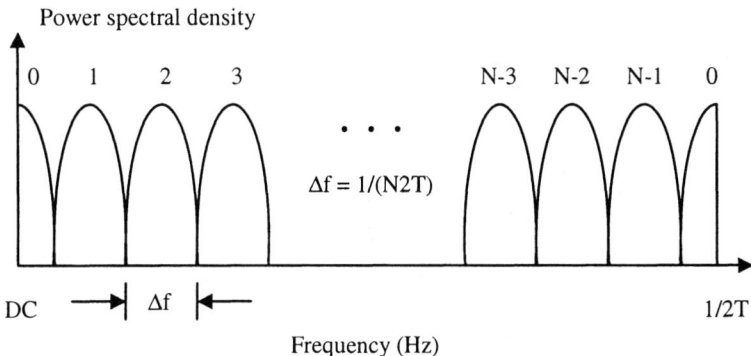

Figure 9.2 Discrete multitone modulation in the frequency domain.

serial estimated data bits $\hat{b}_n$ by using a symbol demapping and a parallel-to-serial (P/S) converter. In general, the parallel sequences of $\{x[0], x[1], \cdots, x[M-1]\}$ and $\{\hat{x}[0], \hat{x}[1], \cdots, \hat{x}[M-1]\}$ are orthogonal to each other since the DMT modulation usually uses the inverse orthonormal transform and the orthonormal transform at the transmitter and the receiver, respectively.

Viewed in the frequency domain, the DMT modulation divides the channel of the transmission data into a fixed number of N parallel, complex, and independent subchannels, as shown in Figure 9.2. Each of the subchannels is referred to as a "tone," with a width of $\Delta f = \frac{1}{T}$ in the frequency domain, where T is the period of the multicarrier symbol. Figure 9.3 shows an arbitrary channel curve of power spectral density (PSD) with vertical "bins" or "tones." In this case, if N is sufficiently large, the channel PSD curve will be approximately flat within each of the subchannels. It is clear that this assumption is valid in the infinite case. This is because we let $N \to \infty$ and then $\Delta f \to 0$, equivalently.

The DMT system continuously uses the concept of multitone transmission methods to achieve the highest levels of performance. The multitone transmission of DMT system partitions the

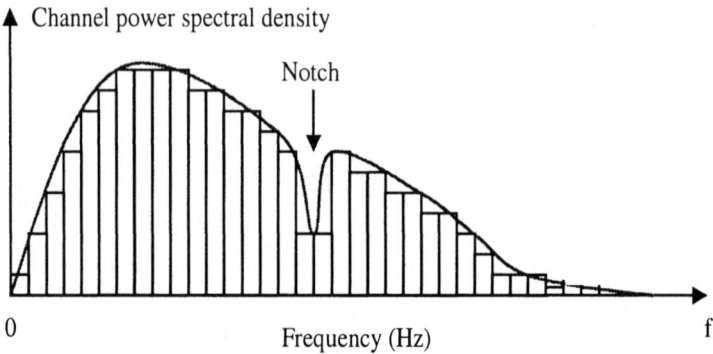

Figure 9.3 An arbitrary channel curve of power spectral density with vertical bins or tones.

transmission channel into a large number of narrowband additive white Gaussian noise (AWGN) subchannels. Those channels usually correspond to contiguous disjoint frequency bands. Thus, the multitone transmission is referred to as multicarrier or multichannel transmission. If the multitone subchannels have a sufficiently narrow bandwidth, then each subchannel has little or even no ISI. In addition, each subchannel independently approximates an AWGN. Furthermore, the capacity of such parallel independent subchannels is the sum of the individual capacities, thereby achieving theoretical maximum data rates.

Note that each of the vertical tones in Figure 9.3 corresponds to a subchannel with its own carrier. Within each of subchannels, the PSD is approximately flat. The height of each vertical tone represents the approximate amount of information, which is transmitted over the partial subchannel. Therefore, for the channel curve of PSD shown in Figure 9.3, we transmit very little data information over the subchannels at DC and high frequency as well as less data information over the subchannels at the notch frequency, where channel attenuation is most severe. As a result, a different number of bits can be conveniently assigned to different subchannels, depending on the SNR in each of the subchannels. Better subchannels with a higher SNR can transmit more information, while poor subchannels transmit little or even no data information. In this way, the overall performance of DMT transmission can be achieved in an optimal sense.

9.2.2 Geometric SNR

For an AWGN channel, the capacity of a transmission channel is a theoretical upper bound on the data rate that can be reliably transmitted. It has been shown in [2] that each of the AWGN subchannels in a multitone transmission carrying the maximum data rate or capacity is given by

$$\bar{b}_n = \frac{1}{2} \log_2 \left(1 + \frac{\text{SNR}_n}{G}\right), \quad (9.1)$$

where G is a gap that quantifies the effective loss in SNR_n given fixed probability of symbol error, P_e. In other words, any reliable and implementable communication system must transmit at a data rate below the capacity given by (9.1).

The mathematical function of gap G is given by [2, 19]

$$G = \frac{\gamma_m}{3 C_{cg}} \left[\text{erfc}^{-1}\left(\frac{P_e}{4}\right)\right]^2, \quad (9.2)$$

where γ_m is the margin (decibels) required to achieve the data and error rates with all anticipated crosstalk and noise levels (for ADSL and VDSL, $\gamma_m = 6$ dB is the accepted value), C_{cg} is the coding gain (decibels) and measures the excess SNR for that given data rate, and erfc(x) is the *complementary error function*,

$$\text{erfc}(x) = \frac{2}{\sqrt{\pi}} \int_x^\infty e^{-y^2} dy. \quad (9.3)$$

Assume that all of subchannels have the same P_e in a multichannel transmission system. This assumption of constant P_e is valid if all of the subchannels use the same class of codes with a constant gap G. In this case, a single performance measure can be used as characteristic of the multichannel transmission system. Thus, for a set of N parallel subchannels, the mean number of bits, $\bar{B}$, can be expressed as the sum of the number of bits carried on each subchannel divided by the total number of subchannels, N, as follows:

$$\begin{aligned}
\bar{B} &= \frac{1}{N} \sum_{n=1}^{N} \bar{b}_n \\
&= \frac{1}{N} \sum_{n=1}^{N} \left[\frac{1}{2} \log_2 \left(1 + \frac{\text{SNR}_n}{G} \right) \right] \\
&= \frac{1}{2} \log_2 \left[\prod_{n=1}^{N} \left(1 + \frac{\text{SNR}_n}{G} \right) \right]^{1/N} \\
&= \frac{1}{2} \log_2 \left(1 + \frac{\text{SNR}_{geo}}{G} \right),
\end{aligned} \qquad (9.4)$$

where the SNR_{geo} is referred to as the *geometric signal-to-noise ratio*, and is given by

$$\text{SNR}_{geo} = G \left\{ \left[\prod_{n=1}^{N} \left(1 + \frac{\text{SNR}_n}{G} \right) \right]^{1/N} - 1 \right\}. \qquad (9.5)$$

In this case, all the N parallel independent subchannels can be treated as one AWGN channel with SNR_{geo}. If the SNR_{geo} in (9.5) improves, then the SNR_{geo} increases. Moreover, the SNR_{geo} can be improved when the available energy is distributed nonuniformly over all or a subset of the parallel subchannels, thereby allowing a higher performance in DMT transmission, such as ADSL, VDSL, and LDSL systems [2, 19, 20]. This leads to a processing of optimizing the bit and energy distribution over a set of N parallel subchannels. This processing is known as the *bit loading algorithm* or referred to as the *water-filling algorithm*, described in Section 9.2.3.

9.2.3 Optimum of Energy Minimum and Bit Loading Maximum

One of the crucial designs in the DMT transmission system is an accurate and efficient bit loading algorithm that will maximize the transmission capacity on any given loop topology. The water filling algorithm is a well-known optimal-power distribution used for parallel communication channels [21]. It can provide a basis scheme for the power minimization and bit loading maximum in a DMT transmission system, especially in a multiuser communication environment. The bit loading algorithm often makes use of (9.1) on a subchannel-by-subchannel basis in an iterative manner to reduce the convergence (or training) time. In this case, the water filling algorithm is called the *iterative water filling algorithm*. The iterative water filling algorithm can also be considered a formulating power allocation in the multiuser interference network in which each user wants to maximize his or her own data rate over the transmission subchannels with respect to crosstalk interference from other users and noise. The iterative water filling algorithm usually converges to an optimal equilibrium point that is referred to as a stationary point in an optimal sense.

Maximization of the Data Rate

In order to maximize the transmission data rate R over a set of N parallel subchannels,

$$R = \frac{\bar{B}}{T}, \qquad (9.6)$$

where $\frac{1}{T}$ is the fixed symbol rate. This requires obtaining a maximization of the achievable mean number of bits $\bar{B}$. The maximum mean number of bits, which can be transmitted over those N parallel subchannels, must maximize the mean number of bits

$$\bar{B} = \frac{1}{2N} \sum_{n=1}^{N} \left[\log_2 \left(1 + \frac{\text{SNR}_n}{G} \right) \right], \qquad (9.7)$$

where SNR_n on an AWGN subchannel can be expressed by

$$\text{SNR}_n = \frac{E_n |H_n|^2}{\sigma_n^2}, \qquad (9.8)$$

where H_n is the subchannel gain, E_n is the energy of a signal constellation over the subchannel, and σ_n^2 is the noise PSD. Note that $\frac{|H_n|^2}{\sigma_n^2}$ is a fixed function of the subchannel, but E_n can be varied to maximize the mean number of bits $\bar{B}$, subject to an energy constraint such that an average energy of a signal constellation is

$$\bar{E}_x = \frac{1}{N} \sum_{n=1}^{N} E_n. \tag{9.9}$$

Using the Lagrange multiplier method, we set the cost function that maximizes the mean number of bits $\bar{B}$ in (9.7) subject to the constraint in (9.8) as follows:

$$L = \frac{1}{2N} \sum_{n=1}^{N} \left[\log_2 \left(1 + \frac{E_n |H_n|^2}{G \sigma_n^2} \right) \right] + \lambda \left(\bar{E}_x - \frac{1}{N} \sum_{n=1}^{N} E_n \right), \tag{9.10}$$

where λ is a constant to be determined. Using the mathematical formula $\log_a y = \frac{\ln y}{\ln a}$, (9.10) can be rewritten as

$$L = \frac{1}{2N \ln(2)} \sum_{n=1}^{N} \left[\ln \left(1 + \frac{E_n |H_n|^2}{G \sigma_n^2} \right) \right] + \lambda \left(\bar{E}_x - \frac{1}{N} \sum_{n=1}^{N} E_n \right), \tag{9.11}$$

Setting the derivative in (9.11) with respect to E_n to zero, we obtain,

$$\frac{\partial L}{\partial E_n} = \frac{1}{2N \ln(2)} \sum_{n=1}^{N} \left(\frac{1}{E_n + G \sigma_n^2 / |H_n|^2} \right) - \lambda = 0. \tag{9.12}$$

Thus, (9.12) yields

$$\lambda = \frac{1}{2N \ln(2)} \sum_{n=1}^{N} \left(\frac{1}{E_n + G \sigma_n^2 / |H_n|^2} \right). \tag{9.13}$$

Since λ is the constant, (9.7) is maximized subject to (9.9) when

$$E_n + \frac{G \sigma_n^2}{|H_n|^2} = K, \tag{9.14}$$

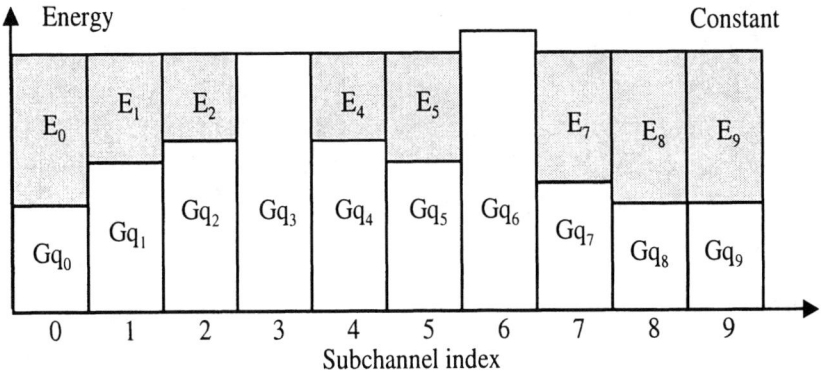

Figure 9.4 Discrete-time water filling solution for 10 subchannels.

where K is the constant. Equation (9.13) can then be rewritten as

$$\lambda = \frac{1}{2K \ln(2)}. \quad (9.15)$$

The solution of (9.14) is referred to as the "water filling" optimization. In other words, the transmit energies with the optimum water filling should satisfy the condition in (9.14) for the DMT transmission. When $G = 1$ (0 dB), we achieve the maximum data rate for a set of parallel channels.

The water fill solution for the set of parallel subchannels can be illustrated by using a graph with the curve of an inverted channel SNR filled with energy (or water) to a constant value. For example, Figure 9.4 shows a discrete-time equivalent of the water filling solution for 10 subchannels with a noise-to-signal ratio (NSR) of $q_n = \frac{\sigma_n^2}{|H_n|^2}$. Note that 8 of the 10 subchannels have positive energies, while 2 subchannels are eliminated due to zero (subchannel index at 3) and negative energy (subchannel index at 6), respectively. Thus, the 8 used subchannels have energy that makes the total of normalized noise and transmit energy constant for the DMT transmission.

The concept of a water filling term arises from the analog of the curve of Gq_n, as shown in Figure 9.4. Water (or energy) is poured into a bowl, thereby filling the bowl until there is no more water to use. In this case, the water (or energy) will gradually rise to reach a constant of a flat level of the bowl. In any subchannel, the amount of water (or energy) is the depth of the water at the corresponding point at Gq_n in the bowl. This process is also referred to as the *water filling optimization* for the DMT transmission.

Minimization of Energy

It is also clear that the energy minimization is equivalent to maximize the data rate. In other words, we want to minimize the total energy

$$\min_{E_n}\{\bar{E}_x\} = \frac{1}{N} \sum_{n=1}^{N} E_n, \qquad (9.16)$$

subject to the data rate being fixed according to

$$\bar{B} = \frac{1}{2N} \sum_{n=1}^{N} \left[\log_2 \left(1 + \frac{E_n |H_n|^2}{G\sigma_n^2} \right) \right]. \qquad (9.17)$$

Using the methods of the Lagrange multiplier and differentiation, the optimum solution of the energy minimization is again the optimization of the water filling solution given by (9.14). In this case, the water (or energy) is poured on until the number of bits per symbol is equal to the given fixed data rate in (9.17). This method is sometimes referred to as *iterative water filling* in the multiuser environment of parallel subchannels.

Adaptive Bit Loading

Adaptive bit loading based on the optimization of the water filling solution in (9.14) and the energy constraint in (9.9) can be computed by using the set of linear equations as follows:

$$E_1 + \frac{G\sigma_1^2}{|H_1|^2} = K$$

$$E_2 + \frac{G\sigma_2^2}{|H_2|^2} = K$$

$$\vdots = \vdots$$

$$E_n + \frac{G\sigma_n^2}{|H_n|^2} = K$$

$$\frac{1}{N}(E_1 + E_2 + \cdots + E_n) = \bar{E}_x. \qquad (9.18)$$

As can be seen, there are $(N+1)$ equations with $(N+1)$ unknown parameters. Those unknown parameters are E_n $(n = 1, 2, \cdots, N)$ and the constant value K. Equations (9.18) can then be expressed in a matrix form as follows:

$$\begin{bmatrix} 1 & 0 & 0 & \cdots & 0 & -1 \\ 0 & 1 & 0 & \cdots & 0 & -1 \\ \vdots & \vdots & \vdots & \ddots & \vdots & \vdots \\ 0 & 0 & 0 & \cdots & 1 & -1 \\ \frac{1}{N} & \frac{1}{N} & \frac{1}{N} & \cdots & \frac{1}{N} & 0 \end{bmatrix} \begin{bmatrix} E_1 \\ E_2 \\ \vdots \\ E_n \\ K \end{bmatrix} = - \begin{bmatrix} \frac{G\sigma_1^2}{|H_1|^2} \\ \frac{G\sigma_2^2}{|H_2|^2} \\ \vdots \\ \frac{G\sigma_n^2}{|H_n|^2} \\ -\bar{E}_n \end{bmatrix}. \qquad (9.19)$$

Equation (9.19) can be solved by using matrix inversion. The energies must be positive values for all of the subchannels. However, the energy solutions of the matrix equation in (9.19) may produce negative values for some subchannels. If this is the case, the largest $\frac{\sigma_N^2}{|H_n|^2}$ should be eliminated, and the corresponding energy E_n should be zero. The matrix equation in (9.19) can be solved iteratively until a solution with nonnegative energy occurs.

An intuitive equation obtaining the constant K can be yielded by summing the first N equations in (9.18),

$$\begin{aligned} K &= \frac{1}{N}\left(\sum_{n=1}^{N} E_n + G\sum_{n=1}^{N} \frac{\sigma_n^2}{|H_n|^2}\right) \\ &= \frac{1}{N}\left(N\bar{E}_x + G\sum_{n=1}^{N} \frac{\sigma_n^2}{|H_n|^2}\right) \\ &= \bar{E}_x + \frac{G}{N}\sum_{n=1}^{N} \frac{\sigma_n^2}{|H_n|^2}. \qquad (9.20) \end{aligned}$$

The corresponding energy E_n can then be obtained by

$$E_n = K - \frac{G\sigma_n^2}{|H_n|^2}, \quad n = 1, 2, \cdots, N. \tag{9.21}$$

Note that if one or more of the energy parameters $E_n < 0$, then the most negative parameter is eliminated first. Then, we can again solve (9.20) and (9.21), but do so by using $N - i$ ($i = 1, 2, \cdots, M$, and $M < N$) to substitute for N in (9.20) for each iteration and eliminating the corresponding term of $\frac{\sigma_n^2}{|H_n|^2}$.

Example 9.1

Assume that a discrete-time channel frequency response is given by

$$H(e^{j\omega}) = 1 + 0.5 e^{-j\omega}, \tag{9.22}$$

the constant noise power is 0.25, and $\bar{E}_n = 1$. We want to determine the water filling solution with $G = 1$ for four subchannels.

The squared magnitude of discrete-time channel is characterized by

$$\begin{aligned}|H(e^{j\omega})|^2 &= H(e^{j\omega})H^*(e^{j\omega}) \\ &= (1 + 0.5e^{-j\omega})(1 + 0.5e^{j\omega}) \\ &= 1.25 + 0.5(e^{-j\omega} + e^{j\omega}) \\ &= 1.25 + \cos\omega. \end{aligned} \tag{9.23}$$

Thus, the characteristics of four subchannels for the water filling solution are obtained by

$$\begin{aligned}\frac{\sigma_1^2}{|H_1|^2} &= \frac{0.25}{1.96 \times (\pi/4 - 0)} = 0.16 \\ \frac{\sigma_2^2}{|H_2|^2} &= \frac{0.25}{1.25 \times (\pi/2 - \pi/4)} = 0.25 \\ \frac{\sigma_3^2}{|H_3|^2} &= \frac{0.25}{0.54 \times (3\pi/4 - \pi/2)} = 0.59 \\ \frac{\sigma_4^2}{|H_4|^2} &= \frac{0.25}{0.25 \times (\pi - 3\pi/4)} = 1.27. \end{aligned} \tag{9.24}$$

Using (9.20) with all of the subchannels, we obtain

$$\begin{aligned} K &= \bar{E}_x + \frac{G}{N} \sum_{n=1}^{N} \frac{\sigma_n^2}{|H_n|^2} \\ &= 1 + \frac{1}{4}(0.16 + 0.25 + 0.59 + 1.27) \\ &= 1.5675. \end{aligned} \qquad (9.25)$$

The corresponding subchannel energies E_n for $n = 1, 2, 3, 4$ are 1.4075, 1.3175, 0.9775, and 0.2975, respectively, which are all positive. Thereby, the water filling solution has been found.

9.3 FFT-Based OFDM

In theory, OFDM is the same as DMT because both of them use the same channel partitioning, multicarrier, and multichannel modulations. However, OFDM differs from DMT in terms of the water filling optimization and the corresponding adaptive bit loading algorithm. Generally, OFDM puts equal bits on all of subchannels used in wireless time-varying channels that have time-varying notches in fading due to multipath propagation, the speed of the mobile and surrounding objects, and the transmission bandwidth of the signal, while DMT uses the adaptive bit loading algorithm based on optimizing a number of bits b_n and the energy E_n for each subchannel used in wired slowly time-varying channels. In this section, we introduce OFDM modulation based on the DFT or FFT (see DFT in Appendix C and FFT in Appendix D).

9.3.1 OFDM System

The key idea behind OFDM technologies is the partitioning of a transmission channel into a set of orthogonal subchannels, each with approximately flat transfer function and AWGN. The transmission data is then transmitted in parallel on all of the subchannels, each of which is completely independent. In other words, the basic idea of OFDM is to transmit blocks of symbols in parallel by using a large number of orthogonal subcarriers.

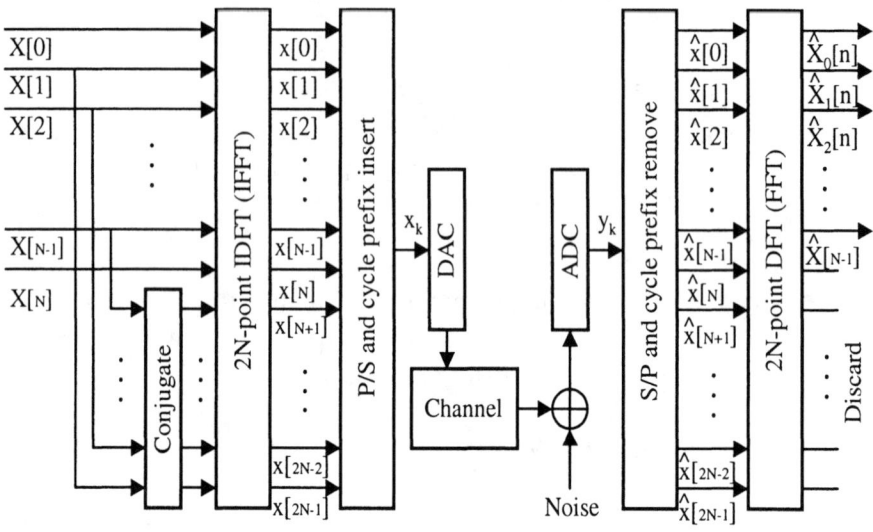

Figure 9.5 The basic structure of the OFDM system.

Figure 9.5 shows structure of an OFDM system, including the modulation, the digital-to-analog (D/A) converter, channel and noise, the analog-to-digital (A/D) converter, and demodulation. With a block transmission, a block of serial data bits b_i ($i = 1, 2, \cdots, N$), each having a bit period of T, is divided into a block of N parallel modulated symbols, $\{X[0], X[1], \cdots, X[N-1]\}$ in the frequency domain, each having a symbol period of $T_s = NT$. The block length of N is usually chosen so that $NT \gg \sigma_c$ where σ_c is the root-mean-square (RMS) delay spread of the channel. The $\{X[0], X[1], \cdots, X[N-1]\}$ are in parallel modulated by using N separate subcarriers, with $X[i]$ symbols modulated by the ith subcarrier. The modulation process is accomplished by using either a $2N$-point *inverse discrete Fourier transform* (IDFT) operation or the more computationally efficient $2N$-point *inverse fast Fourier transform* (IFFT). The parallel output of the $2N$-point IDFT is converted back into a digital serial form, $\{x_k\}$, with the appropriate cyclic prefix attached before passing through the D/A converter, which operates at a sampling frequency rate of $F_s = \frac{2N+\nu}{T}$. The resulting analog waveform is sent through the channel with noise. At the receiver, the received analog signal is converted into digital form by using A/D converter operating at the same sampling frequency rate of F_s. The resulting digital serial sequence, $\{y_k\}$, is converted back to the parallel digital received sequence. The cyclic prefix is then removed to generate the parallel digital received sequence, $\{\hat{x}[0], \hat{x}[1], \cdots, \hat{x}[2N-2], \hat{x}[2N-1]\}$, which is passed to the $2N$-point DFT (or $2N$-point FFT) demodulator. The $2N$-point DFT demodulator converts the discrete-time digital sequence, $\{\hat{x}[0], \hat{x}[1], \cdots, \hat{x}[2N-2], \hat{x}[2N-1]\}$, back to the parallel symbol sequence, $\{\hat{X}[0], \hat{X}[1], \cdots, \hat{x}[N-1]\}$, in the frequency domain.

Note that the symbol rate on each subcarrier of the OFDM system is much less than the serial bit rate. Thus, the effects of delay spread are greatly reduced, thereby reducing or even eliminating the need of equalization at the receiver. However, the channel dispersion will still cause consecutive blocks to overlay, which is referred to as *intercarrier interference* (ICI). This happens due to the effective length of the channel impulse response or the channel constraint length ν. As a result, the tail of the previous block multicarrier symbol will corrupt the beginning of the current block multicarrier symbol. This leads to the conclusion that the subchannels are not strictly independent of each other in the frequency domain. This results in some kind of residual ISI that will degrade the performance. In order to mitigate the effect of ICI, we can eliminate the residual ISI by a guard interval filled with a cyclic prefix between the blocks that is at least the same as the effective channel impulse response ν. In practice, the length of cyclic prefix is selected to be about 25% of the block length of N. However, it is also clear that using the cyclic prefix is at the expense of the channel capacity.

9.3.2 OFDM Modulation by IFFT

Referring to Figure 9.5, a block of N parallel complex symbols is $\{X[0], X[1], \cdots, X[N-1]\}$ in the frequency domain. In order to obtain $2N$-point time-domain values that are real valued after the $2N$-point IFFT operation, the block of N parallel complex input symbols $\{X[0], X[1], \cdots, X[N-1]\}$ is expanded to create $2N$ parallel complex symbols by using Hermitian symmetry,

$$X[k] = \text{conj}\{X[2N-k]\}, \quad k = N+1, N+2, \cdots, 2N-1. \tag{9.26}$$

Note that the subchannel at DC (or the subcarrier index at 0) should not contain energy, $X[0] = 0$, and the subchannel centered at the Nyquist frequency is not used for user data. That subchannel should have a real value or $X[N] = 0$.

For OFDM modulation, the $2N$-point IFFT operation at the transmitter can be implemented by using the N-point IFFT. The idea of reducing computation complexities is to make a new complex sequence $Y[k]$ with a N-point operation rather than a $2N$-point operation. For $k = 0, 1, \cdots, N-1$, we let

$$G[k] = X[k] + \text{conj}\{X[2N-k]\}, \tag{9.27}$$

and

$$H[k] = [X[k] + \text{conj}\{X[2N-k]\}] W_N^{-k}, \tag{9.28}$$

The new complex sequence $Y[k]$ is then formed by

$$Y[k] = G[k] + jH[k], \quad k = 0, 1, \cdots, N-1. \tag{9.29}$$

Thus, the N-point IFFT operation for the new complex sequence $Y[k]$ is as follows:

$$y[n] = \frac{1}{N} \sum_{k=0}^{N-1} Y[k] W_N^{-kn}, \quad n = 0, 1, \cdots, N-1. \tag{9.30}$$

The output $y[n]$ in (9.30) is a complex sequence given by

$$y[n] = g[n] + jh[n], \quad n = 0, 1, \cdots, N-1. \tag{9.31}$$

Therefore, the $2N$-point real sequence $x[n]$ can be formed by examining

$$x[2n] = g[n], \quad n = 0, 1, \cdots, N-1, \tag{9.32}$$

and
$$x[2n+1] = h[n], \quad n = 0, 1, \cdots, N-1. \tag{9.33}$$

The process achieves for the $2N$-point IFFT operation by using the N-point IFFT operation for the OFDM modulation at the transmitter.

9.3.3 OFDM Demodulation by FFT

In an OFDM demodulation, we have $2N$-point real values as an input sequence $\hat{x}[n]$ ($n = 0, 1, \cdots, 2N-1$) to produce an N-point complex value after the FFT operation. If we use the $2N$-point FFT implementation directly, it will cost huge computation complexities for the OFDM demodulation.

Reducing the computation complexities is possible since the input signal sequence $\hat{x}[n]$ is real valued in the discrete-time domain. We split the input sequence $\hat{x}[n]$ into the even-indexed points $\hat{x}_e[n]$,

$$\hat{x}_e[n] = \hat{x}[2n], \quad n = 0, 1, 2, \cdots, N-1, \tag{9.34}$$

and the odd-indexed points $\hat{x}_o[n]$,

$$\hat{x}_o[n] = \hat{x}[2n+1], \quad n = 0, 1, 2, \cdots, N-1. \tag{9.35}$$

Then we can convert the $2N$-point FFT into the N-point FFT ($k = 0, 1, \cdots, N-1$) as follows:

$$\begin{aligned}
\hat{X}[k] &= \sum_{n=0}^{2N-1} \hat{x}[n] W_{2N}^{nk} \\
&= \sum_{n=0}^{N-1} \hat{x}[2n] W_{2N}^{2nk} + \sum_{n=0}^{N-1} \hat{x}[2n+1] W_{2N}^{(2n+1)k} \\
&= \sum_{m=0}^{N-1} \hat{x}_e[m] W_N^{mk} + W_{2N}^k \sum_{m=0}^{N-1} \hat{x}_o[m] W_N^{mk} \\
&= FFT\{x_e[m]\} + W_{2N}^k \left[FFT\{x_o[m]\} \right],
\end{aligned} \tag{9.36}$$

Note that the FFTs in (9.36) are N-point FFT operations. Hence, we can calculate the $2N$-point FFT for the OFDM demodulation by using two N-point FFTs of real signals.

Example 9.2

WLAN 802.11a [9] is the IEEE standard for wireless local area network, which mainly uses an OFDM system. The radio frequency of WLAN 802.11a is initially operated for the 5.15–5.25, 5.25–5.35, and 5.725–5.825 GHz with unlicensed national information structure (U-NII) bands. The OFDM system provides communication data rate capabilities of 6, 9, 12, 18, 24, 36, 48, and 54 Mbps for the WLAN. The OFDM, which is implemented by using the 64-point IFFT/FFT, uses 52 subcarriers with modulations, including binary or BPSK, QPSK, 16-QAM, or 64-QAM. A convolutional encoding of the forward error correction (FEC) is used with a coding rate of $\frac{1}{2}$, $\frac{2}{3}$, or $\frac{3}{4}$. Among the 52 subcarriers, 48 subcarriers N_{SD} are used for data transmission, and 4 subcarriers N_{SP} are used for pilot information. Each channel has 20-MHz frequency bands. Thus, the subcarrier frequency spacing Δ_F is obtained by $\Delta_F = \frac{20}{64} = 0.3125$ MHz. The corresponding IFFT/FFT period T_{FFT} is equal to $\frac{1}{\Delta_F} = 3.2$ μs. The cyclic prefix duration T_{cp} is 25% of T_{FFT}, $T_{cp} = 0.8$ μs. Hence, the duration of the signal OFDM symbol with the BPSK modulation T_{signal} is equal to $T_{cp} + T_{FFT} = 4.0$ μs. The corresponding repetition frequency of the signal F_{signal} is $F_{signal} = \frac{1}{T_{signal}} = 250$ kHz.

We now can determine the transmission data rate. For example, with 64-QAM and a $\frac{3}{4}$ coding rate, we solve $2^b = 64$, where b is the number of bits, and obtain $b = 6$ for coded bits per subcarrier. The coded bits per OFDM symbol N_{CBPS} is equal to $N_{SD} \times b = 288$, and the data bits per OFDM symbol N_{DBPS} is equal to $288 \times \frac{3}{4} = 216$. Therefore, the data rate is equal to $N_{DBPS} \cdot F_{signal} = 216 \times 250$ kHz $= 54$ Mbps.

9.3.4 ADC Resolution for the OFDM Modulation

One of the major concerns for OFDM implementation is the analog front-end design, especially the A/D converter (or D/A converter) whose resolution plays an important role in OFDM system performance. Thus, developing an analytical form that includes all relevant system parameters of the OFDM system is important. In this section, we derive a closed-form expression of the A/D converter resolution for the OFDM system performance.

In uniform quantization, the quantization step size of the A/D converter or the resolution of the quantization is given by [22],

$$\Delta = \frac{2X_{clip}}{2^B}, \qquad (9.37)$$

where B is the number of bits and X_{clip} is known as *clipping* or *hard limiting*, in which the peak signal amplitude will exceed the full-scale amplitude of the A/D converter. The corresponding variance of the quantization error is then

$$\begin{aligned} \sigma_e^2 &= \frac{\Delta^2}{12} \\ &= \frac{2^{-2B} X_{clip}^2}{3}. \end{aligned} \qquad (9.38)$$

The *signal-to-quantization noise ratio* (SQNR), which is a common measure of the fidelity of A/D converters, is defined as the ratio of signal power (variance) σ_x^2 to the quantization noise power (variance) σ_e^2 as follows:

$$\begin{aligned} \text{SQNR} &= \frac{\sigma_x^2}{\sigma_e^2} \\ &= \frac{3 \cdot 2^{2B} \sigma_x^2}{X_{clip}^2} \\ &= \frac{3 \cdot 2^{2B}}{\mu^2}, \end{aligned} \qquad (9.39)$$

where $\mu = \frac{X_{clip}}{\sigma_x}$ is called the *clipping factor* or referred to as the *peak-to-average ratio* (PAR). Equation (9.39) can also be expressed in decibels (dBs)

$$\text{SQNR} = 6.0206B + 4.7712 + 20\log_{10}\left(\frac{\sigma_x}{X_{clip}}\right). \qquad (9.40)$$

Note that the third term in (9.40) suggests that the *root mean square* (RMS) value σ_x of the signal amplitude should be less than the peak amplitude of the signal. If the RMS value σ_x is too large, the peak signal amplitude will exceed the full-scale amplitude of the A/D converter. In this case, (9.40) has a severe distortion known as *clipping* or *hard limiting*.

In order to develop the overall SNR, we use the assumption given by Mestdagh [23], which is that the overall quantization noise σ_e^2 is a factor α smaller than AWGN σ_{AWGN}^2. With $\sigma_{AWGN}^2 = \alpha \sigma_e^2$, we then derive the overall SNR,

$$\begin{aligned} \text{SNR} &= \frac{\sigma_x^2}{\sigma_e^2 + \sigma_{AWGN}^2} \\ &= \frac{\sigma_x^2}{\sigma_e^2 + \alpha \sigma_e^2} \\ &= \frac{1}{1+\alpha}\left(\frac{\sigma_x^2}{\sigma_e^2}\right). \end{aligned} \quad (9.41)$$

Substituting (9.39) into (9.41) yields

$$\text{SNR} = \frac{1}{1+\alpha}\left(\frac{3 \cdot 2^{2B}}{\mu^2}\right). \quad (9.42)$$

By using $\log_2$ on both sides of (9.43), we obtain

$$\log_2 \text{SNR} = \log_2\left(\frac{1}{1+\alpha}\right) + \log_2\left(\frac{3}{\mu^2}\right) + 2B. \quad (9.43)$$

We then obtain

$$\begin{aligned} B &= \frac{\log_2 \text{SNR} + \log_2(1+\alpha) - \log_2(3) + \log_2 \mu^2}{2} \\ &= \frac{10 \log_{10} \text{SNR} + 10\log_{10}(1+\alpha) - 10\log_{10} 3 + 20\log_{10}\mu}{20\log_{10}(2)}. \end{aligned}$$

$$(9.44)$$

The corresponding clipping factor μ can be obtained by solving the probability of clipping given by [24]

$$P_{X_{clip}} = 2\int_{X_{clip}}^{+\infty} \frac{1}{\sigma\sqrt{2\pi}} \exp\left\{-\frac{x^2}{2\sigma^2}\right\} dx. \quad (9.45)$$

Let $y = \frac{x}{\sqrt{2}\sigma}$ and $dx = \sqrt{2}\sigma dy$; we then rewrite (9.45) as

$$\begin{aligned}P_{X_{clip}} &= 2\int_{\frac{X_{clip}}{\sqrt{2}\sigma}}^{+\infty} \frac{1}{\sigma\sqrt{2\pi}} e^{-y^2}(\sqrt{2}\sigma)dy \\ &= \frac{2}{\sqrt{\pi}}\int_{\frac{\mu}{\sqrt{2}}}^{+\infty} e^{-y^2} dy \\ &= \operatorname{erfc}\left(\frac{\mu}{\sqrt{2}}\right) \\ &= 1 - \operatorname{erf}\left(\frac{\mu}{\sqrt{2}}\right), \end{aligned} \qquad (9.46)$$

where $\operatorname{erfc}(x)$ is the complementary error function given by (9.3)

$$\begin{aligned}\operatorname{erfc}(x) &= \frac{2}{\sqrt{\pi}}\int_x^{+\infty} e^{-x^2} dx \\ &= 1 - \operatorname{erf}(x),\end{aligned} \qquad (9.47)$$

where $\operatorname{erf}(x)$ is the error function defined by

$$\operatorname{erf}(x) = \frac{2}{\sqrt{\pi}}\int_0^x e^{-x^2} dx. \qquad (9.48)$$

A relationship between the normal Gaussian function $\Phi_{0,1}(z)$ and the error function $\operatorname{erf}(x)$ is given by

$$\begin{aligned}\Phi_{0,1}(x) &= \int_{-\infty}^x \frac{1}{\sqrt{2\pi}} e^{-\frac{z^2}{2}} dz \\ &= \frac{1}{2} + \frac{1}{2}\operatorname{erf}\left(\frac{x}{\sqrt{2}}\right).\end{aligned} \qquad (9.49)$$

Substituting (9.49) into (9.46), we obtain

$$\Phi_{0,1}(\mu) = 1 - \frac{P_{X_{clip}}}{2}. \qquad (9.50)$$

If we assume $P_{X_{clip}} \leq 10^{-7}$, using a table of the normal function $\Phi_{0,1}(\mu)$ given by [25], we obtain $\mu \approx 5.0$.

With the factor $\alpha = 2$ and $\mu = 5.0$, (9.44) can be approximately rewritten as

$$B \approx \frac{10 \log_{10} \text{SNR}}{6} + 2.322. \qquad (9.51)$$

Equation (9.51) can be further simplified in terms of the maximum number of bits b_{max} carried by each of subchannels. In order to obtain a bit error rate (BER) lower than a specified value P_e, the term of $10 \log_{10} \text{SNR}$ in (9.51) must satisfy the relation given by [23, 26]

$$10 \log_{10} \text{SNR} \geq 10 \log_{10} \left\{ \frac{M-1}{3} \left[\Phi_{0,1}^{-1} \left(\frac{P_e}{4} \frac{1}{1 - 1/\sqrt{M}} \right) \right]^2 \right\}$$
$$+ \gamma_m - \gamma_c, \qquad (9.52)$$

where $\Phi_{0,1}^{-1}(x)$ is the inverse normal Gaussian function of $\Phi_{0,1}(x)$ given by (9.49), M is the QAM constellation size, γ_m is the noise margin in decibels, and γ_c is the coding gain in decibels. Assume that all of subchannels in the OFDM modulation have the same power. Thus, the resolution of the A/D converter, B, is determined by the largest possible QAM constellation size, $M = 2^{b_{max}}$, from (9.51) and (9.52).

If $P_e \leq 10^{-8}$, then

$$\Phi_{0,1}^{-1} \left(\frac{P_e}{4} \frac{1}{1 - 1/\sqrt{M}} \right) \approx 6, \qquad (9.53)$$

and (9.52) can be rewritten as

$$10 \log_{10} \text{SNR} \approx 10 \log_{10}(2^{b_{max}} \cdot 12) + \gamma_m - \gamma_c$$
$$\approx 3 b_{max} + 10.79 + \gamma_m - \gamma_c. \qquad (9.54)$$

Substituting (9.54) into (9.51), we obtain the required resolution for the A/D converter

$$B \approx \frac{b_{max}}{2} + 4.12 + \frac{\gamma_m - \gamma_c}{6}. \qquad (9.55)$$

For a general case, with the noise margin $\gamma_m = 6$ dB and the coding gain $\gamma_c = 2$ dB (e.g., for the forward error correction alone), (9.55) can be simplified to

$$B \approx \frac{b_{max}}{2} + 5. \qquad (9.56)$$

As a result, (9.55) and (9.56) provide a closed-form expression for the required A/D converter resolution for OFDM demodulation.

9.3.5 Equalized OFDM

In order to combat ISI due to a dispersive transmission channel, applying equalization in an OFDM or DMT demodulation system is usually required. An elegant equalization method relies on inserting a *guard time* (or *cyclic prefix*) between transmitted symbols after modulation with an IFFT at the transmitter. If the cyclic prefix is longer than the channel impulse response, demodulation can be implemented by the means of an FFT at the receiver, followed by using an equalizer with a single complex coefficient. This equalization technology works well under the condition that the duration of the guard time is longer than the channel impulse response [27, 28]. However, a long cyclic prefix results in a large overhead with respect to the transmission data rate. In order to reduce the guard time, a common solution is to shorten the channel impulse response by using a time-domain equalizer (TEQ) before the FFT demodulation. However, imperfectly shortened channel impulse responses lead to ISI between two successive symbols and intercarrier interference (ICI) between different carriers. Thus, in this section, we present an overview of the different time-domain and frequency-domain equalization techniques that can be used for the OFDM or DMT demodulation system when the channel impulse response is longer than the cyclic prefix.

Time-Domain Equalization

A transmission channel with severe ISI has a long channel impulse response. A TEQ is usually needed to handle the situation of the severe ISI, especially if the cyclic prefix is significantly shorter than the channel impulse response. Figure 9.6 shows the TEQ in the signal path before the OFDM or DMT demodulation (or a $2N$-point

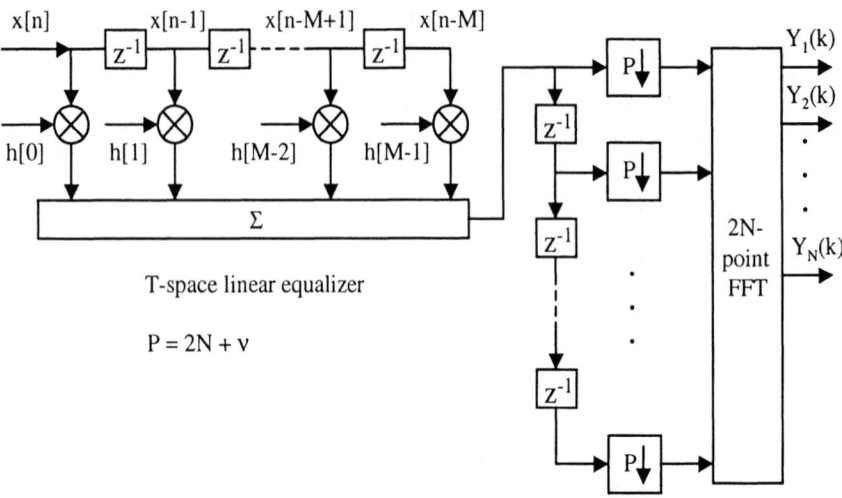

Figure 9.6 A time-domain linear equalized OFDM system.

FFT function), where v denotes the length of the cyclic prefix and $P = 2N + v$ denotes the length of a symbol including cyclic prefix. Assume that $C(z)$ is the channel and $D(z)$ is the time-domain linear equalizer as given in Figure 9.6. The goal here is to design a simple FIR or infinite impulse response (IIR) equalizer $D(z)$ so that the product $D(z)C(z)$ is a good approximation of an FIR filter with a small filter length L. This process is referred to as the *channel shortening* for the OFDM or DMT demodulation system.

The receiver based on the time-domain linear equalized OFDM in Figure 9.6 can be expressed in a matrix form:

$$\mathbf{Y} = \mathbf{WXh}, \tag{9.57}$$

where $\mathbf{h}$ is the M-tap of the time-domain linear equalizer given by

$$\mathbf{h} = \left\{ \begin{array}{cccc} h[0] & h[1] & \cdots & h[M-1] & h[M] \end{array} \right\}^T, \tag{9.58}$$

$\mathbf{X}$ is the $2N \times M$ Toeplitz matrix of the input signals $\mathbf{x}$ as kth symbol given by

$$\mathbf{X} = \begin{bmatrix} x[s_1+v+1] & x[s_1+v] & \cdots & x[s_1+v-M+2] \\ x[s_1+v+2] & x[s_1+v+1] & \cdots & x[s_1+v-M+3] \\ \vdots & \vdots & \ddots & \vdots \\ x[s_2] & x[s_2-1] & \cdots & x[s_2-M+1] \end{bmatrix}, \quad (9.59)$$

where $s_1 = k(2N+v)$ and $s_2 = (k+1)(2N+v)$, and $\mathbf{W}$ is the $2N \times 2N$ FFT-matrix given by

$$\mathbf{W} = \begin{bmatrix} 1 & 1 & 1 & \cdots & 1 \\ 1 & W & W^2 & \cdots & W^{2N-1} \\ 1 & W^2 & W^4 & \cdots & W^{2(2N-1)} \\ \vdots & \vdots & \vdots & \ddots & \vdots \\ 1 & W^{2N-1} & W^{2(2N-1)} & \cdots & W^{(2N-1)^2} \end{bmatrix}, \quad (9.60)$$

where $W = e^{\frac{-j2\pi}{2N}}$ and $\mathbf{Y}$ is the output of N-point FFT operation,

$$\mathbf{Y} = \left\{ \begin{array}{cccc} Y_1[k] & Y_2[k] & \cdots & Y_{N-1}[k] & Y_N[k] \end{array} \right\}^T. \quad (9.61)$$

Note that (9.57) does not consider a synchronization delay δ. In order to synchronize, the delay δ needs to be added into the matrix of $\mathbf{X}$ in (9.59).

Also note that the z^{-1} units and decimator by P units form the polyphase decomposition for the decimator after the TEQ and before the $2N$-point FFT. The polyphase decomposition will provide an efficient way for decimation implementation. However, the TEQ still operates at a normal sampling rate since the TEQ is located before the polyphase decomposition. In other words, the computational complexities of the TEQ cannot be saved while the $2N$-point FFT operates on downsampling rate by P.

Frequency-Domain Equalization

A frequency-domain equalization (FEQ), which is an alternative receiver structure as shown in Figure 9.7, is developed based on a separate MSE optimization for each tone or for each carrier.

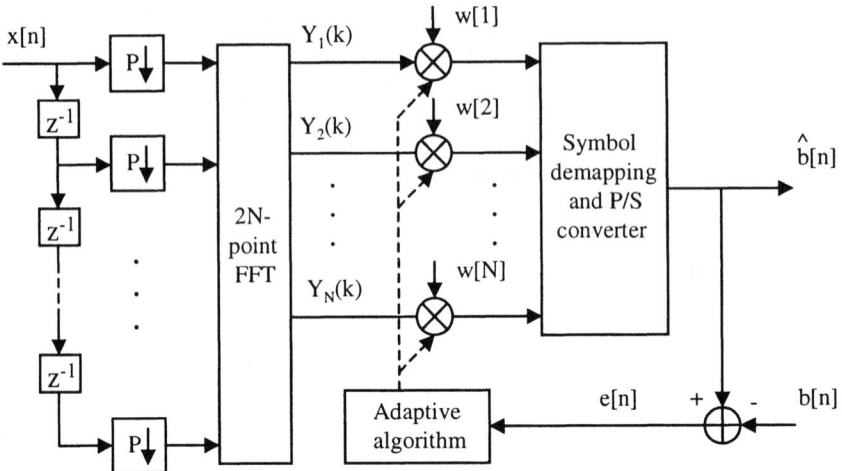

Figure 9.7 A frequency-domain equalized OFDM system.

A detection and decoding system in the OFDM demodulation cannot be simplified if all of the subcarriers do not have the same attenuation and phase. Thus, a single-tap FEQ per tone can be used after the $2N$-point FFT demodulation to correct for the attenuation and phase rotation. In other words, it multiplies the $2N$-point FFT demodulation output by using a diagonal matrix whose elements are one complex multiplication per tone and are the inverse of the transform of the shortened channel impulse response.

Note that the z^{-1} units and decimator by P units form the polyphase decomposition for the decimator before the $2N$-point FFT. The polyphase decomposition will provide an efficient way for decimation implementation for the $2N$-point FFT, which operates at a rate downsampled by P. Furthermore, the FEQ operates on the output of the N-point FFT. Thus, the computational complexities of the FEQ can be reduced since it performs at the downsampled rate. The single-tap based FEQ per tone can be updated by using an adaptive algorithm according to the error signal $e[n] = \hat{b}[n] - b[n]$, where $\hat{b}[n]$ is the output sequence of the symbol demapping and parallel-to-serial (P/S) converter and $b[n]$ is a desired symbol sequence or a training symbol sequence.

A small MSE of a special tone in the OFDM demodulation system generally corresponds to a large SNR and a large transmission capacity for the tone, thereby leading to improved and more predictable performance. Furthermore, the computational complexities and memory requirements can be reduced for the OFDM demodulation system.

9.4 Filter Bank–Based OFDM

In this section, we discuss filter bank approaches for the OFDM system with an emphasis on FFT transmitter filter banks and receiver filter banks, perfect reconstruction, and efficient implementation of polyphase decomposition.

9.4.1 Filter Bank Transmultiplexer

A schematic structure of a filter bank transmultiplexer is shown in Figure 9.8. It is also called a digital transmultiplexer [29–31]. It was initially intended to convert transmission data between *time division multiplexed* (TDM) and *frequency division multiplexed* (FDM). Note that the $H_k(z)$ in Figure 9.8 are referred to as transmitter filters or transmitter interpolation filters. The output of kth transmitter filter is expressed as

$$y_k[n] = \sum_{i=-\infty}^{\infty} x_k[i] h_k[n - iN], \qquad (9.62)$$

where $y_k[n]$ is an interpolated version of $x_k[n]$ and has N times a higher data rate. The 0th transmitter filter $H_0(z)$ is assumed to be a lowpass filter, while other transmitter filters $H_k(z)$, $k = 1, 2, \cdots, N - 1$, are bandpass filters. However, all of the transmitter filters $H_k(z)$, $k = 0, 1, 2, \cdots, N - 1$, usually cover different uniform frequency bands. The interpolated signals $y_k[n]$ are analogous to modulated versions of baseband signals $x_k[n]$. This is because the bandwidth is shifted to the passband of the transmitter filters $H_k(z)$. These are packed into N adjacent frequency bands and added together to obtain the composite signal $y[n]$. If the transmitter

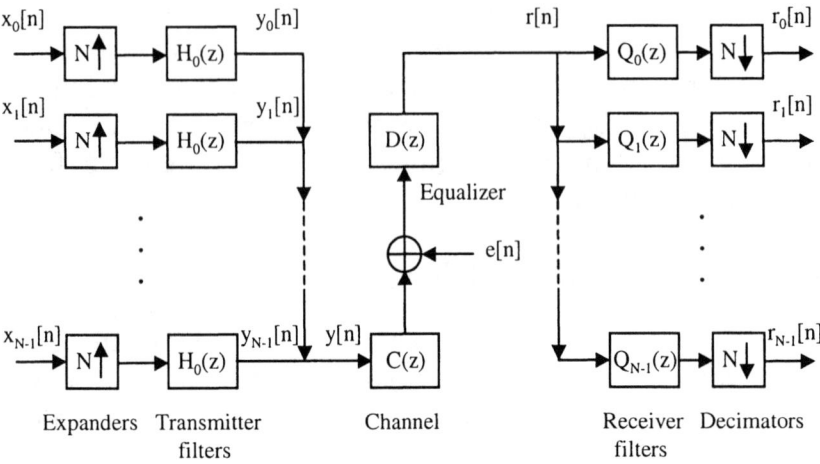

Figure 9.8 A schematic structure of a filter bank transmultiplexer.

filters $H_k(z)$ are chosen as good bandpass filters, we refer to the composite signal $y[n]$ as a FDM version of the separate signals $x_k[n]$, $k = 0, 1, 2, \cdots, N - 1$. On the other hand, if the transmitter filters $H_k(z)$ are just chosen as delay elements z^{-k}, then the composite signal $y[n]$ is the TDM version of the N signals $x_k[n]$.

At the receiver, the receiver filters $Q_k(z)$, $k = 0, 1, 2, \cdots, N - 1$, separate the received signal $r[n]$ into the N parallel signals $r_k[n]$ that are distorted and noisy versions of the symbols $x_k[n]$. Thus, the goal is to detect the symbols $x_k[n]$ from $r_k[n]$ within an acceptable probability of error, thereby making $r_k[n]$ resemble $x_k[n]$, $k = 0, 1, 2, \cdots, N - 1$.

Assume that a transfer function $G_{km}(z)$ is the decimated version of the product filter function $H_k(z)C(z)D(z)Q_m(z)$. If $G_{km}(z)$ is not a zero for $k \neq m$, then the received symbol $r_m[n]$ is affected by the transmitted symbol $x_k[i]$, $i \neq n$. This results in *interband interference*. If $G_{kk}(z)$ is not a constant, then the received symbol $r_k[n]$ is affected by the transmitted symbol $x_k[i]$, $i \neq n$, because of the filtering effect of $G_{kk}(z)$. This leads to *intraband interference*. However, if interband and intraband interferences are cancelled, then the filter bank system is called ISI free.

Further assume that the transmitter filters $H_k(z)$ and the receiver filters $Q_k(z)$ are ideal nonoverlapping bandpass filters. Thus, there is no interband interference. In addition, if the channel $C(z)$ is completely equalized with the inverse filter or equalizer $D(z) = \frac{1}{C(z)}$, then the filter bank system is ISI free and $r_k[n] = x_k[n]$ for all k in the condition of absence of noise. In this case, we have the perfect symbol recovery. In other words, the filter bank system has perfect symbol recovery if and only if the transmitter filters $H_k(z)$ and the receiver filters $Q_k(z)$ satisfy the condition given by

$$H_k(z)Q_n(z)\mid_{\downarrow N} = \delta[k-n]. \tag{9.63}$$

In the time domain, this means that the product filter $P_{kn}(z) = H_k(z)Q_n(z)$ has the zero-crossing property

$$p_{kn}[Mn] = \begin{cases} 0, & k \neq n \\ \delta[n], & \text{otherwise.} \end{cases} \tag{9.64}$$

Equation (9.63) or (9.64) is called the *biorthogonality property*.

9.4.2 The DFT Filter Bank

The DFT can be presented by using the matrix form of the DFT as discussed in Appendix C. The $(M \times M)$ matrix $\mathbf{W}$ of the DFT has $(M \times M)$ elements $[\mathbf{W}]_{km} = W^{km}$ given by (C.9) in Appendix C, where $W = e^{\frac{-j2\pi}{M}}$. Note that $\mathbf{W}^T = \mathbf{W}$ since the column vectors $\mathbf{w}_i$ of matrix $\mathbf{W}$ of the DFT are orthonormal. Then, the quantity $\mathbf{W}^*$ is equal to $\mathbf{W}$, where the matrix $\mathbf{W}^*$ is a conjugate of the matrix $\mathbf{W}$.

By using the definition of the matrix $\mathbf{W}$ and the relationship between input and output given by (C.8) in Appendix C, we can obtain

$$R_k[n] = \sum_{i=0}^{M-1} r[n-i] W^{-ki}, \quad k = 0, 1, 2, ..., M-1. \tag{9.65}$$

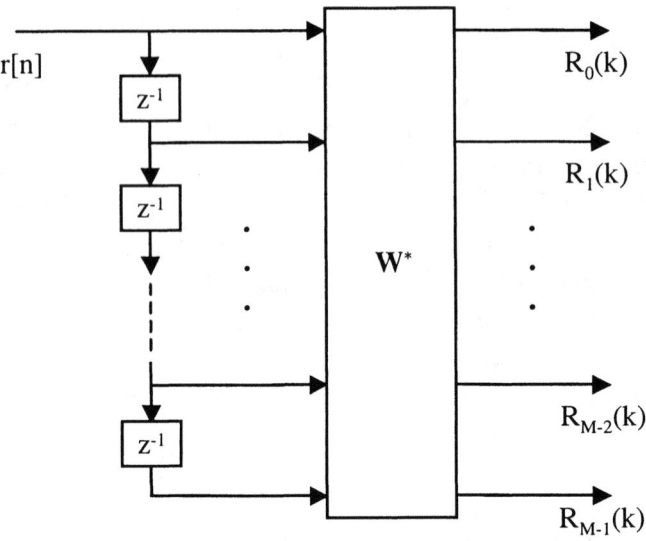

Figure 9.9 The receiver DFT filter bank for the OFDM demodulation.

Equation (9.65) indicates that we need to compute a set of M signals $R_k[n]$ from a set of M signals $r[n - i]$ for every discrete-time index n as shown in Figure 9.9. In the z-transform domain, we can rewrite (9.65) in the following form:

$$\begin{aligned} R_k(z) &= \sum_{i=0}^{M-1} z^{-i} W^{-ki} R(z) \\ &= \sum_{i=0}^{M-1} \left(zW^k \right)^{-i} R(z). \end{aligned} \quad (9.66)$$

Thus, we can represent (9.66) as follows:

$$R_k(z) = F_k(z) R(z), \quad (9.67)$$

where $F_k(z)$ is expressed as

$$F_k(z) = F_0(zW^k), \quad (9.68)$$

and
$$F_0(z) = \sum_{i=0}^{M-1} z^{-i}. \quad (9.69)$$

Equation (9.69) is referred to as a *prototype filter*, and (9.67) is called the receiver filter bank for the OFDM demodulation system. Using the result given by [22], we obtain the frequency response $F_0(e^{j\omega})$ in (9.69) as follows:

$$F_0(e^{j\omega}) = e^{-j\omega(M-1)/2} \left[\frac{\sin(M\omega/2)}{\sin(\omega/2)} \right], \quad (9.70)$$

and the corresponding magnitude response $|H_0(e^{j\omega})|$ is obtained by

$$|F_0(e^{j\omega})| = \left| \frac{\sin(M\omega/2)}{\sin(\omega/2)} \right|. \quad (9.71)$$

Thus, the kth receiver filter bank $F_k(z)$ has the frequency response

$$F_k(e^{j\omega}) = F_0(e^{j(\omega - \frac{2\pi k}{M})}). \quad (9.72)$$

Note that (9.72) is a uniformly shifted version of the prototype filter $F_0(e^{j\omega})$. Thus, the receiver filter bank contains M receiver filters for the OFDM demodulation system. Equation (9.68) or (9.72) is also referred to as a *uniform DFT receiver filter bank*.

Miao and Clements [22] indicate that the receiver filter bank consisting of M filters is obtained from a single prototype filter $F_0(z)$ by uniformly shifting the frequency response based on the relation in (9.72). Those filters have wide transition bands and small stopband attenuation. Furthermore, the magnitude response of those filters has a large amount of overlap because the prototype filter $F_0(z)$ in (9.69) is a very simple filter. However, those problems can be improved when we introduce the polyphase-based DFT filter banks discussed in the next section.

9.4.3 Polyphase-Based DFT Filter Bank

A method of polyphase decomposition can be used to implement the DFT receiver filter bank efficiently. Given the prototype filter $F_0(z)$

in (9.69), the kth receiver filter in (9.68) can be rewritten into the M-channel receiver filter bank as follows:

$$F_k(z) = F_0(zW^k)$$
$$= \sum_{i=0}^{M-1} (z^{-1}W^{-k})^i D_i(z^M), \qquad (9.73)$$

since $(zW^k)^M = z^M$. Substituting (9.73) into (9.67), we obtain

$$R_k(z) = \sum_{i=0}^{M-1} W^{-ki} \left[z^{-i} D_i(z^M) R(z) \right]. \qquad (9.74)$$

By using the method of polyphase decomposition in (9.74), an implementation structure of the DFT receiver filter bank for the OFDM demodulation system can be shown in Figure 9.10. Note that if $D_i(z^M) = 1$ for all i in (9.74), then (9.74) is equivalent to (9.66). This means that Figure 9.9 is a special case of Figure 9.10 when $D_i(z^M) = 1$ for all i. However, the presence of $D_i(z^M)$ in Figure 9.10 allows for increasing the length of the prototype filter $F_0(z)$. As a result, the DFT receiver filter bank, including M filters, can have a sharper cutoff and a higher stopband attenuation for the OFDM demodulation system.

9.4.4 Maximally Decimated DFT Transmitter Filter Bank

Decimating outputs of the receiver filter bank in the DFT receiver filter bank by the decimation factor of M is interesting in the OFDM demodulation system. This can be accomplished because each of these receiver filter outputs has a bandwidth, which is approximately M times narrower than the bandwidth of the input signal $r[n]$. Thus, by using the method of noble identity [22], the polyphase decomposition of the DFT receiver filter bank with a decimation function of M factor is shown in Figure 9.11. As can be seen, this polyphase decomposition structure of the DFT receiver filter bank is an efficient implementation of OFDM demodulation due to the decimation operation before the receiver filters $D_i(z^M)$ for $i = 1, 2, \cdots, M - 1$.

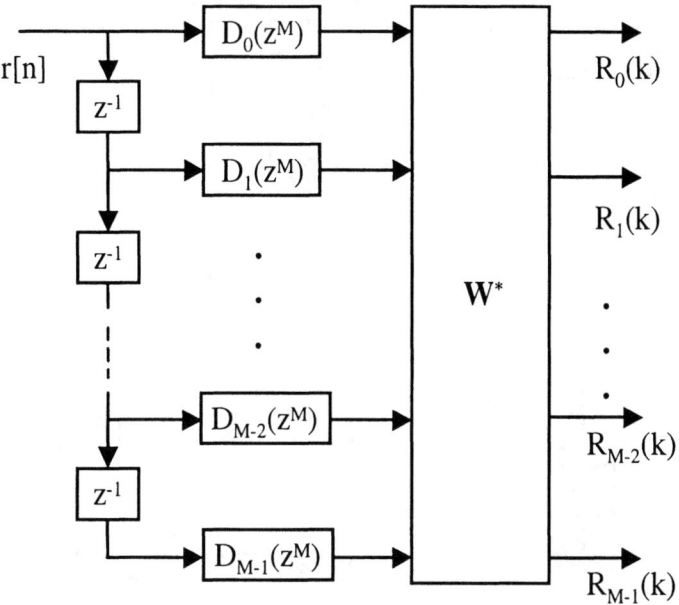

Figure 9.10 Polyphase decomposition of the DFT receiver filter bank for the OFDM demodulation.

9.4.5 Perfect Reconstruction of the DFT Filter Bank

In this section, we discuss the perfect reconstruction of the DFT filter bank for the OFDM modulation and demodulation system based on the fundmental *quadrature mirror filter* (QMF). Figure 9.12 shows a detailed structure for the perfect reconstruction of the DFT transmitter and receiver filter banks for the OFDM modulation and demodulation system. By using the prototype filter in (9.68) and (9.69), we recall that the kth receiver filter $Q_k(z)$ in the receiver filter bank for the OFDM demodulation system is given by

$$Q_k(z) = Q_0(zW^k). \tag{9.75}$$

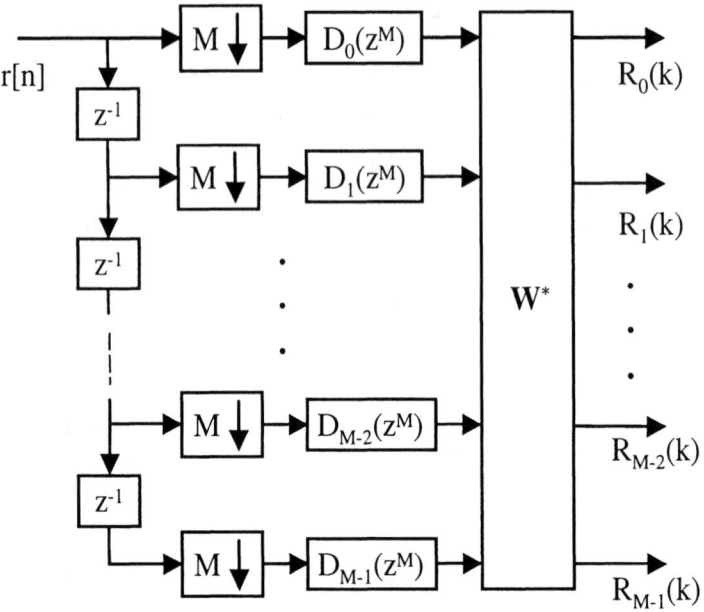

Figure 9.11 Polyphase decomposition of the DFT receiver filter bank for the OFDM modulation with decimation by a factor of M.

The kth transmitter filter $H_k(z)$ in the transmitter filter bank for the OFDM modulation is obtained by

$$H_k(z) = W^{-k} H_0(zW^k), \qquad (9.76)$$

where

$$H_0(z) = Q_0(z). \qquad (9.77)$$

This indicates that each filter in the transmitter filter bank of the OFDM modulation system has precisely the same magnitude response as the corresponding filter in the receiver filter bank of the OFDM demodulator.

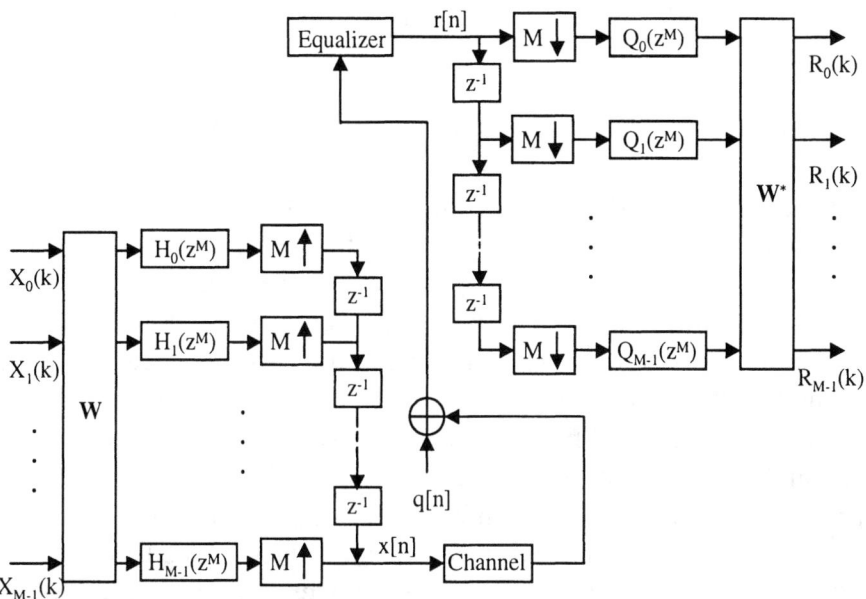

Figure 9.12 A perfect reconstruction of the DFT transmitter and receiver filter banks for the OFDM modulation and demodulation system.

Assume that the channel $C(z)$ is completely equalized with the inverse filter or equalizer $D(z) = \frac{1}{C(z)}$ and is noise free. Under these conditions, the transmitter filters $H_k(z)$ and the receiver filters $Q_k(z)$ satisfy the biorthogonality property given by (9.63) or (9.64). Thus, we obtain the polyphase matrix of the transmitter filter bank given by $\mathbf{WH}(z)$, and the polyphase matrix of the receiver filter bank given by $\mathbf{Q}(z)\mathbf{W}^*$. For the perfect symbol recovery, we multiply the polyphase matrix of the transmitter filters, the channel, the equalizer, and the polyphase matrix of the receiver filters to yield a combined polyphase matrix as follows:

$$\mathbf{WH}(z)C(z)D(z)\mathbf{Q}(z)\mathbf{W}^* = M\mathbf{I}. \qquad (9.78)$$

As a result, the receiver signals $R_i(k)$, $i = 0, 1, 2, \cdots, M - 1$, after the OFDM demodulation satisfies the perfect reconstruction condition by

$$R_i(k) = MX_i(k - M + 1), \quad i = 0, 1, 2, \cdots, M - 1. \quad (9.79)$$

Equation (9.79) indicates that the received symbols $R_i(k)$ have an attenuation of M and a time delay of $M - 1$ that differs from the transmitted symbols $X_i(k)$. However, the ISI, interband, and intraband interferences can be completely canceled because the transmitter filters in the OFDM modulation and the receiver filters in the OFDM demodulation satisfy the condition of biorthogonality.

In this section, we have discussed the DFT filter banks based on a channel frequency response at uniformly spaced points. However, in practice, filter banks at nonuniformly spaced samples of the channel frequency response also exist, especially when spaced in octaves. This is because there are relatively fewer equalizer coefficients with large values for channels with rapidly decaying frequency responses, thereby reducing the noise amplification of the channel at the receiver. On the other hand, the filter banks with nonuniformly spaced samples can be used for a digital channelizer, which is a digital transponder of satellite communications. The digital channelizer is used to decompose an input signal bandwidth into a set of subchannels that are nonuniformly spaced and recombine them to form output signals for satellite communications. The filter banks with nonuniformly spaced samples are referred to as *nonuniform filter banks*. The interested reader may refer to [32, 33].

9.5 Summary

In this chapter, multicarrier modulation, DMT, and OFDM are introduced. We first presented the fundamentals of multicarrier modulation and DMT with an emphasis on multitone transmission, geometric SNR, and optimum energy minima and bit loading maxima. Second, we described the theory of an FFT-based OFDM system in which we addressed the OFDM modulation and demodulation by IFFT and FFT operations, the determination of the A/D converter resolution, and an equalized OFDM system. We then introduced the filter bank–based OFDM modulation and demodulation system, focusing on the DFT (FFT) filter bank and its corresponding efficient implementation forms including the polyphase, maximally decimated, and perfect reconstruction structures at the transmitter and receiver.

DMT has been used in wired, slowly time-varying channels, especially in telephone lines with NEXT and FEXT impairments and different loop topologies, while OFDM has been applied in wireless, time-varying channels with multipath and fading. However, the fundamentals of DMT and OFDM both belong to the same multicarrier modulation, and their multicarrier spacing is selected to ensure that each of the subcarriers is orthogonal to the other subcarriers. In addition, because of using an FFT for the multicarrier modulation, the symbol rate on each subcarrier of the DMT and OFDM system is much less than the serial bit data rate. Thus, the effects of delay spread in the the DMT and OFDM modulation are greatly reduced. DMT usually uses a dynamic loading algorithm to assign bit information to subchannel signals based on a geometric SNR of subchannel output, thereby leading to a different number of bits assigned in each of subchannels for the DMT transmitter. On the other hand, OFDM assigns a fixed number of bits to all subchannels at the transmitter. Furthermore, both DMT and OFDM systems are mainly developed based on the DFT (FFT). This is because the DFT (FFT) can provide efficient implementation for the DMT and OFDM systems at the transmitter and receiver. However, the channel dispersion will cause consecutive blocks to overlay in DMT and OFDM systems, which leads to ICI and ISI. This results in the degradation of performance. In practice, the DMT and OFDM systems both use cyclic prefixes (or guard intervals) between the consecutive blocks to mitigate the effects of ICI and ISI. However, using the cyclic prefix wastes a fraction of the channel capacity for the communication system.

The filter bank–based DMT and/or OFDM system is developed based on the theory of QMF in which it decomposes the spectral frequency into subbands in such a way that all the aliasing incurred in the transmitter is eliminated at the receiver. The approach requires designing transmitter filter bank and receiver filter bank to meet the properties of perfect reconstruction, thereby allowing elimination of amplitude and phase distortion completely at the receiver. Efficient implementation of the filter banks based on the DMT and/or OFDM modulation and demodulation system exists by using the polyphase decomposition along with the FFT structures. However, this approach usually requires a TEQ before the DMT and/or OFDM demodulation to compensate the channel dispersion.

References

[1] ANSI, "Asymmetric Digital Subscriber Line (ADSL) Metallic Interface," ANSI Standard T1.413, 1995.

[2] Starr, T., J. M. Cioffi, and P. J. Silverman, *Understanding Digital Subscriber Line Technology*, Prentice Hall, Upper Saddle River, New Jersey, 1998.

[3] T1E1.4, "Very-High-Bit-Rate Digital Subscriber Line (VDSL) Metallic Interface, Part I: Functional Requirement and Common Specification," November 2000.

[4] Miao, G. J., "Dual-Mode Wireless/Wired Power Line Communications," U.S. patent pending, June 2004.

[5] ETSI DAB Std, http://portal.etsi.org/radio/DigitalAudioBroadcasting/DAB.asp.

[6] ETSI DVB Std, http://portal.etsi.org/radio/DigitalVideoBroadcasting/DVB.asp.

[7] Zhou, W. Y., and Y. Wu, "COFDM: An Overview," *IEEE Trans. on Broadcasting*, Vol. 41, No. 1, pp. 1–9, March 1995.

[8] Yoshida, J. "Chip Makers Place Their Bets on Mobile-TV Silicon," *EE Times*, p. 31, August 29, 2005 (http://www.eetimes.com).

[9] IEEE Std 802.11a, "Part 11: Wireless LAN Medium Access Control (MAC) and Physical Layer (PHY) Specifications: Higher-Speed Physical Layer Extension in the 5 GHz Band," *IEEE-SA Standards Board*, 1999.

[10] IEEE Std 802.11g, "Part 11: Wireless LAN Medium Access Control (MAC) and Physical Layer (PHY) Specifications: Further Higher Data Rate Extension in the 2.4 GHz Band," *IEEE-SA Standards Board*, June 2003.

[11] IEEE Draft Std 802.16, "Corrigendum to IEEE for Local and Metropolitan Area Network - Part 16: Air Interface for Fixed Broadband Wireless Access Systems," *IEEE-SA Standards Board*, August 2005.

[12] Doelz, M. L., E. T. Heald, and D. L. Martin, "Binary Data Transmission Techniques for Linear Systems," *Proc. IRE*, Vol. 45, pp. 656–661, May 1957.

[13] Hirosaki, B., "An Orthogonally Multiplexed QAM System Using the Discrete Fourier Transform," *IEEE Trans. on Communications*, Vol. COM-29, pp. 982–989, July 1981.

[14] Kalet, I., and N. A. Zervos, "Optimized Decision Feedback Equalization Versus Optimized Orthogonal Frequency Division Multiplexing for High-Speed Data Transmission over the Local Cable Network," *IEEE International Conference on Communications Records*, pp. 1080–1085, September 1989.

[15] Bingham, J. A. C., "Multicarrier Modulation for Data Transmission: An Idea Whose Time Has Come," *IEEE Communications Magazine*, pp. 5–14, May 1990.

[16] Kasturia, S., J. Aslanis, and J. M. Cioffi, "Vector Coding for Partial-Response Channels," *IEEE Trans. on Information Theory*, Vol. 36. No. 4, pp. 741–762, July 1990.

[17] Lechleider, J. W., "The Optimum Combination of Block Codes and Receivers for Arbitrary Channels," *IEEE Trans. on Communications*, Vol. 38, No. 5, pp. 615–621, May 1990.

[18] Ruiz, A., J. M. Cioffi, and S. Kasturia, "Discrete Multiple Tone Modulation with Coset Coding for the Spectrally Shaped Channel," *IEEE Trans. on Communications*, Vol. 40, No. 6, pp. 1012–1029, June 1992.

[19] Bingham, J. A. C., *ADSL, VDSL, and Multicarrier Modulation*, John Wiley & Sons, New York, 2000.

[20] Miao, G. J., "Introduction to Long-Range Extended DSL System: LDSL," *White Paper*, pp. 1–21, December 2002.

[21] Cover, T. M., and J. A. Thomas, *Elements of Information Theory*, John Wiley & Sons, New York, 1991.

[22] Miao, G. J., and M. A. Clements, *Digital Signal Processing and Statistical Classification*, Artech House, Norwood, Massachusetts, 2002.

[23] Mestdagh, D. J. G., "Calculation of ADC Resolution for DMT Modulation," *Electronics Letters*, Vol. 31, No. 16, p. 1315, August 1995.

[24] Mestdagh, D. J. G., P. M. P. Spruyt, and B. Biran, "Effect of Amplitude Clipping in DMT-ADSL Transceivers," *Electronics Letters*, Vol. 29, No. 15, pp. 1354–1355, 1993.

[25] Editor Group, *Mathematical Handbook*, Advanced Edcuation Press, Bejing, China, 1977.

[26] Proakis, J. G., *Digital Communications*, 2nd ed., McGraw-Hill Book Company, New York, 1989.

[27] Pollet, T., et al., "Equalization for DMT-Based Broadband Modems," *IEEE Communications Magazine*, pp. 106–113, May 2000.

[28] Acker, K. V., et al., "Per Tone Equalization for DMT-Based Systems," *IEEE Trans. on Communications*, Vol. 49, No. 1, pp. 109–119, January 2001.

[29] Vetterli, M., "Perfect Transmultiplexers," *Proc. of IEEE Int. Conf. on Acoustic, Speech, and Signal Processing*, pp. 2567–2570, April 1986.

[30] Akansu, A. N., et al., "Orthogonal Transmultiplexers in Communications: A Review," *IEEE Trans. on Signal Processing*, Vol. 46, No. 4, pp. 979–995, April 1998.

[31] Vaidyanathan, P. P., "Filter Banks in Digital Communications," *IEEE Circuits and Systems Magazine*, Vol. 1, No. 2, pp. 4–24, 2001.

[32] Chen, T., L. Qiu, and E. Bai, "General Multirate Building Structures with Application to Nonuniform Filter Banks," *IEEE Trans. Circuits and Systems II: Analog and Digital Signal Processing*, Vol. 45, No. 8, pp. 948–958, August 1998.

[33] Vaidyanathan, P. P., and B. Vrcelj, "Filter Banks for Cyclic-Prefixing the Nonuniform DMT System," *Proceedings 36th Asilomar Conference on Signal, Systems and Computers*, Pacific Grove, California, pp. 1–5, November 2002.

10

Discrete-Time Synchronization

10.1 Introduction

In the previous chapters, the analysis of theory and algorithm development presumed that a transmitter (or modulator) and receiver (or demodulator) in the wireless and wireline communication systems are synchronized already. In other words, both the transmitter and the receiver know the exact symbol rate, phase, and timing, and where appropriate, both the transmitter and the receiver may also know the exact carrier frequency and phase. However, in practice, the common knowledge of the exact transmission frequency and/or phase, the same time, and the carrier clocks are not known unless some information is provided so that the receiver can synchronize with the transmitter. Coherent demodulation requires the use of a properly phased carrier at the receiver. The receiver must estimate the carrier phase from the incoming signal. Furthermore, symbol timing must be derived from the received signal for both the coherent and noncoherent demodulation in order to synchronously sample the output of the matched filters at the proper time instant. Therefore, the recovery of the transmitted data must include a method for synchronizing the received data with the transmitted data. This typically involves a phase detector, a loop filter, a *voltage-controlled oscillator* (VCO), and a feedback controlled sampler in which the frequency and phase are slowly adjusted in response to synchronization errors in received samples, thereby setting estimate values close to their correct values. Such a device for synchronization is often referred to as a *phase locked loop* (PLL) and is the essential function for phase and frequency recovery in many communication applications.

Accurate synchronization is an important consideration in both wireless and wireline communication system performance, including digital data storage systems. This is because the performance of the receiver is critically dependent on the quality of the synchronization via the PLL. Improper synchronization can have an adverse impact on the quality of the sample values that are then used for either symbol-by-symbol or sequence detection in digital communication receivers. Thus, using robust signal processing techniques is necessary to provide an optimum sampling instant for every received signal sample.

In general practice, the PLL will have some finite delay so the regenerated local phase will have to predict the incoming phase and then assess how well that prediction did in quantitative form in terms of phase error. The more quickly the PLL tracks phase deviations in phase, the more susceptible it will be to random noise and other imperfections. Thus, the communication system designer must trade these two competing effects appropriately when designing a synchronization system. Meanwhile, the design of the transmitted signals can facilitate or complicate such trade-off analysis for the receiver system.

In digital communications systems, using completely analog continuous-time synchronization is uncommon. Most synchronization approaches used are either hybrid analog and digital or mixed continuous- and discrete-time. The discrete-time synchronization has become important for the communications system designer to implement because of recent advances in VLSI and digital signal processing (DSP) technologies, especially in software defined radio (SDR). Therefore, in this chapter, we consider the theory of signal processing techniques for symbol and carrier synchronization, which derive the symbol timing and carrier phase from the information-bearing signal, and focus on the implementation based on the all discrete-time synchronization approaches.

This chapter is organized as follows. In this section, a short background and the importance of symbol and carrier recovery for wireless and wireline communications are presented. In Section 10.2, we introduce discrete-time PLL functions and characteristics that are an integral part of most discrete-time synchronizations for symbol timing and carrier recovery. This section is focused on a discrete-time loop filter, a phase detector, and a *numerically controlled oscillator* (NCO). In Section 10.3, we describe timing recovery methods, including early-late gate synchronizers, bandedge timing, decision-directed timing, multirate timing, polyphase filter bank timing recovery, and multicarrier modulation timing recoveries. Subsequently, the fundamentals of discrete-time carrier recovery are given in Section 10.4, with emphases on carrier phase error, open-loop carrier recovery, carrier recovery for multiple phase signals, and decision-directed carrier recovery. Finally, a brief summary of this chapter is provided in Section 10.5.

10.2 Discrete-Time Phase Locked Loop

The PLL is an integral part of most synchronization schemes for considering the problems of carrier and symbol synchronization and timing recovery. The basic theory of PLL schemes has been extensively studied by Proakis [1], Lee and Messerschmitt [2] Waggener [3], Razavi [4], and Starr et al. [5], especially in analog and mixed continuous- and discrete-time domains. Recently advanced developments in VLSI and DSP technologies allow implementation of all discrete-time PLL schemes for digital communications systems to become possible eventually. Thus, in this section, we focus on reviewing the fundamental concepts of the discrete-time PLL as shown in Figure 10.1, thereby forming a basis

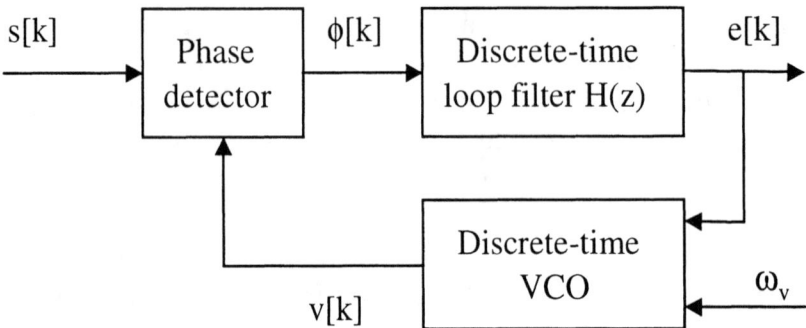

Figure 10.1 A block diagram of the discrete-time phase locked loop.

for discrete-time synchronization.

Assume that the received discrete-time input signal $s[k]$ is given by

$$s[k] = A_s \cos(\omega_v kT + \theta_k), \qquad (10.1)$$

where A_s is called the constant amplitude response, ω_v is called the constant common phase reference, T is the symbol interval ($1/T$ is the symbol rate), and θ_k is the phase of the received discrete-time input signal.

The discrete-time VCO function expression can be described as

$$\begin{aligned} \hat{\theta}_{k+1} &= \hat{\theta}_k + C_{vco} e[k] \\ &= \hat{\theta}_k + C_{vco} h[n] * \phi_k, \end{aligned} \qquad (10.2)$$

where C_{vco} is the constant, $h[n]$ is the discrete-time loop filter, and ϕ_k is the phase error, $\phi_k = \theta_k - \hat{\theta}_k$, which represents the local phase at sampling time instant k. The discrete-time VCO also produces the estimated discrete-time output signal $v[k]$, given by

$$v[k] = A_{vco} \cos(\omega_v kT + \hat{\theta}_k), \qquad (10.3)$$

where A_{vco} is the constant value.

Taking the z-transform (see Appendix A) of both sides of (10.2) obtains

$$z^{-1}\hat{\Theta}(z) = \hat{\Theta}(z) + C_{vco}H(z)\Phi(z). \tag{10.4}$$

where the z-transform of the phase error, $\Phi(z)$, is

$$\Phi(z) = \Theta(z) - \hat{\Theta}(z). \tag{10.5}$$

Substituting (10.5) into (10.4), we get

$$z^{-1}\hat{\Theta}(z) = [1 - C_{vco}H(z)]\hat{\Theta}(z) + C_{vco}H(z)\Theta(z). \tag{10.6}$$

Thus, we obtain a phase transfer function from the input phase to the estimated output phase based on (10.6) as

$$\frac{\hat{\Theta}(z)}{\Theta(z)} = \frac{C_{vco}H(z)z}{1 - [1 - C_{vco}H(z)]z}. \tag{10.7}$$

By evaluating (10.7) at $z = 1$, we notice that the discrete-time PLL has unity gain to DC phase input.

According to (10.4) and (10.5), we then obtain the transfer function between the phase error $\Phi(z)$ and the input phase $\Theta(z)$ as

$$\frac{\Phi(z)}{\Theta(z)} = \frac{1 - z}{1 - z + C_{vco}H(z)z}, \tag{10.8}$$

where $H(z)$ is the discrete-time loop filter that determines the Nth-order discrete-time PLL.

10.2.1 Discrete-Time Loop Filter

In this section, we discuss in detail the principles of the first- and second-order discrete-time PLLs, which have been used extensively in common digital communications systems.

First-Order Discrete-Time PLL

For the first-order discrete-time PLL, the discrete-time loop filter is a constant value, K. In other words, the discrete-time loop filter is frequency independent, that is,

$$C_{vco}H(z) = K. \tag{10.9}$$

Substituting (10.9) into (10.7) yields

$$\frac{\hat{\Theta}(z)}{\Theta(z)} = \frac{Kz}{1 - [1 - K]z}. \tag{10.10}$$

In order to have a stability of (10.10), we need to have $|1 - K| < 1$. This is the same as

$$0 \leq K < 2. \tag{10.11}$$

As can be seen, when K approaches 2, the bandwidth of the overall loop filter from $\Theta(z)$ to $\hat{\Theta}(z)$ is wide and can distort the estimated phase if noise is on the input sinusoid.

Given certain inputs, a steady-state operating point of discrete-time PLL is often useful to know precisely. The steady-state phase error ξ_{ϕ_k} is defined to be

$$\xi_{\phi_k} = \lim_{k \to \infty} \phi_k. \tag{10.12}$$

If the discrete-time PLL does not achieve phase lock, then $\xi_{\phi_k} \neq 0$. On the other hand, if $\xi_{\phi_k} = 0$ for $k < 0$, then we can determine ξ_{ϕ_k} by using the *final value theorem* for z-transform

$$\xi_{\phi_k} = \lim_{z \to 1}(1 - z)\Phi(z). \tag{10.13}$$

If we assume that the input phase is

$$\theta_k = \theta_0 u_k, \quad k \geq 0, \tag{10.14}$$

where θ_0 is a constant and u_k is the unit step, then (10.14) has a z-transform function,

$$\Theta(z) = \frac{\theta_0 z}{z - 1}. \tag{10.15}$$

Substituting (10.15) and (10.9) into (10.8), the transform form of the phase error can be obtained as follows

$$\Phi(z) = \frac{(1-z)\Theta(z)}{1-z+C_{vco}H(z)z}$$
$$= \left(\frac{1-z}{1-z+Kz}\right)\left(\frac{\theta_0 z}{z-1}\right)$$
$$= \frac{\theta_0 z}{(1-K)z-1}. \qquad (10.16)$$

Hence, the steady-state phase error ξ_{ϕ_k} can be obtained by using (10.13)

$$\xi_{\phi_k} = \lim_{z\to 1}(1-z)\Phi(z)$$
$$= \lim_{z\to 1}\left[\frac{\theta_0(1-z)z}{(1-K)z-1}\right]$$
$$= 0. \qquad (10.17)$$

Thus, (10.17) indicates that the first-order phase lock loop can track and decay to zero for any phase difference between a constant θ_0 and an estimated $\hat{\theta}_k$.

The first-order phase lock loop usually suffers a constant phase offset. To illustrate this effect, we assume that the input phase has frequency offset introduced at time $k = 0$,

$$\theta_k = \omega_0 T k u_k, \quad k \geq 0, \qquad (10.18)$$

where ω_0 is the frequency offset and $1/T$ is the sampling rate. The z-transform of (10.18) is given by

$$\Theta(z) = \frac{\omega_0 T z}{(z-1)^2}. \qquad (10.19)$$

Then, substituting (10.19) and (10.9) into (10.8), the transform form of the phase error $\Phi(z)$ can be obtained in the following,

$$\Phi(z) = \frac{(1-z)\Theta(z)}{1-z+C_{vco}H(z)z}$$

$$= \left(\frac{1-z}{1-z+Kz}\right)\left[\frac{\omega_0 Tz}{(z-1)^2}\right]$$
$$= \frac{\omega_0 Tz}{[1-(1-K)z](1-z)}. \qquad (10.20)$$

Therefore, the steady-state phase error ξ_{ϕ_k} for the frequency offset ω_0 case can also be computed by the final value theorem

$$\begin{aligned}\xi_{\phi_k} &= \lim_{z\to 1}(1-z)\Phi(z) \\ &= \lim_{z\to 1}\frac{(1-z)\omega_0 Tz}{[1-(1-K)z](1-z)} \\ &= \frac{\omega_0 T}{K}. \end{aligned} \qquad (10.21)$$

As can be seen, (10.21) shows that the first-order phase lock loop cannot track a nonzero frequency offset $\omega_0 T$ so that the phase error will not decay to zero. If K satisfies the condition in (10.11), for very small frequency offsets, the first-order phase lock loop will incur only a very small distortion of the phase error. After the first-order phase lock loop has converged, the magnitude of the frequency offset $|\omega_0|$ must be less than $K\pi/T$ to focus the phase within the linear phase of the modulo-2π phase detector, thereby avoiding cycle slips. Further note that either increasing K at fixed sampling rate $1/T$ or increasing the sampling rate $1/T$ will increase the bandwidth of the first-order phase lock loop. As a result, the phase lock loop filter will filter less noise on the incoming phase, resulting in a lower quality estimate of the phase.

Second-Order Discrete-Time PLL

In a second-order discrete-time PLL, the phase lock loop filter has a z-transform form,

$$C_{vco}H(z) = \alpha + \frac{\beta}{1-z^{-1}}, \qquad (10.22)$$

where α and β are the proportional and integral step sizes, respectively. In other words, the offset estimate is found by

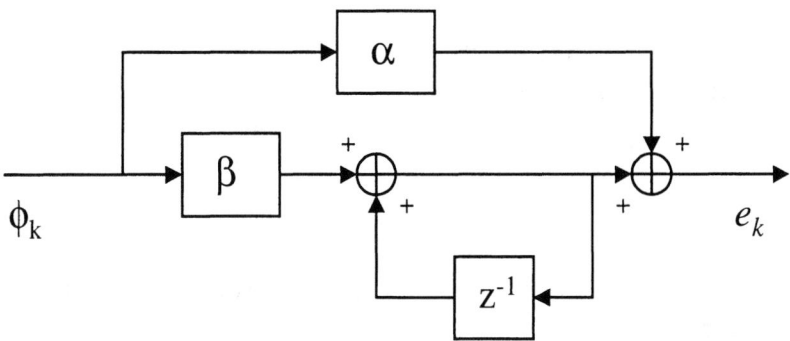

Figure 10.2 A block diagram of the second-order discrete-time phase locked loop structure.

accumulating the output of the phase lock loop filter with transfer function in (10.22) and input estimated phase error θ_k. A block diagram of the second-order discrete-time PLL structure is shown in Figure 10.2.

Substituting (10.22) into (10.7) yields

$$\frac{\hat{\Theta}(z)}{\Theta(z)} = \frac{(\alpha + \beta)z^2 - \alpha z}{(\alpha + \beta - 1)z^2 + (2 - \alpha)z - 1}. \qquad (10.23)$$

Consider the case if $\beta = 0$; thus, (10.23) reduces to the first-order discrete-time PLL in (10.10). Equation (10.23) has poles

$$z_{1,2} = \frac{\alpha - 2 \pm \sqrt{(2-\alpha)^2 + 4(\alpha + \beta - 1)}}{2(\alpha + \beta - 1)}. \qquad (10.24)$$

For stability, α and β must satisfy the following conditions: $0 \leq \alpha < 2$ and $\beta > 1$.

The second-order phase lock loop can track for any frequency offset between θ_k and the estimated $\hat{\theta}_k$. To illustrate this effect, we again assume that the input phase has a frequency offset introduced at time $k = 0$,

$$\theta_k = \omega_0 T k u_k, \quad k \geq 0, \qquad (10.25)$$

where ω_0 is the frequency offset and $1/T$ is the sampling rate. The z-transform of (10.25) is

$$\Theta(z) = \frac{\omega_0 T z}{(z-1)^2}. \qquad (10.26)$$

Thus, substituting (10.26) and (10.22) into (10.8), the transform form of the phase error $\Phi(z)$ for the second-order phase lock loop is obtained by

$$\begin{aligned}
\Phi(z) &= \frac{(1-z)\Theta(z)}{1-z+C_{vco}H(z)z} \\
&= \left(\frac{1-z}{1-z+[\alpha+\beta/(1-z^{-1})]z}\right)\left[\frac{\omega_0 T z}{(z-1)^2}\right] \\
&= \frac{\omega_0 T z}{1-(2-\alpha)z-(\alpha+\beta-1)z^2}. \qquad (10.27)
\end{aligned}$$

Therefore, the steady-state phase error ξ_{ϕ_k} of the second-order discrete-time PLL for the frequency offset ω_0 can be calculated by the final value theorem

$$\begin{aligned}
\xi_{\phi_k} &= \lim_{z \to 1}(1-z)\Phi(z) \\
&= \lim_{z \to 1}\left[\frac{(1-z)\omega_0 T z}{1-(2-\alpha)z-(\alpha+\beta-1)z^2}\right] \\
&= 0. \qquad (10.28)
\end{aligned}$$

As can be seen, as long as the parameters of α and β are selected within the stability ranges, the second-order phase lock loop should be able to track any constant phase or frequency offset. However, choosing α and β to reject noise, based on making the second-order phase lock loop too sharp or narrow in bandwidth, may make its initial convergence to steady-state slow. Therefore, a trade-off analysis is needed to evaluate the second-order discrete-time PLL in any particular application.

10.2.2 Phase Detector

A wide variety of phase detectors, ranging from simple to complicated, exist. Much design effort is needed in the design of phase detectors for carrier and time recovery. In this section, we discuss some variations of phase detectors based on discrete-time domain operation.

Ideal and Modulo-2π Phase Detectors

A device that can calculate exactly the difference between the input phase θ_k and the estimated phase $\hat{\theta}_k$ at the time index k is referred to as an *ideal discrete-time phase detector*, $\phi_k = \theta_k - \hat{\theta}_k$.

Given the discrete-time input signal $s[k]$ in (10.1) and the estimated discrete-time output signal $v[k]$ in (10.3) from the discrete-time VCO, a seemingly straightforward method to compute the phase error θ_k would then be to compute θ_k and $\hat{\theta}_k$, according to

$$\theta_k = \pm \arccos\{s[k]\} - \omega_v kT, \qquad (10.29)$$

and

$$\hat{\theta}_k = \pm \arccos\{v[k]\} - \omega_v kT. \qquad (10.30)$$

Hence, the phase error is $\phi_k = \theta_k - \hat{\theta}_k$. However, for a reasonable implementation of the arccos function, angles can only be produced between 0 and π in such a way that ϕ_k would then always lie between $-\pi$ and π. Any difference of magnitude greater than π would be therefore effectively computed through a modulo operation of $(-\pi, \pi)$. As a result, the arccos function can be implemented by using a look-up table. We refer to the arccos function look-up table implementation of the phase detector as a *modulo-2π phase detector*.

The comparison of characteristics of the ideal phase detector and the modulo-2π phase detector is shown in Figure 10.3. The large difference will not be shown in the phase error even if the phase difference does exceed $|\pi|$ in magnitude. This phenomenon is known as a *cycle slip* in which the phase detection missed or added an entire period of the input sinusoid. In most applications,

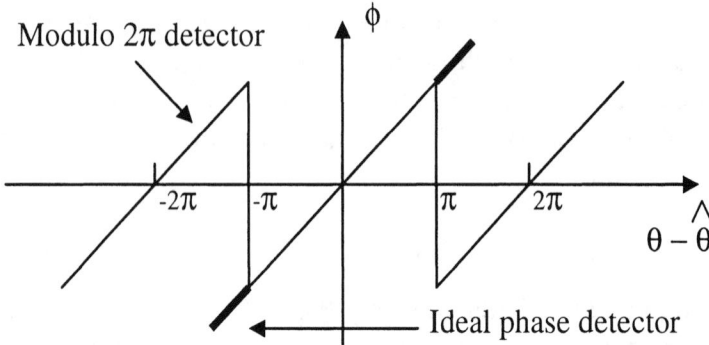

Figure 10.3 Comparison characteristics of ideal and modulo-2π phase detectors.

this is an unwelcome phenomenon. Thus, one tries to ensure that the phase error ϕ_k does not exceed π after the modulo-2π phase detector has converged. Therefore, the phase error ϕ_k should be kept as close to zero as possible so that the modulo-2π phase detector can be operated correctly under the necessary condition of small phase error.

Demodulation Phase Detector

A block diagram of a *demodulation phase detector* is shown in Figure 10.4. The discrete-time input signal is given by

$$s[n] = A_s \cos(\omega_v kT + \theta_k), \qquad (10.31)$$

and the estimated discrete-time output signal from the discrete-time VCO is given by

$$\hat{s}[n] = A_{vco} \sin(\omega_v kT + \hat{\theta}_k). \qquad (10.32)$$

The digital lowpass filter $h[n]$ is cascaded with the phase error processing in the phase lock loop. The phase error ϕ_k is obtained

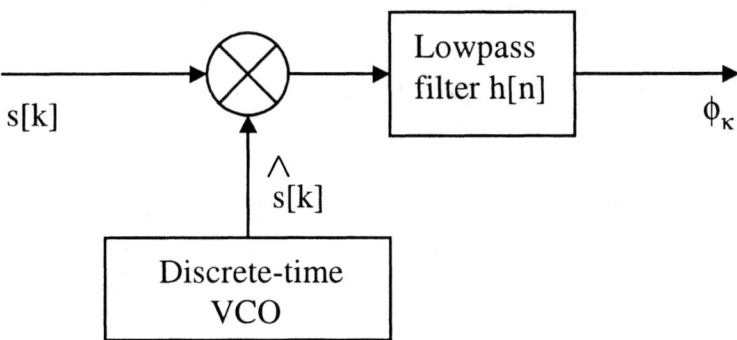

Figure 10.4 A block diagram of a demodulation phase detector.

by

$$\phi_k = h[n] * \left[A_{vco} \sin(\omega_v kT + \hat{\theta}_k) A_s \cos(\omega_v kT + \theta_k) \right], \quad (10.33)$$

where "*" is a convolution operation.

Since $\sin(x)\cos(y) = \frac{1}{2}[\sin(x+y) + \sin(x-y)]$, (10.33) can be rewritten as

$$\phi_k = \frac{A_{vco} A_s}{2} h[n] * \left[\sin(2\omega_v kT + \theta_k + \hat{\theta}_k) + \sin(\theta_k - \hat{\theta}_k) \right]. \quad (10.34)$$

Assuming the first term is removed by the digital lowpass filter $h[n]$, the phase error ϕ_k is obtained

$$\phi_k = \frac{A_{vco} A_s}{2} \sin(\theta_k - \hat{\theta}_k). \quad (10.35)$$

If ϕ_k is small, the generation of the phase error does not require the arcsin function. Since $\sin(x) \approx x$ when x is very small, when θ_k is close to $\hat{\theta}_k$, (10.35) is approximately equal to

$$\phi_k \approx \frac{A_{vco} A_s}{2} (\theta_k - \hat{\theta}_k), \quad (10.36)$$

which is an approximately linear characteristic. Now, ϕ_k is a much more reasonable estimate of the phase error.

10.2.3 Discrete-Time VCO

An ideal discrete-time VCO has a discrete-time output sinusoid with an estimated phase $\hat{\theta}_k$ that is determined by using a discrete-time input error signal $e[k]$ approximately according to

$$\hat{\theta}_{k+1} = \hat{\theta}_k + C_{vco}e[k], \qquad (10.37)$$

where C_{vco} is a constant. In other words, the discrete-time VCO regenerates the local phase information from the processed phase error ϕ_k in order to match the incoming phase θ_k. Thus, the phase reconstruction attempts to force

$$\phi_k = \theta_k - \hat{\theta}_k \Longrightarrow 0, \qquad (10.38)$$

by producing a local phase $\hat{\theta}_k$ so that the estimated discrete-time output signal $v[n]$ equals the discrete-time input signal $s[n]$ as shown in Figure 10.1.

According to (10.37), since the $e[n]$ is the discrete-time signal, the discrete-time VCO can be implemented by using a look-up table and adder whose output is used to generate the discrete-time sinusoidal signal $v[n]$ in (10.3). Such an implementation of the discrete-time VCO is often referred to as an NCO in the literature.

When the discrete-time VCO output frequency is not equal to the input frequency, but is related by a fixed rational multiple, it is possible to design a discrete-time PLL to maintain phase lock. Thus, when the discrete-time PLL is phase locked as

$$\frac{F_{in}}{M} = \frac{F_{out}}{N}, \qquad (10.39)$$

then

$$F_{out} = \frac{N}{M} F_{in}, \qquad (10.40)$$

where F_{in} is the input frequency and F_{out} is the output frequency and M and N are integers as shown in Figure 10.5. Thus, the frequency synthesizer produces an output signal with frequency F_{out} equal to $\frac{N}{M}$ times the input frequency F_{in}. The discrete-time PLL is called a *frequency synthesizer*.

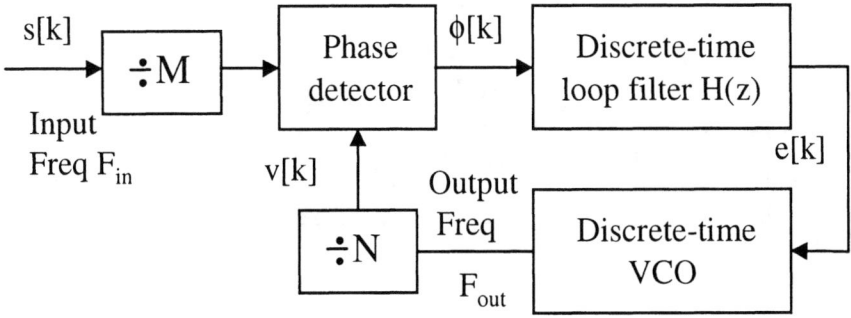

Figure 10.5 A discrete-time PLL-based frequency synthesizer.

Example 10.1

Assuming that an input clock is at 49.408 MHz, how can we select M and N such that an output clock is at 65.536 MHz for a discrete-time PLL-based frequency synthesizer? Since F_{in} = 49,408 kHz,

$$F_{out} = 49{,}408 \frac{N}{M}. \qquad (10.41)$$

In order to generate F_{out} = 65.536 MHz, we could simply set M = 49,408 and N = 65,536, and the inputs to the phase detector would be on the order of 1 kHz. Thus, the discrete-time loop filter $H(z)$ has to have sufficiently narrow bandwidth to remove this fundamental and its harmonics. In addition, the narrow bandwidth means that the discrete-time PLL-based frequency synthesizer responds slowly to changing conditions. On the other hand, if we note that 49,408 = 193×256 and 65,536 = 256 × 256, then we can use M = 193 and N = 256. This results in inputs to the phase detector on the order of 256 kHz, so the bandwidth of the discrete-time loop filter $H(z)$ can be made larger. Thus, the design significantly improves the discrete-time loop filter $H(z)$ specifications.

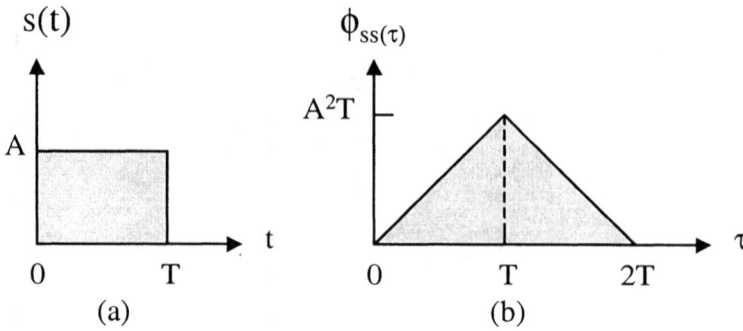

Figure 10.6 (a) Rectangular pulse and (b) its autocorrelation function.

10.3 Timing Recovery

Symbol timing or synchronization is concerned with the problem of determining a clock for periodically sampling the output of matched filters for the purpose of recovering the transmitted information. One method for the solution of the problem is to transmit a clock signal along with the information signal in multiplexed form to the digital communications receiver. Another method is to derive the clock signal from the information signal at the digital communications receiver. In the digital communications receiver, an analog waveform must be sampled, usually by using an analog-to-digital (A/D) converter. Sampling at the right times is critical to achieving good overall performance for the digital communications receiver. Such a process of synchronizing the sampler (or A/D converter) with the pulses of the received analog waveform, along with the PLL, is known as *timing recovery*.

10.3.1 Early-Late Gate Synchronizer

Suppose that an analog signal pulse $s(t)$, $0 \leq t \leq T$, is a rectangular pulse as shown in Figure 10.6(a). The autocorrelation function of the rectangular pulse by passing the rectangular pulse through its matched filter is a triangular pulse, as shown in Figure 10.6(b). In other words, when the analog signal pulse $s(t)$ is passed through a filter matched to it, the output of the filter has a maximum value of $t = T$. Thus, the proper time is to sample the output of the matched filter for a maximum value output at $t = T$, which is the peak of the autocorrelation function. However, in practice, if we sample the output of the matched filter early at $t = T - \delta$ or late at $t = T + \delta$, the sampled values will be smaller than the peak value at $t = T$. Note that the autocorrelation function is even symmetric relative to $t = T$. Hence, the values of the autocorrelation function at $t = T - \delta$ and $t = T + \delta$ are equal in magnitude, that is,

$$|\phi_{ss(T-\delta)}| = |\phi_{ss(T+\delta)}|. \tag{10.42}$$

This is to say that the difference in the magnitude of these sampled values is zero. Therefore, in this situation, the proper time to sample is at the midpoint between $t = T - \delta$ and $t = T + \delta$, for example, at $t = T$. This simple condition is the fundamental theory for a commonly used synchronizer referred to as an *early-late gate synchronizer*.

A block diagram of the early-late gate synchronizer, as shown in Figure 10.7, is realized by using autocorrelators, A/D converters, square-law devices ($|s_1[n]|^2$ and $|s_2[n]|^2$), and discrete-time PLL. The received signal $s(t)$ is multiplied by using a carrier signal,

$$c(t) = \cos(2\pi f_c t + \hat{\phi}), \tag{10.43}$$

to translate in frequency to lowpass and passed through two autocorrelators. The discrete-time error signal $e[n]$ between the outputs of two autocorrelators should be zero because the autocorrelation function of any signal pulse is even symmetric. Thus, if the sampling timing is off at an optimal position, the discrete-time error signal $e[n]$ will not be zero,

$$e[n] = r_1[n] - r_2[n] \neq 0. \tag{10.44}$$

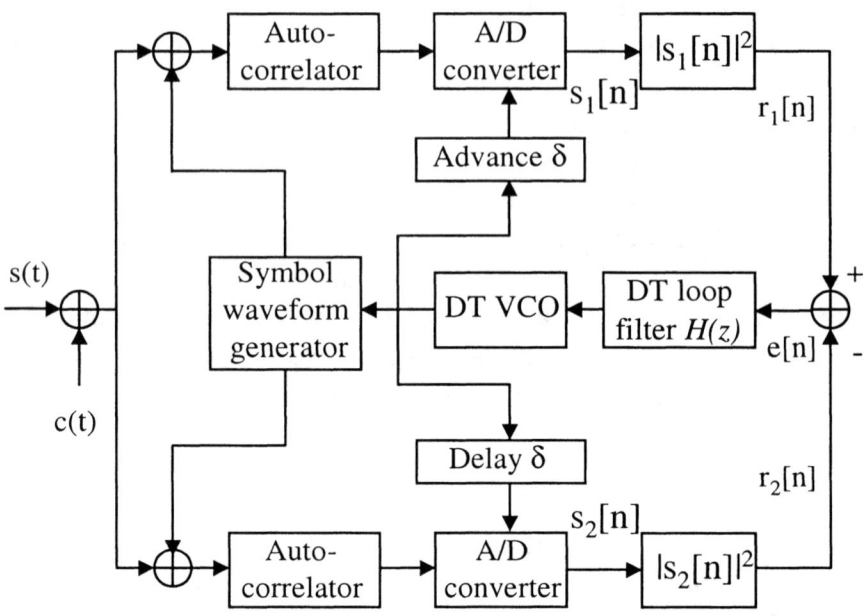

Figure 10.7 A block diagram of an early-late gate synchronizer, where "DT" means discrete-time.

In order to smooth the error signal $e[n]$, it is passed through a discrete-time lowpass loop filter $H(z)$, whose output is the control voltage for a discrete-time VCO. The output of the discrete-time VCO is used to control a symbol waveform generator that feeds to the autocorrelators. In addition, the clock from the discrete-time VCO is advanced or delayed by δ and these clock signals then are used to control the A/D converters to sample the outputs of the autocorrelators. Note that there is no need to use the symbol waveform generator if the received signal $s(t)$ is a rectangular pulse. In this case, the autocorrelators become integrate-and-dump filters. The early-late gate synchronizer is especially useful for phase-shift keying (PSK) and pulse amplitude modulation (PAM) signals.

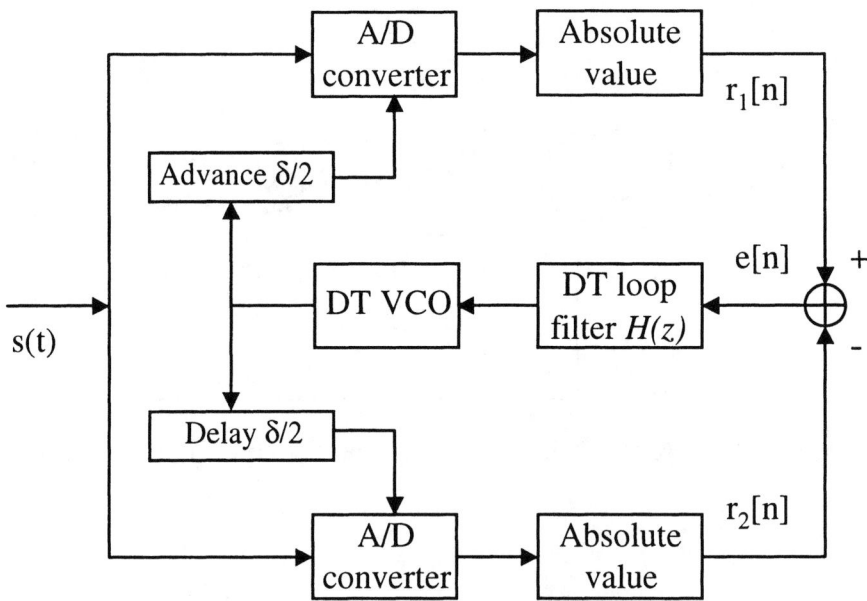

Figure 10.8 A block diagram of an early-late gate synchronizer, an alternative form.

A similar realization of the early-late gate synchronizer that is somewhat easier to implement in the discrete-time domain is shown in Figure 10.8. In this case, the received signal $s(t)$ is sampled two extra times by using A/D converters, once prior to the sampling instant by an amount of $\frac{\delta}{2}$ and once after the sampling instant by the same amount of $\frac{\delta}{2}$. The error signal $e[n]$, $e[n] = r_1[n] - r_2[n]$, is passed through the discrete-time loop filter whose output is used to control the discrete-time VCO. The clock from the discrete-time VCO is advanced or delayed by $\frac{\delta}{2}$, and these clock signals are then used to control the A/D converters of sampling the received signal $s(t)$. The sampling instant for the A/D converters is adjusted until the two extra samples are equal.

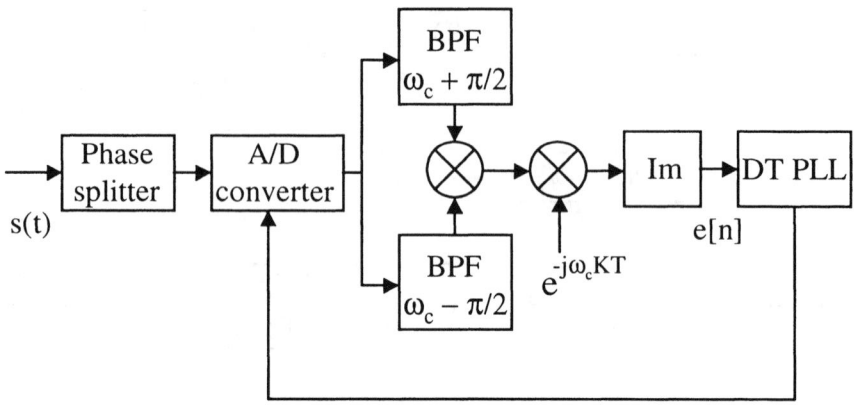

Figure 10.9 A block diagram of the bandedge timing recovery.

10.3.2 Bandedge Timing Recovery

For quadrature modulated data transmission, one widely used method that does not depend on knowing a correct carrier frequency is the so-called *bandedge timing recovery*, as shown in Figure 10.9. In this case, the two bandpass filters are narrowband and identical. If the correct timing phase is selected, an approximate maximum of the energy should be within the two bandedges. The outputs of the two narrowband bandpass filters are then multiplied to produce the maximum value. This means that the output of the multiplier should be a real value. Thus, at this timing phase, the discrete-time PLL uses the error signal $e[n]$ from the samples of the imaginary part of the multiplier output for the phase-lock loop, which is used to adjust the timing phase for the A/D converter. Although the carrier frequency is assumed to be known in the design of the narrowband bandpass filters, this information has little effect as long as the carrier is close to the correct value. This is typically true in most quadrature modulated data transmission methods.

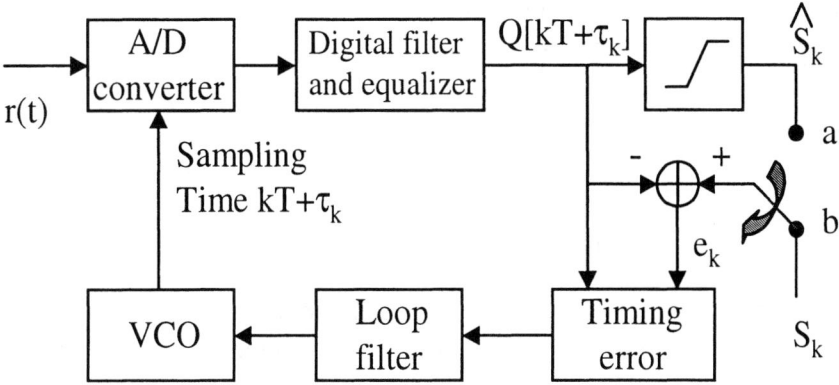

Figure 10.10 A block diagram of the decision-directed timing recovery.

10.3.3 Decision-Directed Timing Recovery

The basic idea of decision-directed timing recovery is to minimize the mean square error (MSE) over the sampling time phase, between a digital matched filter output and a decision output. Decision-directed timing recovery is also referred to as the minimum mean square error (MMSE) timing recovery. Figure 10.10 shows a block diagram of the decision-directed timing recovery. In this figure, the received signal $r(t)$ is assumed to be the baseband signal and is sampled at times $kT + \tau_k$, where T is the symbol interval and thus τ_k represents the timing error in the kth sample. After the digital filter and equalizer processing, the output of the kth sample is $Q[kT + \tau_k]$, which is dependent on the timing phase. Ideally, τ_k is a constant that corresponds to the best sampling phase. However, in practice, τ_k has a time variation, which is referred to as the *timing jitter*. In this case, the MMSE timing recovery adjusts τ_k to minimize MSE between the equalizer output $Q[kT + \tau_k]$ and the decision output $\hat{S}_k$ at the point "a" or the correct symbols S_k (such as a training sequence during a training mode) at the point "b." That is, to minimize the expected squared error between the input to the decision $Q[kT + \tau_k]$ and the output of the decision $\hat{S}_k$,

$$\begin{aligned} \xi(\tau_k) &= E\{|E_k(\tau_k)|^2\} \\ &= E\{|Q[kT + \tau_k] - \hat{S}_k|^2\}, \end{aligned} \quad (10.45)$$

where $Q[kT + \tau_k]$ is the equalizer output at sampling time k corresponding to sampling phase τ_k. In order to minimize the expected square error, the update uses a stochastic-gradient estimate of τ in the opposite direction of the unaveraged derivative of $\xi(\tau_k)$ with respect to τ_k. The derivative is as follows:

$$\begin{aligned} \frac{d\xi(\tau_k)}{d\tau_k} &= \frac{dE\{|E_k(\tau_k)|^2\}}{d\tau_k} \\ &= E\left[\frac{d|E_k(\tau_k)|^2}{d\tau_k}\right] \\ &= 2E\left\{\operatorname{Re}\left[E_k^*(\tau_k)\frac{dE_k(\tau_k)}{d\tau_k}\right]\right\}. \end{aligned} \quad (10.46)$$

Note that $\hat{S}_k$ does not depend on τ_k in (10.45). Thus, we obtain

$$\frac{dE_k(\tau_k)}{d\tau_k} = \frac{dQ[kT + \tau_k]}{d\tau_k}. \quad (10.47)$$

Hence, we use the stochastic gradient algorithm by adjusting the timing phase in the direction opposite the gradient,

$$\begin{aligned} \tau_{k+1} &= \tau_k - \beta\left\{\operatorname{Re}\left[E_k^*(\tau_k)\frac{dQ[kT + \tau_k]}{d\tau_k}\right]\right\} \\ &= \tau_k - \beta\left\{\operatorname{Re}\left[(Q[kT + \tau_k] - \hat{S}_k)^*\frac{dQ[kT + \tau_k]}{d\tau_k}\right]\right\}, \end{aligned} \quad (10.48)$$

where β is a step size that is used to empirically ensure stability, minimize timing jitter, and ensure adequate tracking ability. As a

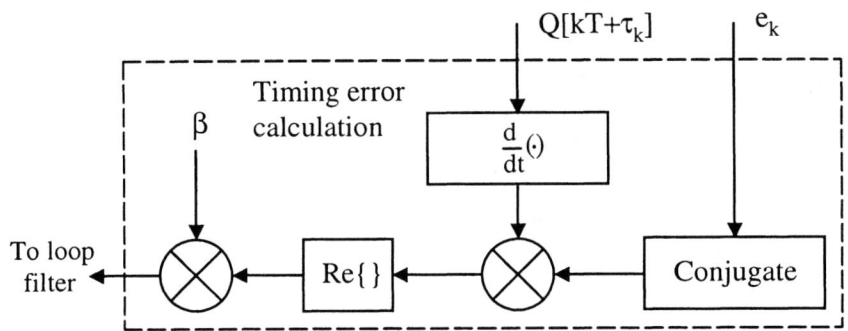

Figure 10.11 The detailed structure of the timing error block in the decision-directed timing recovery design.

result, the block diagram of the timing error in Figure 10.10 can be redrawn in detail as shown in Figure 10.11.

Further note that the stochastic gradient algorithm in (10.48) does not guarantee to converge to the optimal timing phase. This is because $E_k(\tau_k)$ is not a linear function of τ_k. This indicates that $|E_k(\tau_k)|^2$ is not a quadratic function of τ_k.

Using an approximate MMSE technique, we can express the discrete-time derivative as follows:

$$\frac{dQ[kT+\tau_k]}{d\tau_k} = Q[kT+\tau_k] * h[k], \qquad (10.49)$$

where "$*$" is a convolutional operation, and $h[n]$ is a discrete-time FIR filter, with impulse response $h[k] \approx (\delta_{k+1} - \delta_{k-1})/T$. Thus, we obtain an approximation

$$\frac{dQ[kT+\tau_k]}{d\tau_k} \approx \frac{Q[(k+1)T+\tau_{k+1}] - Q[(k-1)T+\tau_{k-1}]}{T}$$

$$= h_{\tau_k}[\tau_k]. \qquad (10.50)$$

Substituting (10.50) into (10.48), adjusting the timing phase in the discrete-time domain is approximately given by

$$\tau_{k+1} = \tau_k - \beta z_k[\tau_k], \quad (10.51)$$

where

$$z_k[\tau_k] = \text{Re}\left\{(Q[kT + \tau_k] - \hat{S}_k)^* h_{\tau_k}[\tau_k]\right\}. \quad (10.52)$$

The discrete-time derivative can also be found by approximately taking each sample either one sample ahead or one sample behind the current estimate τ_k of the timing phase. By alternating the two phases, the difference between the error at even samples and the error at odd samples is an indication of the derivative of the error with respect to the timing phase. This leads to a popular *timing error detector* (TED) that generates the timing error signal $\hat{\epsilon}$ developed by Mueller and Müller [6],

$$\hat{\epsilon} = Q[kT + \tau_k]\hat{S}_{k-1} - Q[(k-1)T + \tau_{k-1}]\hat{S}_k. \quad (10.53)$$

Note that $\hat{S}_k$ in (10.53) is the output of the decision device as shown in Figure 10.10. The decision device that generates these decisions $\hat{S}_k$ will invariably introduce a processing delay. Therefore, the performance of the decision-directed PLL depends critically on two factors: (1) reliability of the decisions, and (2) processing delay of the decision device. The reliability of the decision device is so important because incorrect decisions will drive the timing estimates away from their optimal values. On the other hand, the importance of delay is also obvious since a processing delay in a feedback loop of the discrete-time loop filter in (10.22) can lead to instability. Increasing the delay moves the closed-loop poles closer to the unit circle so that the parameter α in (10.22) must be decreased to maintain stability. However, decreasing α makes the system less agile to time-varying timing offsets. Therefore, there is a fundamental trade-off between these two parameters.

10.3.4 Multirate Timing Recovery

A timing recovery that is operated based on multirate signal processing is called a *multirate timing recovery*. Figure 10.12

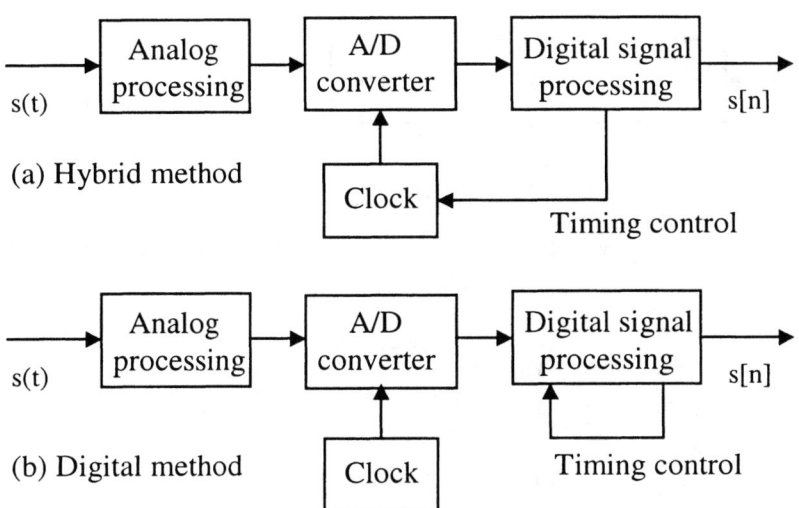

Figure 10.12 Multirate timing recovery: (a) hybrid and (b) digital methods.

shows a basic block diagram of multirate timing recovery. The implementation of communication systems using digital techniques introduces an oversampling of the analog signal, which is a topic of intense present activity. In some circumstances, the oversampling can be synchronized to the symbol rate of the incoming signal as shown in Figure 10.12(a). In this case, the timing phase can be recovered in a synchronous digital system in much the same way as is familiar from analog practice. In other circumstances, the oversampling cannot be synchronized to the incoming signal. Figure 10.12(b) shows an asynchronized structure of timing recovery with oversampling according to Gardner [7] and Erup et al. [8]. These examples of the asynchronized structure of timing recovery include digital signal processing of asynchronized frequency-multiplexed signals, asynchronized digital capture and subsequent postprocessing of a signal, or avoiding the problem of

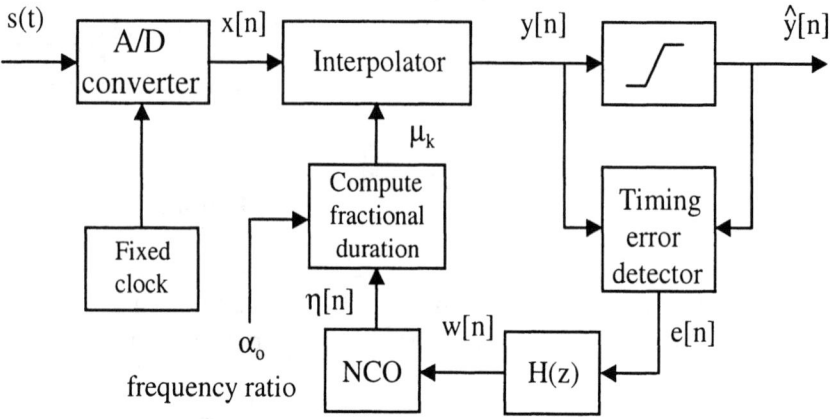

Figure 10.13 An interpolation-based multirate timing recovery.

detection delay processing. Instead of using the discrete-time VCO of the discrete-time PLL to control the sampling times, we could instead sample asynchronously and uniformly with a free-running clock rate at a rate fast enough to avoid aliasing. This may be the baud rate, a few percent higher than the baud rate, twice the baud rate, or oversampling the baud rate, depending on the amount of excess bandwidth and the severity of the worst-case frequency offset. In any case, for one reason or another, the oversampling clock must remain independent of the symbol timing in the *asynchronous configuration*.

Figure 10.13 shows an interpolation-based multirate timing recovery. Assume that $s(t)$ is a bandlimited analog signal, such as a symbol pulse with uniformly spaced at intervals T. The bandlimited analog signal $s(t)$ can be sampled at a sampling frequency rate $F_s = \frac{1}{T_s}$ without aliasing by using an A/D converter. The output of the A/D converter is $x[n] = s[nT_s]$, which is taken at uniform intervals T_s. Also assume that the ratio of $\frac{T}{T_s}$ is irrational because the symbol timing is derived from a source that is independent of the sampling clock. The discrete-time sampled signals $x[n]$ are then applied to the interpolator, which computes interpolations $y[n] = y[nT_i]$ at time T_i, where $T_i = \frac{T}{K}$ and K is a small integer.

The TED is characterized by using a decision-directed approach based on the number samples $y[n]$ per symbol and estimated samples $\hat{y}[n]$ after the slice as shown in Figure 10.13. There are different TED approaches as follows:

one-sample Mueller and Müller TED [6]:

$$e[n] = \hat{y}[n]y[n-1] - y[n]\hat{y}[n-1]. \qquad (10.54)$$

two-sample and two-point Gardner TED [9]:

$$e[n] = (y[n] - y[n-2])\hat{y}[n-1]. \qquad (10.55)$$

two-sample crossing-point Gardner TED [9]:

$$e[n] = (\hat{y}[n] - \hat{y}[n-2])y[n-1]. \qquad (10.56)$$

In order to adjust the timing for interpolation, a fractional interval μ_k must be calculated based on the output $\eta[n]$ of the NCO at the nth clock tick. The NCO is controlled by a discrete-time input signal $w[n]$. The NCO difference equation between $\eta[n]$ and $w[n]$ is given by [7]

$$\eta[n] = (\eta[n-1] - w[n-1]) \bmod\text{-}1. \qquad (10.57)$$

Equation (10.57) is a modulo-1 operation and denotes the NCO. The relationship between the output signal μ_k and the input signal $\eta[n]$ at the kth interpolation time is given by

$$\frac{\mu_k T_s}{\eta[n_k]} = \frac{(1-\mu_k)T_s}{1-\mu[n_k+1]}. \qquad (10.58)$$

Then, μ_k can be obtained by solving (10.58) along with (10.57)

$$\begin{aligned}\mu_k &= \frac{\eta[n_k]}{1-\eta[n_k+1]+\eta[n_k]} \\ &= \frac{\eta[n_k]}{w[n_k]}.\end{aligned} \qquad (10.59)$$

Thus, (10.59) indicates that an estimate of μ_k can be obtained by performing a division of $\eta[n_k]$ by $w[n_k]$, which are both available from the input and output of the NCO. Furthermore, to avoid division, note that a period of the NCO is

$$T_i = \frac{T_s}{w[n]}, \qquad (10.60)$$

which leads to a normalized frequency ratio

$$\frac{1}{w[n_k]} = \frac{T_i}{T_s} = \alpha_0. \qquad (10.61)$$

Therefore, the fractional interval in (10.59) can be approximately obtained by

$$\mu_k = \alpha_0 \eta[n_k], \qquad (10.62)$$

where α_0 is uniformly distributed. If α_0 is too large, then a first-order correction equation can be used

$$\mu_k = \alpha_0 \eta[n_k](2 - \alpha_0 w[n_k - 1]). \qquad (10.63)$$

In this method, timing error cannot be accumulated because of multiplying by using α_0 in (10.62) instead of dividing by using $w[n]$ in (10.59). Thus, the discrete-time feedback loop filter $H(z)$ in Figure 10.13 is able to remove any constant error.

10.3.5 Polyphase Filter Bank Timing Recovery

In Section 10.3.4, we introduced the multirate timing recovery based on the multirate signal processing techniques. The multirate signal processing techniques offer flexibility for symbol timing recovery and synchronization. Using multirate approaches to symbol timing recovery and synchronization leads to straightforward implementation, especially in asynchronous modes. In this section, we extend the multirate signal processing techniques into multirate filtering based on polyphase filtering, thereby leading to the efficiently parallel architecture of *polyphase filter bank timing recovery*.

For an interpolation filter with upsampling by L, a direct implementation of the polyphase filter bank for the interpolation filter and upsampling of L requires the operation of the L polyphase filters, which operate in efficiently parallel architecture. However, for timing recovery, we only need to select a single stage filter with L filter coefficients out of the L polyphase filters under control of the discrete-time PLL. Harris and Rice [10] have proposed several approaches based on the polyphase filter bank for timing recovery. The polyphase filter bank is applied to symbol timing recovery by using a polyphase decomposition of the matched filter to realize interpolation instead of a separate polynomial interpolation filter. Thus, interpolation and matched filtering are integrated into a single filtering operation. The polyphase filter bank incorporates the timing phase detector in a very efficient manner, thereby making the polyphase filter bank timing recovery an especially attractive approach for DSP and FPGA implementation platforms.

Figure 10.14 shows an early-late gate approach for a timing phase error detector by using a polyphase filter bank implementation. The interpolation filters corresponding to the early and late gate approaches are the polyphase filters $H_{k-1}(z)$ and $H_{k+1}(z)$ at the polyphase segments $(k-1)$ and $(k+1)$. The received signal $r[n]$ is filtered with the polyphase filters $H_{k-1}(z)$ and $H_{k+1}(z)$ that are passed through absolute value functions to form corresponding output signals $p[n]$ and $q[n]$, respectively, where

$$\begin{aligned} p[n] &= |r[n] * h_{k-1}[n]| \\ &= |r[n] * h[Ln + k - 1]|, \end{aligned} \qquad (10.64)$$

and

$$\begin{aligned} q[n] &= |r[n] * h_{k+1}[n]| \\ &= |r[n] * h[Ln + k + 1]|. \end{aligned} \qquad (10.65)$$

Thus, the timing phase error detector $e[n]$ can be formed as follows:

$$\begin{aligned} e[n] &= q[n] - p[n] \\ &= |r[n] * h[Ln + k + 1]| - |r[n] * h[Ln + k - 1]|. \end{aligned} \qquad (10.66)$$

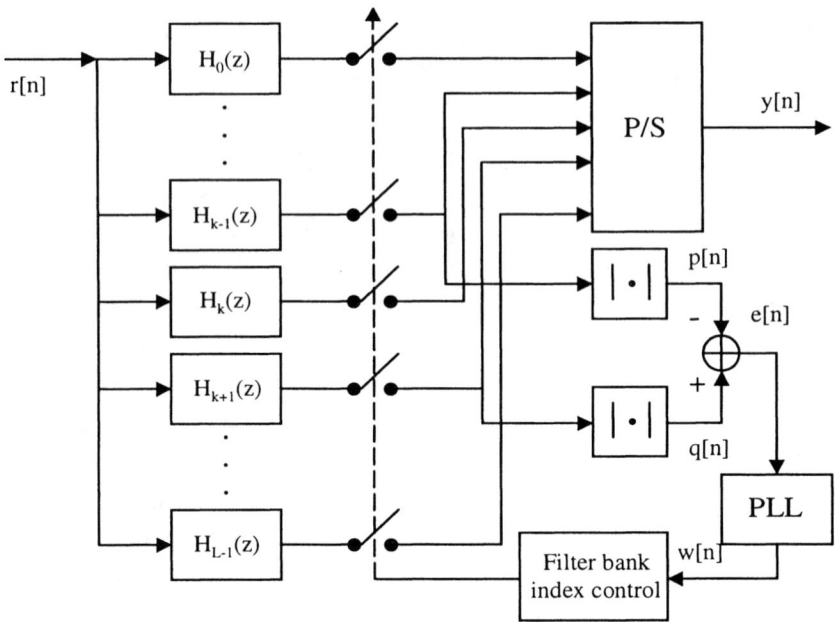

Figure 10.14 An early-late gate approach for a timing error phase detector based on a polyphase filter bank.

Equation (10.66) is passed through a discrete-time PLL where output will control a filter bank index control for timing phase selection.

Figure 10.15 shows another maximum likelihood approach for timing phase error detectors by using two polyphase filters $H_{k-1}(z)$ and $H_{k+1}(z)$ to compute the discrete-time derivative. The discrete-time derivative $d[n]$ is formed by using the output $p_1[n]$ from the polyphase filter $H_{k-1}(z)$ and the output $p_2[n]$ from the polyphase filter $H_{k-2}(z)$,

$$\begin{aligned} d[n] &= p_2[n] - p_1[n] \\ &= r[n] * h_{k+1}[n] - r[n] * h_{k-1}[n] \\ &= r[n] * (h_{k+1}[n] - h_{k-1}[n]) \end{aligned}$$

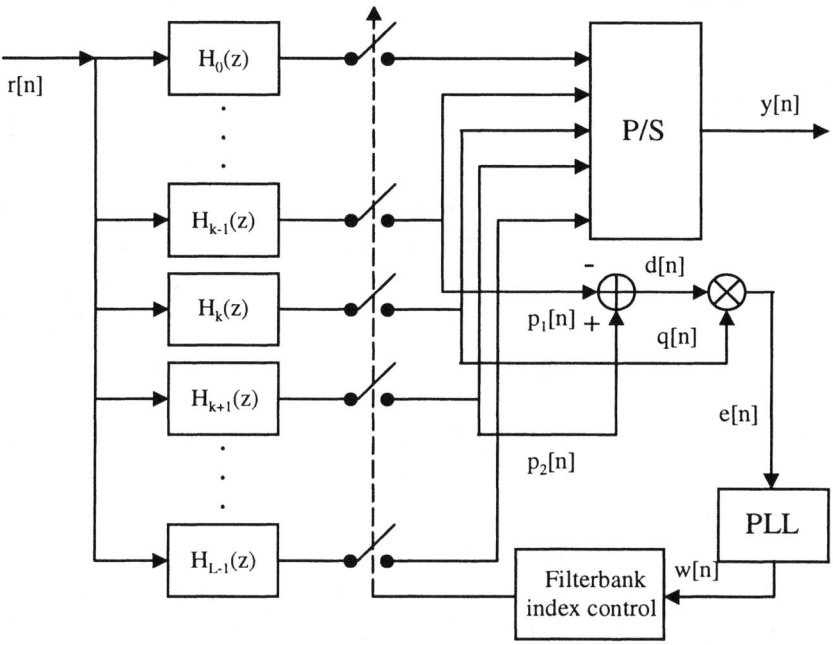

Figure 10.15 A maximum likelihood timing phase error detector based on a polyphase filter bank.

$$= r[n] * (h[Ln + k + 1] - h[Ln + k - 1]) . \quad (10.67)$$

Then, the maximum likelihood timing phase error detector $e[n]$ is obtained by

$$\begin{aligned} e[n] &= d[n]q[n] \\ &= r[n] * (h[Ln + k + 1] - h[Ln + k - 1]) \cdot \\ & \quad (r[n] * h[Ln + k]) \\ &= (r[n])^2 * (h[Ln + k + 1] - h[Ln + k - 1]) * h[Ln + k]. \end{aligned}$$
$$(10.68)$$

Equation (10.68) is passed through the discrete-time PLL to control the filter bank index control for time phase selection.

10.3.6 Multicarrier Modulation Timing Recovery

Multicarrier modulation, including DMT and OFDM, has shown great interest with applications in wireless and wireline personal broadband communication systems, which are expected to provide ubiquitous, high-quality, and high-rate multimedia transmission. This is because of its various advantages in the severe effects of frequency selective fading wireless channels and in the severe effects of near-end crosstalk (NEXT) and far-end crosstalk (FEXT) on wireline channels. However, the multicarrier modulation (DMT and OFDM) systems are vulnerable to synchronization error, including carrier frequency offsets and symbol timing recovery. The carrier frequency offsets are usually caused by the inherent instabilities of the transmitter and receiver carrier frequency oscillators. They can lead to severe multicarrier modulation system degradation because of intercarrier interference (ICI) [11]. On the other hand, symbol timing recovery must be achieved in order to avoid ISI [12].

Timing recovery for the multicarrier modulation systems can be divided into symbol synchronization and sampling clock synchronization. The purpose of symbol synchronization is to find the correct position for the fast Fourier transform (FFT) frame that is needed to satisfy the synchronization requirement in coherent DMT and OFDM systems. In contrast to the symbol synchronization, the purpose of sampling clock synchronization is to align the receiver sampling clock frequency to that of the transmitter. The sampling clock frequency error can not only cause ICI but also can result in a drift in the symbol timing and even further worsen the symbol synchronization.

Multicarrier modulation timing recovery has been extensively reported based on the symbol time offset [13], timing error [14], and frequency synchronization [15]. Sands and Jacobsen [16] have researched a pilotless timing recovery for multicarrier modulation, for which we focus on introducing these results in the following sections.

The Effect of Sampling Shift

Assume that the system sampling rate is F_s and the length of the cyclic prefix is v samples. Also let the number of subchannels be $N/2$ where N is a relatively large number. Then, each transmitted symbol X_k is composed of $N/2$ complex variables and in the time domain has a duration of $\frac{N+v}{F_s}$ second. The ith subcarrier has a frequency $f_i = \frac{iF_s}{N}$ for $i = 0, 1, \cdots, \frac{N}{2} - 1$. Considering the system in the frequency domain, if a sequence of subsymbols is transmitted over a channel with a transfer function $H(z)$ with a combined noise and interference $n_{i,k}$, the received subsymbols from the channel output are

$$\begin{aligned} Y_{i,k} &= X_{i,k} H(f)|_{f=\frac{iF_s}{N}} + n_{i,k} \\ &= X_{i,k} H_i + n_{i,k}. \end{aligned} \quad (10.69)$$

If the sampling frequency error is Δf Hz, the fractional frequency error is defined as $\varepsilon = \frac{\Delta f}{F_s}$. The sampling instant shifts by $\frac{\varepsilon}{F_s}$ second in the time domain for each sample. This leads to shift in time $\frac{(N+v)\varepsilon}{F_s}$ for each symbol. Therefore, for the kth symbol, the sampling instant shift is $\frac{k(N+v)\varepsilon}{F_s}$. Using the time-shift theorem of Fourier theory given by [17], which states that a time shift τ has a phase rotation in the frequency domain of $2\pi\tau f$, the ith subcarrier during the kth symbol is accordingly rotated by an angle

$$\begin{aligned} \phi_{i,k} &= 2\pi\tau f \\ &= \frac{2\pi\tau i F_s}{N} \\ &= \frac{2\pi i \varepsilon k (N+v)}{N}. \end{aligned} \quad (10.70)$$

Thus, the effect of the phase shift on the received subsymbols from the channel output is expressed as follows:

$$Y_{i,k} = X_{i,k} H_i e^{-j\left[\frac{2\pi i \varepsilon k(N+v)}{N}\right]} + n_{i,k}, \quad (10.71)$$

where the phase shift on the noise term in (10.71) is neglected.

The Estimate of Frequency Error

Assume that one symbol X_p is transmitted repeatedly during an acquisition stage. It is not necessary to know the exact signal at the receiver. However, it is preferred that its power spectrum is a uniform distribution over the used signal frequency band. Further assume that the channel insertion loss remains constant and a frequency rotation on any tone is entirely due to frequency error from symbol to symbol.

It is also assumed that the noise terms in (10.71) are uncorrelated, $E\{n_{i,k}n_{i,k-1}\} = 0$, and the noise term has zero mean, $E\{n_{i,k}\} = 0$. Thus, the estimate of the frequency error can be obtained from any subchannel by calculating the expected value of the product of the current subsymbol $Y_{i,k}$ and the complex conjugate of the previous subsymbol $Y^*_{i,k-1}$. Under these assumptions, we obtain the expected value $E\{Y_{i,k}Y^*_{i,k-1}\}$ as follows:

$$\begin{aligned}
E\{Y_{i,k}Y^*_{i,k-1}\} &= E\left\{\left[X_p H_i e^{-j\frac{2\pi i\varepsilon k(N+v)}{N}} + n_{i,k}\right] \cdot \right. \\
&\quad \left. \left[X_p H_i e^{j\frac{2\pi i\varepsilon(k-1)(N+v)}{N}} + n_{i,k-1}\right]\right\} \\
&= |X_p|^2 |H_i|^2 e^{-j\frac{2\pi i\varepsilon(N+v)}{N}} \\
&= |X_p|^2 |H_i|^2 \left\{\cos\left[\frac{2\pi i\varepsilon(N+v)}{N}\right] \right.\\
&\quad \left. + j\sin\left[\frac{2\pi i\varepsilon(N+v)}{N}\right]\right\}.
\end{aligned} \quad (10.72)$$

Since the acquisition symbol X_p is repeatly transmitted, a frame misalignment of m samples can be treated as a time shift of $\frac{m}{F_s}$ second. In other words, for the ith subcarrier in the frequency domain, this is a constant phase rotation as follows:

$$\frac{2\pi f_i m}{F_s} = \frac{2\pi i m}{N}. \quad (10.73)$$

In order to eliminate the frequency error in (10.72), the terms of $Y_{i,k}$ and $Y^*_{i,k-1}$ would be multiplied by using $e^{-j2\pi i m/N}$ and $e^{j2\pi i m/N}$, respectively. Thereby, the effect in (10.72) disappears.

Example 10.2

In this example, for a DMT system in VDSL applications, the expected frequency error ε is less than 100 ppm (10^{-4}). Assume that the DMT system has 512 subcarriers and the cyclic prefix is about 40 samples. Thus, for the ith subcarrier during the kth symbol, the average rotated phase shift is obtained by

$$\frac{2\pi i \varepsilon (N+v)}{N} = \frac{2\pi \times 255 \times 10^{-4} \times (512+40)}{512} < 0.1727. \tag{10.74}$$

Given the small phase shift in (10.74), using (10.72) obtains

$$E\{\text{Re}(Y_{i,k} Y^*_{i,k-1})\} \approx |X_p|^2 |H_i|^2 \tag{10.75}$$

and

$$E\{Y_{i,k} Y^*_{i,k-1}\} \approx |X_p|^2 |H_i|^2 \left[\frac{2\pi i \varepsilon (N+v)}{N}\right]. \tag{10.76}$$

Based on (10.72), we also note that the phase rotation caused by the frequency error is proportional to the subchannel frequency index i. Thus, each subchannel can provide an estimate given by Sands and Jacobsen [16] as follows:

$$\hat{\varepsilon}_{i,k} = \frac{N[\text{Im}(Y_{i,k} Y^*_{i,k-1})]}{2\pi i \varepsilon (N+v) \text{Re}(Y_{i,k} Y^*_{i,k-1})}, \tag{10.77}$$

where $\hat{\varepsilon}_{i,k}$ is the estimate of the fractional frequency error of the sampling clock at the ith subcarrier.

Assume that the channel transfer function H_i and noise variance σ_i^2 of each subchannel are known or can be estimated. If the transmission power on each subchannel is P_t, the signal-to-noise ratio (SNR) of the subchannels is given by

$$\text{SNR}_i = \frac{P_t |H_i|^2}{\sigma_i^2}. \tag{10.78}$$

The radial component of the noise will have no effect on the phase angle of $Y_{i,k}$. The noise in the tangential direction is about $\frac{\sigma_i^2}{2}$, and the angular noise variance is obtained by

$$\sigma_{\phi,i}^2 = \frac{\sigma_i^2}{2P_t|H_i|^2}$$
$$= \frac{1}{2\text{SNR}_i}, \quad (\text{radian}^2). \qquad (10.79)$$

Translating (10.79) into the frequency error by using (10.70) is given by

$$\sigma_{\varepsilon,i}^2 = \frac{N^2}{8[\pi i(N+v)]^2 \text{SNR}_i}. \qquad (10.80)$$

To estimate the frequency error, we need to determine a weighted sum of these estimates. Using the method of MSE, optimal weights based on (10.80) are obtained by [16],

$$w_{i,opt} = \frac{1}{\sigma_{\varepsilon,i}^2 \sum_i \frac{1}{\sigma_{\varepsilon,i}^2}}$$
$$= \frac{i^2 \text{SNR}_i}{\sum_i i^2 \text{SNR}_i}, \qquad (10.81)$$

and the corresponding minimum error variance ξ_{min}^2 of the estimate is given by

$$\xi_{min}^2 = \frac{1}{\sum_i \frac{1}{\sigma_{\varepsilon,i}^2}}$$
$$= \frac{1}{8}\left[\frac{N}{\pi(N+v)}\right]^2 \left(\frac{1}{\sum_i i^2 \text{SNR}_i}\right). \qquad (10.82)$$

Therefore, the estimate of the frequency error is obtained by

$$\hat{\varepsilon}_k = \sum_i w_{i,opt} \hat{\varepsilon}_{i,k}$$
$$= \left[\frac{N}{2\pi(N+v)}\right]\left[\sum_i \frac{i\,\text{SNR}_i \text{Im}(Y_{i,k}Y_{i,k-1}^*)}{\text{Re}(Y_{i,k}Y_{i,k-1}^*)}\right]\left(\frac{1}{\sum_i i^2 \text{SNR}_i}\right). \qquad (10.83)$$

Assuming knowledge of the subchannel SNR_i, (10.81), (10.82), and (10.83) provide the best performance.

The frequency estimate of (10.83) can be used as the input to a discrete-time loop filter, for example, with the transfer function in the z-domain given by

$$H(z) = \frac{Kz^{-1}}{1-z^{-1}}\left(1 + \frac{\beta z^{-1}}{1-z^{-1}}\right), \qquad (10.84)$$

where K is a constant of loop filter gain of the system, which is related to the filter bandwidth and settle time, and β is used to control the damping of the step response. Thus, PLL transient behavior can be controlled during acquisition.

The Analysis of Timing Jitter

The rotation error will be susceptible to the constellation points, especially at the edges of square or cross constellations. This is because the angular separation between points is the smallest. A square constellation with loading b_i bits with minimum separation of points has edges of distance $2^{b_i/2} - 1$ from the axes [16]. The angular separation between points at the edges of a square constellation is about $\frac{2}{2^{\lceil b_i/2 \rceil}-1}$ radians, where $\lceil \cdot \rceil$ denotes the ceiling function.

Note that the sampling phase error τ leads to a phase rotation at f_i of $2\pi f_i \tau$; accordingly, the minimum angular separation of each constellation is reduced by this amount. Assume that, in DMT and OFDM, bits are loaded onto all subcarriers in use, with equal detection SNR, and that the nominal constellation point separation is 2. Then the detection SNR is given by

$$\text{SNR}_{det} = \frac{2^2}{2\sigma^2}. \qquad (10.85)$$

If the sampling clock jitter is present, the minimum distance between the constellation points is reduced and the detection SNR becomes

$$\text{SNR}_{det-jitter} = \frac{\left[2 - 2\pi\sigma_j f_i(2^{\lceil b_i/2 \rceil} - 1)\right]^2}{2\sigma^2}. \qquad (10.86)$$

In order to determine the noise margin reduction, the SNR reduction can be expressed as follows:

$$\frac{\text{SNR}_{det-jitter}}{\text{SNR}_{det}} = \left[1 - \pi\sigma_j f_i(2^{\lceil b_i/2 \rceil} - 1)\right]^2. \qquad (10.87)$$

Equation (10.87) can be expressed in decibels as

$$\begin{aligned}\Re_i &= -10\log_{10}\left(\frac{\text{SNR}_{det-jitter}}{\text{SNR}_{det}}\right) \\ &= -20\log_{10}\left[1 - \pi\sigma_j f_i(2^{\lceil b_i/2 \rceil} - 1)\right] \text{ (dB)}. \quad (10.88)\end{aligned}$$

On the other hand, because of the timing jitter, the degradation can be used to define the maximum tolerable margin reduction $\Re_{max}$ and to establish a limit of the jitter as follows:

$$\sigma_j < \min_i \left\{\frac{1 - 10^{-\Re_{max}/20}}{\pi f_i(2^{\lceil b_i/2 \rceil} - 1)}\right\}, \qquad (10.89)$$

where σ_j is the sampling clock jitter and $f_i = \frac{iF_s}{N}$ for $i = 0, 1, \cdots, (N/2) - 1$.

10.4 Carrier Recovery

In the previous section, we derived the symbol timing recovery without knowledge of the carrier frequency and phase. However, coherent demodulation at the receiver requires exactly the known carrier frequency and phase information to perform the demodulation, with an independent timing reference.

Wireline channels, such as telephone lines, often introduce frequency offset. The frequency offset is indistinguishable from using a different carrier frequency. However, the symbol rate cannot be changed from a transmitter to a receiver. This is because the receiver can only receive exactly as many symbols per time unit as the symbols are sent from the transmitter. Thus, the relationship of the symbol rate to carrier frequency at the receiver is dependent on the unknown frequency offset.

In wireless channels, when either the transmitter or the receiver is in motion, the carrier frequency at the receiver is subject to a Doppler shift while the symbol timing is clearly not. Thus, the resulting frequency offset is similar to that found in the telephone channels [2], even though the frequency offset is more likely to be time-varying as the velocity of the receiver changes.

Furthermore, it is also possible for a channel to introduce *phase jitter* that appears as fluctuations in the phase of the carrier. Tracking the phase jitter is necessary to ensure that the phase jitter does not cause the performance degradation of the receiver system. Even in the absence of frequency offset, it is still highly recommended to derive carrier frequency and phase information independently from the symbol timing so that the phase jitter can be tracked.

In this section, we will first assume that symbol timing is known and then derive the carrier frequency and phase directly from the modulated signal, with an assumption that the modulated signal is transmitted via a suppressed carrier.

10.4.1 Carrier Phase Error

Extracting an accurate carrier phase estimate is important for a receiver to make carrier synchronization correctly. In order to emphasize the importance, we consider the effect of a carrier phase error on the demodulation of PAM, QAM, and multiphase phase-shift keying (M-PSK) signals.

Effect of the Carrier Phase Error in a PAM Signal

The demodulation of a PAM signal is given by

$$s(t) = A(t)[\cos(2\pi f_c t + \phi)], \qquad (10.90)$$

where $A(t)$ is the amplitude, f_c is the carrier frequency, and ϕ is the carrier phase. Correspondingly, the discrete-time PAM signal $s[n]$ can be obtained by using an A/D converter, sampling the continuous-time signal $s(t)$ at a sampling rate of $F_s = \frac{1}{T_s}$. We then have the discrete-time PAM signal as follows:

$$s[n] = s(t)|_{t=nT_s}$$

$$= A[n][\cos(2\pi f_c n T_s + \phi)],$$
$$= A[n]\left[\cos\left(2\pi \frac{f_c}{F_s} n + \phi\right)\right], \qquad (10.91)$$

where $A[n] = A(t)|_{t=nT_s}$. Demodulating the discrete-time PAM signal $s[n]$ of (10.91) by multiplying with the discrete-time carrier reference

$$\begin{aligned} c[n] &= c(t)|_{nT_s} \\ &= \cos(2\pi f_c n T_s + \hat{\phi}) \\ &= \cos\left(2\pi \frac{f_c}{F_s} n + \hat{\phi}\right), \end{aligned} \qquad (10.92)$$

we obtain the demodulation of the discrete-time PAM signal

$$\begin{aligned} r[n] &= s[n]c[n] \\ &= \frac{A[n]}{2}\left[\cos(\phi - \hat{\phi})\right] + \frac{A[n]}{2}\left[\cos\left(4\pi \frac{f_c}{F_s} n + \phi + \hat{\phi}\right)\right]. \end{aligned} \qquad (10.93)$$

Note that the second term of (10.93) can be eliminated by using a discrete-time lowpass filter. Thus, filtering produces the discrete-time demodulated PAM signal $r[n]$ as follows:

$$r[n] = \frac{A[n]}{2}\left[\cos(\phi - \hat{\phi})\right], \qquad (10.94)$$

where $\phi - \hat{\phi}$ is called the *carrier phase error*. The effect of the carrier phase error reduces the signal level in voltage and in power by the amount of $\cos(\phi - \hat{\phi})$ and $\cos^2(\phi - \hat{\phi})$, respectively. For example, if the phase error is 15 degrees, this results in a signal power loss of 0.30 dB. If the phase error is 30 degrees, this results in a signal power loss of 1.25 dB.

Effect of the Carrier Phase Error in QAM and M-PSK Signals

Discrete-time QAM and M-PSK signals can be expressed as follows:

$$s[n] = A[n]\left[\cos\left(2\pi\frac{f_c}{F_s}n + \phi\right)\right] - B[n]\left[\sin\left(2\pi\frac{f_c}{F_s}n + \phi\right)\right]. \quad (10.95)$$

The modulated signal of (10.95) is demodulated by using the two discrete-time quadrature carriers as follows:

$$c_I[n] = \cos\left(2\pi\frac{f_c}{F_s}n + \hat{\phi}\right), \quad (10.96)$$

and

$$c_Q[n] = -\sin\left(2\pi\frac{f_c}{F_s}n + \hat{\phi}\right). \quad (10.97)$$

Multiplying $s[n]$ in (10.95) with $c_I[n]$ in (10.96) followed by using a discrete-time lowpass filter yields the in-phase component as follows:

$$\begin{aligned} y_I[n] &= s[n]c_I[n] \\ &= \frac{A[n]}{2}\left[\cos(\phi - \hat{\phi})\right] - \frac{B[n]}{2}\left[\sin(\phi - \hat{\phi})\right]. \end{aligned} \quad (10.98)$$

Similarly, multiplying $s[n]$ in (10.95) with $c_Q[n]$ in (10.97) followed by using a discrete-time lowpass filter yields the quadrature component as follows:

$$\begin{aligned} y_Q[n] &= s[n]c_Q[n] \\ &= \frac{A[n]}{2}\left[\cos(\phi - \hat{\phi})\right] + \frac{B[n]}{2}\left[\sin(\phi - \hat{\phi})\right]. \end{aligned} \quad (10.99)$$

Note that (10.98) and (10.99) show a much more severe effect of phase error in the demodulation of QAM and M-PSK signal than

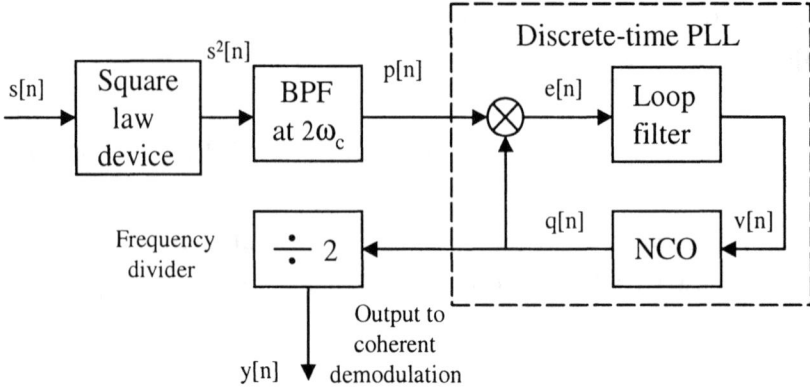

Figure 10.16 An open-loop carrier recovery based on a square-law device.

in the demodulation of a PAM signal given by (10.94). This is because there is not only a reduction in the power of the desired signal component by the factor of $\cos^2(\phi - \hat{\phi})$ but also a crosstalk interference from the in-phase and quadrature components. As can be seen, a small phase error causes a large degradation in performance since the average power levels of amplitudes A and B are similar. Therefore, in general, the phase accuracy requirements for QAM and M-PSK signals are much higher than those for a PAM signal.

10.4.2 Open-Loop Carrier Recovery

In this section, we introduce an open-loop carrier recovery method based on a squaring loop technique. Figure 10.16 shows the block diagram of the open-loop carrier recovery by using a square-law device. Assume that the input signal $s[n]$ is a PAM signal given by (10.91). Generating a carrier from the input signal $s[n]$ is to square the input signal $(s[n])^2$ and to produce a frequency component at $2\omega_c$, where $\omega_c = 2\pi f_c$, by using the square-law device.

The output signal of the square-law device is given by

$$(s[n])^2 = (A[n])^2 \left[\cos^2\left(2\pi\frac{f_c}{F_s}n + \phi\right)\right]$$

$$= \frac{(A[n])^2}{2}\left[1 + \cos\left(4\pi\frac{f_c}{F_s}n + 2\phi\right)\right]. \quad (10.100)$$

Note that the input signal $s[n]$ and the amplitude $A[n]$ have a zero mean, $E\{s[n]\} = E\{A[n]\} = 0$, because the input signal levels are symmetric about zero. Also, since the PAM is a cyclostationary stochastic process [1], the expected value of the output signal $(s[n])^2$ of the square-law device is obtained by

$$E\{(s[n])^2\} = \frac{E\{(A[n])^2\}}{2}\left[1 + \cos\left(4\pi\frac{f_c}{F_s}n + 2\phi\right)\right]. \quad (10.101)$$

Thus, from (10.101), we can clearly see that there is power at the frequency component at $2f_c$ (or $2\omega_c$) that can be used to drive a discrete-time PLL tuned to the frequency at $2f_c$.

Note that the effect of squaring $s[n]$ has removed the sign information contained in the amplitude $A[n]$ and then has resulted in phase coherent frequency components at twice the carrier. As shown in Figure 10.16, the output signal $(s[n])^2$ of the square-law device is passed through a discrete-time bandpass filter (BPF), which can be designed tuned to the double-frequency term given by (10.100). If the discrete-time input output signal $p[n]$ of the discrete-time BPF is the sinusoid expressed as follows:

$$p[n] = \cos\left(4\pi\frac{f_c}{F_s}n + 2\phi\right), \quad (10.102)$$

and the discrete-time output signal of the numerically controlled oscillator (NCO) is

$$q[n] = \sin\left(4\pi\frac{f_c}{F_s}n + 2\hat{\phi}\right), \quad (10.103)$$

where $\hat{\phi}$ denotes the estimate of phase ϕ, the product of these two discrete-time signals $p[n]$ and $q[n]$ is obtained by

$$\begin{aligned} e[n] &= p[n]q[n] \\ &= \cos\left(4\pi\frac{f_c}{F_s}n + 2\phi\right)\sin\left(4\pi\frac{f_c}{F_s}n + 2\hat{\phi}\right) \\ &= \frac{1}{2}\sin\left[2(\phi - \hat{\phi})\right] + \frac{1}{2}\sin\left(8\pi\frac{f_c}{F_s}n + 2\phi + 2\hat{\phi}\right). \end{aligned}$$
(10.104)

The second term with the frequency component at $4f_c$ of (10.104) can be removed by using the loop filter, which is a discrete-time lowpass filter responding to the low frequency component of $\frac{1}{2}\sin[2(\phi - \hat{\phi})]$. Generally, this loop filter can be selected to have a relatively simple transfer function, but a high-order discrete-time lowpass filter can also be used if a better loop response is needed. Thus, the discrete-time PLL can be used to further tune the accuracy of the sinusoid. The output frequency of the discrete-time PLL is double the desired carrier frequency and is divided by 2 to generate the carrier frequency.

10.4.3 Carrier Recovery for Multiple Phase Signals

Consider a digital signal transmitted via an M-phase modulation of a carrier. The discrete-time signal with an M-phase modulation has an expression as follows:

$$s[n] = A[n]\cos\left[2\pi\frac{f_c}{F_s}n + \phi + \frac{2\pi}{M}(m-1)\right], \qquad (10.105)$$

where $m = 1, 2, \cdots, M$, and $\frac{2\pi}{M}(m-1)$ is the information bearing component of the signal phase. The goal of carrier recovery is to remove the information bearing component and then to obtain the unmodulated carrier of $\cos(2\pi\frac{f_c}{F_s}n + \phi)$. Figure 10.17 is the power of M carrier recovery for M-phase modulation signals.

The received signal $s[n]$ is passed through an Mth power law device that produces the signal $(s[n])^M$ with a number of harmonics

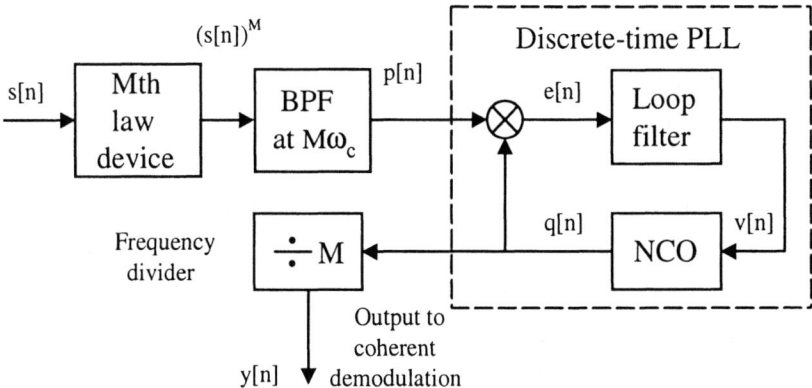

Figure 10.17 A power of M carrier recovery for multiple phase signals.

of ω_c. A discrete-time bandpass filter tuned to the frequency component at $M\omega_c$ is able to generate the discrete-time signal $p[n]$,

$$p[n] = \cos\left(2\pi M \frac{f_c}{F_s} n + M\phi\right). \tag{10.106}$$

Equation (10.106) can drive the discrete-time PLL. Also note that the term

$$\frac{2\pi}{M}(m-1)M = 2\pi(m-1) = 0, \mod 2\pi, \tag{10.107}$$

where $m = 1, 2, \cdots, M$. Therefore, the information of (10.107) has been eliminated. The discrete-time output signal $q[n]$ of the NCO is given by

$$q[n] = \sin\left(2\pi M \frac{f_c}{F_s} n + M\hat{\phi}\right). \tag{10.108}$$

Thus, the product of these two discrete-time signals $p[n]$ and $q[n]$ yields

$$e[n] = p[n]q[n]$$

$$= \cos\left(2\pi M \frac{f_c}{F_s}n + M\phi\right)\sin\left(2\pi M \frac{f_c}{F_s}n + M\hat{\phi}\right)$$
$$= \frac{1}{2}\sin\left[M(\phi - \hat{\phi})\right] + \frac{1}{2}\sin\left(2\pi M \frac{f_c}{F_s}n + M\phi + M\hat{\phi}\right).$$
(10.109)

The loop filter is a discrete-time lowpass filter that responds only to the first term of the low frequency component $\frac{1}{2}\sin\left[M(\phi - \hat{\phi})\right]$ and eliminates the second term of the frequency component at Mf_c. Thus, the output of the NCO is divided in frequency by M to yield $\sin\left(2\pi \frac{f_c}{F_s}n + \hat{\phi}\right)$. Phase-shifting by $\frac{\pi}{2}$ radians yields $\cos\left(2\pi \frac{f_c}{F_s}n + \hat{\phi}\right)$. Then, these components can be fed to the demodulator for the carrier recovery of the M-phase modulation signal.

10.4.4 Decision-Directed Carrier Recovery

Consider that a discrete-time passband PAM signal has an input to the system in Figure 10.18:

$$r[k] = e^{j(\omega_c kT_s + \theta_k)} \sum_{i=-\infty}^{\infty} A_i p[kT_s - iT_s], \qquad (10.110)$$

where ω_c is the carrier frequency, θ_k is the frequency offset and phase jitter, and $p[kT_s]$ represents a transmit filter, a linear distortion in the channel, and a receive filter. If there is a discrete-time demodulation image of the carrier

$$q[k] = e^{-j(\omega_c kT_s + \phi_k)}, \qquad (10.111)$$

where ϕ_k is the receiver estimate of the carrier phase, then we obtain the output of the phase detector

$$y[k] = e^{j(\theta_k - \phi_k)} \sum_{i=-\infty}^{\infty} A_i p[(k-i)T_s]. \qquad (10.112)$$

With any amount of noise and ISI, if the carrier recovery follows using an equalizer, then the equalizer output should approximately

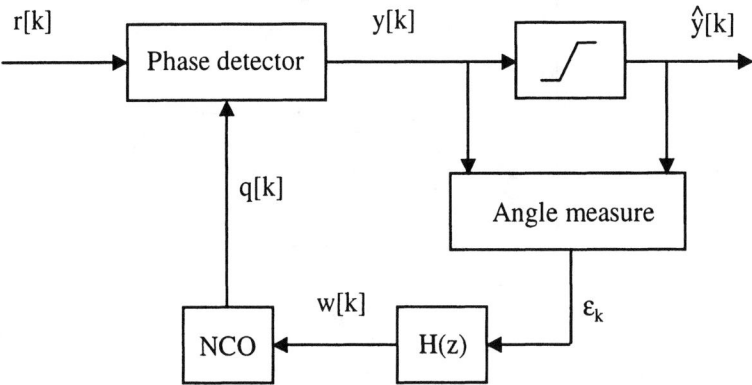

Figure 10.18 A decision-direction carrier recovery based on angular measurement.

satisfy the Nyquist criterion $p[k] = c[k]\delta_k$, where $c[k]$ is the real value and greater than 0 for amplitude errors. In this case, (10.112) can be rewritten as

$$\begin{aligned} y[k] &= c[k]e^{j(\theta_k - \phi_k)}A_k \\ &= c[k]e^{j\varepsilon_k}A_k, \end{aligned} \quad (10.113)$$

where ε_k is the phase errors, which are due to noise, ISI, phase jitter, and frequency offset.

Given (10.113), the phase errors can be obtained by [2]

$$\varepsilon_k = \sin^{-1}\left[\frac{\text{Im}\{y[k]A_k^*\}}{|y[k]||A_k|}\right]. \quad (10.114)$$

Since the symbols A_k are not available in the receiver, we can use decisions $\hat{y}[k]$ instead of the the symbols A_k. Thus, (10.114) can be rewritten as

$$\varepsilon_k = \sin^{-1}\left[\frac{\text{Im}\{y[k]\hat{y}^*[k]\}}{|y[k]||\hat{y}[k]|}\right]. \quad (10.115)$$

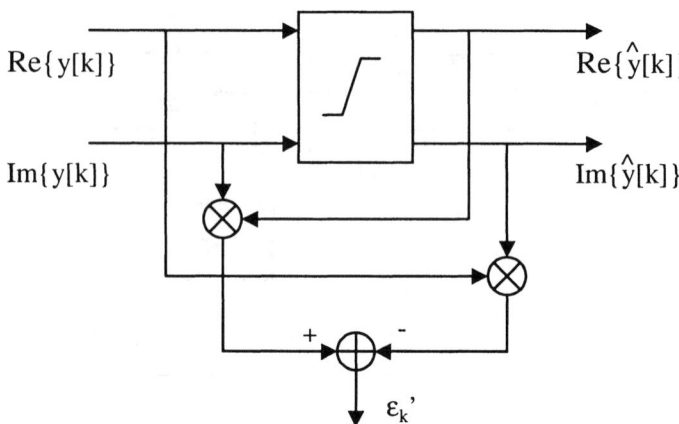

Figure 10.19 An implementation structure of the decision-direction carrier recovery.

Note that for small phase errors, we approximately have

$$\varepsilon_k \approx \sin(\varepsilon_k), \quad (10.116)$$

and then

$$\sin(\varepsilon_k) = \left[\frac{\text{Im}\{y[k]\hat{y}^*[k]\}}{|y[k]||\hat{y}[k]|}\right]. \quad (10.117)$$

With small phase errors, the characteristic becomes approximately linear. Equation (10.117) can be simplified further to omit the denominator

$$\varepsilon_k' = \sin(\varepsilon_k) \approx \text{Im}\{y[k]\hat{y}^*[k]\}. \quad (10.118)$$

Thus, in practice, the estimate of the phase error is obtained by

$$\varepsilon_k' = \text{Re}\{\hat{y}[k]\}\text{Im}\{y[k]\} - \text{Re}\{y[k]\}\text{Im}\{\hat{y}[k]\}. \quad (10.119)$$

The implementation structure of (10.119) is shown in Figure 10.19, which represents the block diagram of the angle measure in Figure 10.18.

10.5 Summary

In this chapter, discrete-time synchronization techniques, including the discrete-time PLL, timing, and carrier recoveries, were introduced. We first described the fundamentals of the discrete-time PLL with an emphasis on the discrete-time loop filter, phase detector, and discrete-time VCO. Second, we presented theories and architectures of the timing recovery, including the early-late gate synchronizer, bandedge timing recovery, decision-directed timing recovery, multirate timing recovery, and polyphase filter bank timing recovery, as well as multicarrier modulation timing recovery. We then focused on introducing the carrier recovery in which we address the carrier phase error, open-loop carrier recovery, carrier recovery for multiple phase signals, and decision-directed carrier recovery.

Timing recovery is used to recover a clock at the symbol rate or a multiple of the symbol rate from the modulated waveform, while carrier recovery is used to find exactly the same carrier frequency and phase to perform coherent demodulation from the data-bearing signal. Timing recovery can be derived without knowledge of the carrier phase, but estimating the carrier phase usually needs to know the symbol timing first. Thus, a receiver should first derive timing, then estimate the carrier phase, and finally adapt other devices, such as an equalizer.

Multirate timing recovery was developed based on oversampling and/or downsampling for the symbol rate of the incoming signal. It is especially useful in the asynchronized mode. Polyphase filter bank timing recovery, which is a special case of the multirate timing recovery, offers flexibility for symbol timing recovery and synchronization. This technique leads to efficient parallel architecture for straightforward implementation.

References

[1] Proakis, J. G., *Digital Communications*, 2nd ed., McGraw-Hill Book Company, New York, 1989.

[2] Lee, E. A., and D. G. Messerschmitt, *Digital Communications*, Kluwer Academic Publishers, Boston, Massachusetts, 1988.

[3] Waggener, W., *Pulse Code Modulation Systems Design*, Artech House, Norwood, Massachusetts, 1999.

[4] Razavi, B., *RF Microelectronics*, Prentice Hall, Upper Saddle River, New Jersey, 1998.

[5] Starr, T., J. M. Cioffi, and P. J. Silverman, *Understanding Digital Subscriber Line Technology*, Prentice Hall, Upper Saddle River, New Jersey, 1998.

[6] Mueller, K. H., and M. Müller, "Timing Recovery in Digital Synchronous Data Receivers," *IEEE Trans. on Communications*, Vol. COM-24, No. 5, pp. 516–531, May 1976.

[7] Gardner, F. M., "Interpolation in Digital Modems: Part I: Fundamentals," *IEEE Trans. on Communications*, Vol. 41, No. 3, pp. 501–507, March 1993.

[8] Erup, L., F. M. Gardner, and R. A. Harris, "Interpolation in Digital Modems: Part II: Implementation and Performance," *IEEE Trans. on Communications*, Vol. 41, No. 6, pp. 998–1,008, June 1993.

[9] Gardner, F. M., "A BPSK/QPSK Timing-Error Detector for Sampled Receivers," *IEEE Trans. on Communications*, Vol. 34, No. 5, pp. 423–429, May 1986.

[10] Harris, F. J., and M. Rice, "Multirate Digital Filters for Symbol Timing Synchronization in Software Defined Radios," *IEEE Journal on Selected Area in Communications*, Vol. 19, No. 12, pp. 2,346–2,357, December 2001.

[11] Moose, P. H., "A Technique for Orthogonal Frequency Division Multiplexing Frequency Offset Correction," *IEEE Trans. on Communications*, Vol. 42, pp. 2,908–2,914, October 1994.

[12] Yang, B., et al., "Timing Recovery for OFDM Transmission," *IEEE Journal on Selected Area in Communications*, Vol. 18, No. 11, pp. 2,278–2,291, November 2000.

[13] Landström, et al., "Symbol Time Offset Estimation in Coherent OFDM Systems," *IEEE Trans. on Communications*, Vol. 50, No. 4, pp. 545–549, April 2002.

[14] Louveauz, J., et al., "Bit-Rate Sensitivity of Filter-Bank-Based VDSL Transmission to Timing Errors," *IEEE Trans. on Communications*, Vol. 49, No. 2, pp. 375–383, February 2001.

[15] Lambrette, U., M. Speth, and H. Meyr, "OFDM Burst Frequency Synchronization by Single Carrier Training Data," *IEEE Communications Letters*, Vol. 1, No. 2, pp. 46–48, March 1997.

[16] Sands, N. P., and K. S. Jacobsen, "Pilotless Timing Recovery for Baseband Multicarrier Modulation," *IEEE Journal on Selected Area in Communications*, Vol. 20, No. 5, pp. 1,047–1,054, June 2002.

[17] Miao, G. J., and M. A. Clements, *Digital Signal Processing and Statistical Classification*, Artech House, Norwood, Massachusetts, 2002.

Appendix A: The z-Transform

A.1 Introduction

In this appendix, we first introduce the z-transform representation, region of convergence, properties, common pairs of the z-transform, and the z-transform of the transfer function and then discuss the corresponding inverse z-transform along with technique methods for finding the inverse z-transform, including contour integration and partial fractions expansion. The relationship between the all-pass and the phase systems is sequentially addressed by using the z-transform, with an emphasis on the all-pass system and the minimum-phase system and their relations in the decomposition phase system, compensation system, and FIR system. Furthermore, the z-transform can be used to represent a linear time-invariant (LTI) system, discrete-time transform, digital filtering, and multirate signal processing, and to allow frequency-domain interpretation. The z-transform allows the interested reader to bring the power of the complex variable theory to support the problem analysis of discrete-time signals and systems.

A.2 The z-Transform

Let $X(z)$ denote the z-transform of a discrete-time signal sequence $x[n]$. The z-transform is then defined as

$$X(z) = \sum_{n=-\infty}^{\infty} x[n] z^{-n}, \quad (A.1)$$

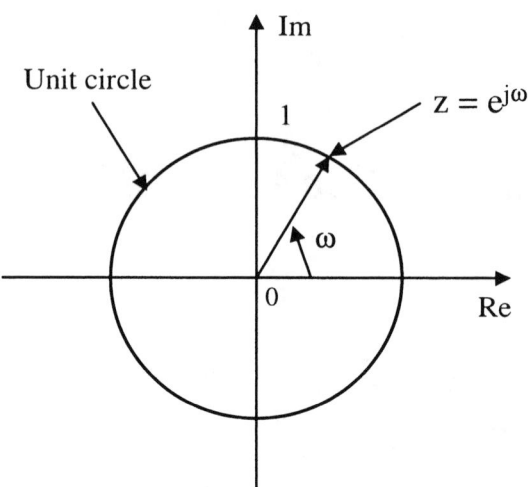

Figure A.1 The complex z-plane of a unit circle with the contour corresponding to $z = 1$ and the ω representing the angle between the vector to a point.

where z is a complex variable. Thus, the z-transform is a function of a complex variable and can be interpreted by using the complex z-plane.

Figure A.1 shows the complex z-plane of a unit circle plot or a contour. The contour corresponding to $|z| = 1$ is a circle of the unit radius and ω is the angle between the vector to a point z on the unit circle and the real axis of the complex z-plane. The unit circle plot is useful for determining the *region of convergence* (ROC) for the z-transform. This is because the z-transform does not converge for all sequences. Thus, for any given discrete-time sequence $x[n]$, all the values of z for which the z-transform converges are called the ROC. In other words, the z-transform of the discrete-time signal sequence $x[n]$ converges absolutely only for values of z in its ROC.

Note that the convergence of (A.1) depends on the magnitude of z. This is because the z-transform $|X(z)| < \infty$ if and only if we hold the following equation

$$\sum_{n=-\infty}^{\infty} |x[n]||z^{-n}| < \infty. \quad (A.2)$$

If this is the case, the ROC will be bounded by circles centered at the origin of the complex z-plane. In general, the ROC of the z-transform can be classified into four configuration regions, including the interior of a circle, the exterior of a circle, the annulus, and the entire z-plane. Figure A.2 shows the four configuration regions.

It is most useful for the z-transform if the infinite sum can be expressed in a closed form of a rational function inside the ROC, for instance,

$$X(z) = \frac{P(z)}{Q(z)}, \quad (A.3)$$

where $P(z)$ and $Q(z)$ are polynomials of z. Note that the values of z are referred to as the *zeros* of $X(z)$ if the z-transform $X(z) = 0$. If $X(z)$ in (A.3) is infinite, then the values of z are referred to as the *poles* of $X(z)$. Furthermore, it is also possible for the poles to take place at $z = 0$ or $z = \infty$. Therefore, in theory, the ROC should not contain any poles because the $X(z)$ becomes infinite at the poles.

Example A.1

Assume that a discrete-time signal sequence is

$$x[n] = \left(\frac{1}{5}\right)^n u[n], \quad (A.4)$$

where

$$u[n] = \begin{cases} 1, & n \geq 0 \\ 0, & n < 0. \end{cases} \quad (A.5)$$

Then the z-transform of $x[n]$ is obtained by using (A.1),

$$X(z) = \sum_{n=-\infty}^{\infty} \left(\frac{1}{5}\right)^n u[n] z^{-n}$$

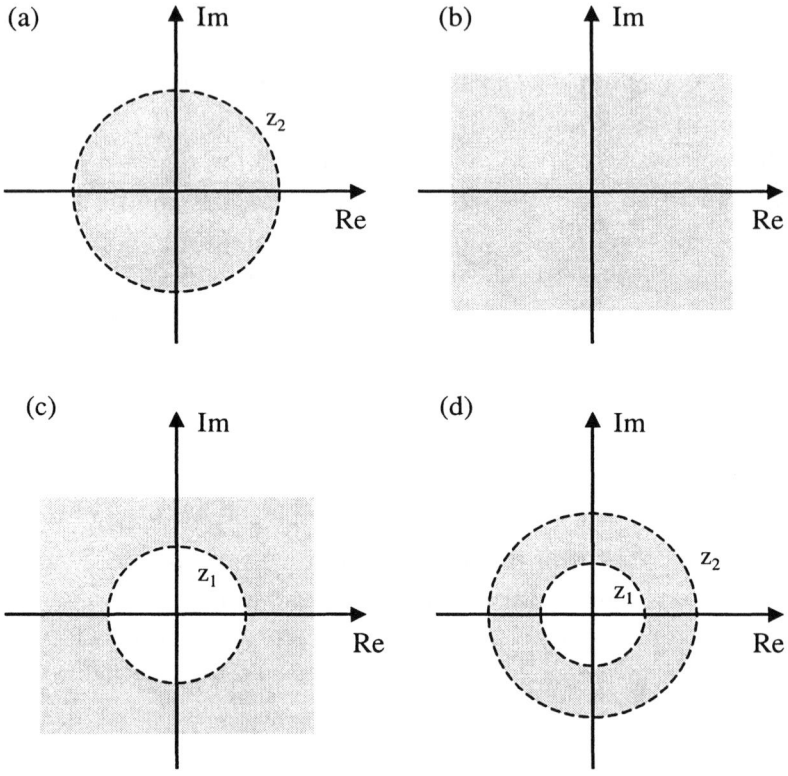

Figure A.2 The configuration regions of the z-transform: (a) the interior of a circle, $|z| < |z_2|$; (b) the entire plane; (c) the exterior of a circle, $|z| > |z_1|$; and (d) an annulus, $|z_1| < |z| < |z_2|$.

$$= \sum_{n=0}^{\infty}\left(\frac{1}{5}z^{-1}\right)^n. \tag{A.6}$$

In order to get the convergence of $X(z)$, we should have

$$\sum_{n=0}^{\infty}\left|\frac{1}{5}z^{-1}\right|^n < \infty. \tag{A.7}$$

Thus, the ROC is the range of values of z for $|z| > \frac{1}{5}$. Within the ROC, the infinite series in (A.6) converges to

$$X(z) = \frac{1}{1 - \frac{1}{5}z^{-1}}, \quad \text{for } |z| > \frac{1}{5}. \tag{A.8}$$

Example A.2

Consider a discrete-time signal sequence with the sum of two real exponentials as follows:

$$x[n] = a^n u[n] + b^n u[n], \tag{A.9}$$

and

$$u[n] = \begin{cases} 1, & n \geq 0 \\ 0, & n < 0. \end{cases} \tag{A.10}$$

where $a < 1$, $b < 1$, and $a \geq b$. The z-transform $X(z)$ is then obtained by

$$\begin{aligned}
X(z) &= \sum_{n=-\infty}^{\infty} \{a^n u[n] + b^n u[n]\} z^{-n} \\
&= \sum_{n=-\infty}^{\infty} a^n u[n] z^{-n} + \sum_{n=-\infty}^{\infty} b^n u[n] z^{-n} \\
&= \sum_{n=0}^{\infty} (az^{-1})^n + \sum_{n=0}^{\infty} (bz^{-1})^n \\
&= \frac{1}{1 - az^{-1}} + \frac{1}{1 - bz^{-1}} \\
&= \frac{2 - (a+b)z^{-1}}{(1 - az^{-1})(1 - bz^{-1})}.
\end{aligned} \tag{A.11}$$

In order to converge, we should require that both $|az^{-1}| < 1$ and $|bz^{-1}| < 1$ at the same time. This is equivalent to $|z| > a$ and $|z| > b$. As a result, the ROC of (A.11) is $|z| > a$, since $a \geq b$.

A.2.1 The z-Transform Properties

The z-transform has many properties that are useful tools to design and analyze the discrete-time signals and systems. These properties are often used to obtain the inverse z-transform of more complicated expressions. The z-transform properties are summarized below. There are many references of the z-transform properties. The interested reader may refer to Miao and Clements [1], Oppenheim et al. [2], Oppenheim and Schafer [3], and D'Azzo and Houpis [4].

To simplify the notation, we denote $X(z)$ as the z-transform of $x[n]$ and R_x as the ROC of $X(z)$ as follows:

$$x[n] \xleftrightarrow{Z} X(z), \qquad (A.12)$$

where the ROC is R_x, which represents a set of values of z with the condition of $r_R < |z| < r_L$.

Linearity

The linearity property states that if

$$x_1[n] \xleftrightarrow{Z} X_1(z), \quad \text{ROC} = R_{x_1}, \qquad (A.13)$$

and

$$x_2[n] \xleftrightarrow{Z} X_2(z), \quad \text{ROC} = R_{x_2}, \qquad (A.14)$$

then

$$ax_1[n] + bx_2[n] \xleftrightarrow{Z} aX_1(z) + bX_2(z), \qquad (A.15)$$

where the ROC includes $R_{x_1} \cap R_{x_2}$.

Time Shifting

Time shifting states that if

$$x[n] \xleftrightarrow{Z} X(z), \quad \text{ROC} = R_x, \qquad (A.16)$$

then
$$x[n-d] \stackrel{Z}{\Longleftrightarrow} z^{-d}X(z) \qquad (A.17)$$

where d is an integer. The discrete-time signal sequence $x[n]$ is shifted right if d is positive. The discrete-time signal sequence $x[n]$ is shifted left if d is negative. Thus, the ROC in (A.17) is R_x except for the possible addition or deletion of $z = 0$ or $z = \infty$.

Frequency Shifting or Modulation

Frequency shifting states if
$$x[n] \stackrel{Z}{\Longleftrightarrow} X(z), \quad \text{ROC} = R_x, \qquad (A.18)$$

then
$$e^{j\omega n}x[n] \stackrel{Z}{\Longleftrightarrow} X(e^{-j\omega}z). \qquad (A.19)$$

Note that the $e^{j\omega n}x[n]$ will not be real if the discrete-time signal sequence $x[n]$ is real unless ω is an integer multiple of π. Furthermore, even if the poles and zeros of $X(z)$ are in complex conjugate pairs, they may not have this symmetry after frequency shifting.

Time Reversal

A signal shows time reversal if
$$x[n] \stackrel{Z}{\Longleftrightarrow} X(z), \quad \text{ROC} = R_x, \qquad (A.20)$$

then
$$x[-n] \stackrel{Z}{\Longleftrightarrow} X(z^{-1}), \qquad (A.21)$$

where the ROC is $\frac{1}{R_x}$. In this case, the ROC means that $R_x(z)$ is inverted. This is to say that if z_0 is in the ROC for $x[n]$, then $1/z_0$ is in the ROC for $x[-n]$.

Convolution of Two Sequences

The convolution of sequences states if
$$x_1[n] \stackrel{Z}{\Longleftrightarrow} X_1(z), \quad \text{ROC} = R_{x_1}, \qquad (A.22)$$

and
$$x_2[n] \xleftrightarrow{Z} X_2(z), \quad \text{ROC} = R_{x_2}, \quad (A.23)$$
then
$$x_1[n] * x_2[n] \xleftrightarrow{Z} X_1(z)X_2(z), \quad (A.24)$$
where the ROC includes $R_{x_1} \cap R_{x_2}$. In other words, the ROC contains the intersection of the regions of convergence of $X_1(z)$ and $X_2(z)$. In addition, the ROC may be larger if a pole that borders on the ROC of one of the z-transforms is eliminated by a zero of the other ROC of the z-transform.

Multiplication of an Exponential Sequence

When a signal is multiplied by an exponential sequence, the multiplication of exponential sequence states that if
$$x[n] \xleftrightarrow{Z} X(z), \quad \text{ROC} = R_x, \quad (A.25)$$
then
$$z_0^n x[n] \xleftrightarrow{Z} X(z_0^{-1}z), \quad (A.26)$$
where the ROC is $|z_0|R_x$. If R_x has a set of values of z with range of $r_R < |z| < r_L$, then $|z_0|R_0$ is the set of values of z such that $|z_0|r_R < |z| < |z_0|r_L$.

Conjugation of a Complex Sequence

The conjugation of a complex sequence states that if
$$x[n] \xleftrightarrow{Z} X(z), \quad \text{ROC} = R_x, \quad (A.27)$$
then
$$x^*[n] \xleftrightarrow{Z} X^*(z^*) \quad (A.28)$$
$$\text{Re}\{x[n]\} \xleftrightarrow{Z} \frac{1}{2}[X(z) + X^*(z^*)] \quad (A.29)$$
$$\text{Im}\{x[n]\} \xleftrightarrow{Z} \frac{1}{2j}[X(z) - X^*(z^*)] \quad (A.30)$$

The ROC is equal to R_x.

Differentiation of X(z)

The differentiation of $X(z)$ states that if

$$x[n] \overset{Z}{\Longleftrightarrow} X(z), \quad \text{ROC} = R_x, \tag{A.31}$$

then

$$nx[n] \overset{Z}{\Longleftrightarrow} -z\frac{dX(z)}{dz}, \tag{A.32}$$

where the ROC is R_x except for the possible addition or deletion of $z = 0$ or $z = \infty$.

Initial Value

The initial value states that if a discrete-time signal sequence $x[n]$ is zero for $n < 0$ and if $\lim_{z \to \infty} X(z)$ exists, then

$$x[0] = \lim_{z \to \infty} X(z). \tag{A.33}$$

Final Value

The final value states that if $X(z)$ converges for $\|z\| > 1$ and all poles of $(1 - z)X(z)$ are inside the unit circle, then

$$x[\infty] = \lim_{z \to 1}(1 - z^{-1})X(z). \tag{A.34}$$

A.2.2 Common Pairs of the z-Transform

There are many common pairs of the z-transform in literature. In this section, we list a number of the basic common z-transform pairs that include values of $x[n]$ only for $n \geq 0$, as shown in Table A.1. These common z-transform pairs are useful for finding the z-transform $X(z)$ if a discrete-time signal sequence $x[n]$ is given. On the other hand, these common pairs of the z-transform are also useful for obtaining the discrete-time signal sequence $x[n]$ corresponding to a given z-transform $X(z)$.

Table A.1 The Common z-Transform Pairs

Sequence	z-Transform	ROC
$\delta[n]$	1	All z
$\delta[n-m], m > 0$	z^{-m}	All z except 0
$u[n]$	$\frac{1}{1-z^{-1}}$	$|z| > 1$
$-u[-n-1]$	$\frac{1}{1-z^{-1}}$	$|z| < 1$
$nu[n]$	$\frac{z}{(z-1)^2}$	$|z| > 1$
$n^2 u[n]$	$\frac{z(z+1)}{(z-1)^3}$	$|z| > 1$
$n^3 u[n]$	$\frac{z(z^2+4z+1)}{(z-1)^4}$	$|z| > 1$
$a^{|n|}$	$\frac{1-a^2}{(1-az^{-1})(1-az)}$	$a < |z| < \frac{1}{a}$
$\frac{a^n}{n} u[n]$	$e^{a/z}$	All z
$a^n u[n]$	$\frac{1}{1-az^{-1}}$	$|z| > |a|$
$n a^n u[n]$	$\frac{az^{-1}}{(1-az^{-1})^2}$	$|z| > |a|$
$n^2 a^n u[n]$	$\frac{az(z+a)}{(z-a)^3}$	$|z| > |a|$
$n^3 a^n u[n]$	$\frac{az(z^2+4az+a^2)}{(z-a)^4}$	$|z| > |a|$
$(n+1) a^n u[n]$	$\frac{z^2}{(z-a)^2}$	$|z| > |a|$
$\frac{(n+1)(n+2)}{2} a^n u[n]$	$\frac{z^3}{(z-a)^3}$	$|z| > |a|$
$\frac{(n+1)(n+2)(n+3)}{3} a^n u[n]$	$\frac{z^4}{(z-a)^4}$	$|z| > |a|$
$-a^n u[-n-1]$	$\frac{1}{1-az^{-1}}$	$|z| < |a|$
$-n a^n u[-n-1]$	$\frac{az^{-1}}{(1-az^{-1})^2}$	$|z| < |a|$
$a^n (u[n] - u[n-N])$	$\frac{1-a^N z^{-N}}{1-az^{-1}}$	$|z| > 0$
$[r^n \cos(\omega_0 n)] u[n]$	$\frac{1-(r\cos\omega_0) z^{-1}}{1-(2r\cos\omega_0) z^{-1} + r^2 z^{-2}}$	$|z| > r$
$[r^n \sin(\omega_0 n)] u[n]$	$\frac{(r\sin\omega_0) z^{-1}}{1-(2r\cos\omega_0) z^{-1} + r^2 z^{-2}}$	$|z| > r$

A.2.3 The z-Transform Transfer Function

An important application of the z-transform is in the analysis and representation of a discrete-time LTI system. This is because we can represent the discrete-time LTI system as follows:

$$Y(z) = H(z)X(z), \qquad (A.35)$$

where $H(z)$, $X(z)$, and $Y(z)$ are the z-transforms of the system impulse response, input, and output, respectively. The $H(z)$ of the z-transform is called the discrete-time system function or the z-transform transfer function.

The z-transform transfer function can be derived from the LTI difference equation as follows:

$$y[n] + \sum_{k=1}^{N-1} a_k y[n-k] = \sum_{k=0}^{M-1} b_k x[n-k]. \qquad (A.36)$$

By taking the z-transform of each term on both sides of (A.36) and using the properties of the linearity and the time-shifting, we then obtain as follows:

$$Y(z) + Y(z)\sum_{k=1}^{N-1} a_k z^{-k} = X(z)\sum_{k=0}^{M-1} b_k z^{-k}. \qquad (A.37)$$

We solve (A.37) for (A.35) and yield the result as follows:

$$\begin{aligned} H(z) &= \frac{Y(z)}{X(z)} \\ &= \frac{\sum_{k=0}^{M-1} b_k z^{-k}}{1 + \sum_{k=1}^{N-1} a_k z^{-k}}. \end{aligned} \qquad (A.38)$$

Equation (A.38) can be further expressed in a factored form

$$H(z) = b_0 \left[\frac{\prod_{k=1}^{M-1}(1 - c_k z^{-1})}{\prod_{k=1}^{N-1}(1 - d_k z^{-1})} \right]. \qquad (A.39)$$

Note that each of the factors $(1 - c_k z^{-1})$ in the numerator contributes a zero at $z = c_k$ and a pole at $z = 0$. Similarly, each of the factors $(1 - d_k z^{-1})$ in the denominator contributes a pole at $z = d_k$ and a zero at $z = 0$.

Example A.3

Consider a discrete-time LTI system with the relationship of the input and output in a difference equation

$$y[n] - \frac{1}{2}y[n-1] = x[n] + \frac{1}{3}x[n-1]. \quad (A.40)$$

By using the z-transform on both sides of (A.40) and the properties of the linearity and time-shifting, we are able to obtain

$$Y(z) - \frac{1}{2}z^{-1}Y(z) = X(z) + \frac{1}{3}z^{-1}X(z). \quad (A.41)$$

Thus, the z-transform transfer function is given by

$$H(z) = \frac{Y(z)}{X(z)} = \frac{1 + \frac{1}{3}z^{-1}}{1 - \frac{1}{2}z^{-1}}. \quad (A.42)$$

Note that the ROC contains two different possibilities: (1) $|z| > 1/2$ is associated with the assumption that the impulse response $h[n]$ is right-sided, and (2) $|z| < 1/2$ is associated with the assumption that the impulse response $h[n]$ is left-sided.

A.3 The Inverse z-Transform

The inverse z-transform is derived using the Cauchy integral theorem contributed by Churchill and Brown [5] and is given by the contour integral theorem as follows:

$$x[n] = \frac{1}{2\pi j} \oint_C X(z)z^{n-1}dz, \quad (A.43)$$

where the integral notation with C is a counterclockwise closed contour in the ROC of $X(z)$ and encircling the origin of the complex z-plane.

Equation (A.43) can be solved by using other approaches, including the contour integration based on the direction calculation, examination method, and partial fractions expansion based on the well-known procedures of containing the common pairs of the z-transform.

A.3.1 The Contour Integration

Equation (A.43) can be directly calculated by using the contour integration. Assume that $X(z)$ is a function of the complex variable z and C is a counterclockwise closed contour in the ROC of $X(z)$. If the $(k+1)$-order derivative $\frac{dX(z)}{dz}$ exists on and inside the contour C, and if $X(z)$ does not have any poles at $z = z_0$, then the inverse z-transform is obtained by

$$\begin{aligned} x[n] &= \frac{1}{2\pi j} \oint_C \frac{X(z)}{(z-z_0)^k} dz \\ &= \begin{cases} \frac{1}{(k-1)!} \left. \frac{d^{k-1} X(z)}{dz^{k-1}} \right|_{z=z_0}, & \text{if } z_0 \text{ is inside } C \\ 0, & \text{otherwise.} \end{cases} \end{aligned} \quad (A.44)$$

Equation (A.44) is the result in terms of the expression of the Cauchy residue theorem.

A.3.2 The Examination Approach

Another solution for solving the inverse z-transform is to use the common pairs z-transform by discovering the relationship between the z-transform $X(z)$ and the discrete-time signal sequence $x[n]$ as shown in Table A.1. This approach is often called the *examination method*.

To illustrate the examination method, we consider an example of the z-transform $X(z)$ given by

$$X(z) = \frac{1}{1+\frac{1}{5}z^{-1}} + \frac{1}{1-\frac{1}{2}z^{-1}}, \quad (A.45)$$

where the ROC is the annular region $\frac{1}{5} < |z| < \frac{1}{2}$. Using the common pairs z-transform, we can directly obtain the discrete-time signal sequences as follows:

$$\left(-\frac{1}{5}\right)^n u[n] \stackrel{Z}{\Longleftrightarrow} \frac{1}{1+\frac{1}{5}z^{-1}}, \quad |z| > \frac{1}{5} \quad (A.46)$$

and

$$-\left(\frac{1}{2}\right)^n u[-n-1] \stackrel{Z}{\Longleftrightarrow} \frac{1}{1-\frac{1}{2}z^{-1}}, \quad |z| < \frac{1}{2}. \quad (A.47)$$

Then, by using the linearity property, we obtain the discrete-time signal sequence,

$$x[n] = \left(-\frac{1}{5}\right)^n u[n] - \left(\frac{1}{2}\right)^n u[-n-1]. \quad (A.48)$$

Note that the discrete-time signal sequence $x[n]$ in this example grows exponentially if n goes to $-\infty$.

A.3.3 The Partial Fraction Expansion

The *partial fraction expansion* is referred to as an alternative expression for the z-transform $X(z)$ as a sum of the terms. Each of the terms is tabulated to identify the discrete-time signal sequence $x[n]$ that corresponds to the individual terms by using the common pairs z-transform in Table A.1.

The partial fraction expansion can be used in a different way that slightly depends on whether the z-transform transfer function has all distinct poles or some multiple poles. In the case of the multiple poles, a procedure of the partial fraction expansion is more complicated. For a mathematical treatment, the interested reader may refer to Oppenheim and Schafer [3]. In this section, we are interested in presenting the procedure of the partial fraction expansion applying the z-transform transfer function in the case of single poles. This is because the most filter designs in practice have this property.

The procedure of the partial fractions expansion for the z-transform transfer function with distinct poles requires factoring the denominator of $H(z)$ as follows:

$$H(z) = \frac{\sum_{k=0}^{M} b_k z^{M-k}}{\prod_{k=1}^{N}(z - d_k)}. \quad (A.49)$$

Then we express the discrete-time signal sequence $h[n]$ in the following form

$$h[n] = c_0 \delta[n] + \sum_{k=1}^{N} c_k (d_k)^n, \quad (A.50)$$

where $h[n]$ is the inverse z-transform of the transfer function $H(z)$.

To obtain the discrete-time signal sequence $h[n]$ in (A.50), we have to determine the coefficients c_k, for $k = 0, 1, \cdots, N$. The first coefficient c_0 is obtained by

$$\begin{aligned} c_0 &= H(z)\,|_{z=0} \\ &= \frac{b_M}{\prod_{k=1}^{N}(-d_k)}. \end{aligned} \qquad \text{(A.51)}$$

The other coefficients c_k for $k = 1, 2, \cdots, N$ can be calculated by using

$$c_k = \left(\frac{z - d_k}{z}\right) H(z)\,|_{z=d_k}. \qquad \text{(A.52)}$$

To illustrate the operation of the inverse z-transform, we show an example next.

Example A.4

Consider the z-transform transfer function given by

$$H(z) = \frac{z^2}{z^2 + z - 12}, \qquad \text{(A.53)}$$

where the ROC is the exterior of a circle region $|z| > 4$. If we factor the denominator of $H(z)$, we obtain

$$H(z) = \frac{z^2}{(z-3)(z+4)}. \qquad \text{(A.54)}$$

The first coefficient c_0 is then determined by

$$c_0 = H(z)|_{z=0} = 0. \qquad \text{(A.55)}$$

The other coefficients c_1 and c_2 are obtained by

$$c_1 = \left[\frac{z-3}{z}\frac{z^2}{(z-3)(z+4)}\right]\bigg|_{z=3} = \frac{3}{7}, \qquad \text{(A.56)}$$

and

$$c_1 = \left[\frac{z+4}{z}\frac{z^2}{(z-3)(z+4)}\right]\bigg|_{z=-4} = \frac{4}{7}. \quad (A.57)$$

Therefore, the inverse z-transform, $h[n]$, is given by

$$h[n] = \frac{3}{7}3^n u[n] + \frac{4}{7}(-4)^n u[n]. \quad (A.58)$$

A.4 The z-Transform All-Pass and Phase Systems

This section presents a theory and a method of all-pass and phase systems based on the z-transform. The all-pass and phase systems have many application values in inverse filtering, compensation for phase distortion, deconvolution, and system identification in the areas of signal processing and digital communications.

A.4.1 All-Pass Systems

Assume that a stable system, denoted by $H_{ap}(z)$, has a z-transform transfer function as follows:

$$H_{ap}(z) = \frac{z^{-1} - a^*}{1 - az^{-1}}, \quad (A.59)$$

where a^* is the complex conjugate of a. Equation (A.59) has a frequency-response magnitude, but it is independent of ω. To see this, we rewrite (A.59) into the following form in the frequency domain by using $z = e^{j\omega}$ and obtain the discrete-time Fourier transform (DTFT) representation,

$$\begin{aligned} H_{ap}(e^{j\omega}) &= \frac{e^{-j\omega} - a^*}{1 - ae^{-j\omega}} \\ &= e^{-j\omega}\left(\frac{1 - a^* e^{j\omega}}{1 - ae^{-j\omega}}\right), \end{aligned} \quad (A.60)$$

where the term $e^{-j\omega}$ has a unity magnitude. Further note that the remaining term's numerator and denominator have the same magnitude because of the complex conjugates of each other. Thus,

$|H_{ap}(e^{j\omega})|$ in (A.60) is equal to unity. In other words, the z-transform transfer function system $H_{ap}(z)$ in (A.59) passes all of the frequency components of its input without changing magnitude gain. Due to this reason, (A.59) is referred to as the *all-pass* or the *all-phase* system.

There are two important properties of an all-phase system, including: (1) the group delay of the all-phase system is always positive, and (2) the phase of the all-phase system is negative within the frequency range of $0 < \omega < \pi$.

A.4.2 Phase Systems

Consider the z-transform of a finite impulse response (FIR) transfer function system $H(z)$ with an arbitrary length as follows:

$$H(z) = C_0(1 - z_1 z^{-1})(1 - z_2 z^{-1}) \cdots (1 - z_M z^{-1}), \quad (A.61)$$

where $z = e^{j\omega}$, z_i ($i = 1, 2, \cdots, M$) denotes the zeros and C_0 is an arbitrary constant. Equation (A.61) is referred to as a *minimum-phase system* if all the zeros are inside the unit circle in the z-plane. Equation (A.61) is called a *maximum-phase system* if all the zeros are outside the unit circle in the z-plane. Furthermore, if the FIR system in (A.61) has some of its zeros inside the unit circle and its remaining zeros outside the unit circle in the z-plane, we then refer to the FIR system $H(z)$ as a *mixed-phase system*.

For the z-transform of an infinite impulse response (IIR) transfer function system given by

$$H(z) = \frac{B(z)}{A(z)}. \quad (A.62)$$

Equation (A.62) is referred to as a minimum-phase system if all the zeros and poles are inside the unit circle in the z-plane. Equation (A.62) is called a maximum-phase system if all the zeros are outside the unit circle and all the poles are inside the unit circle in the z-plane. Moreover, the IIR transfer function system in (A.62) is known as a mixed-phase system if some, but not all, of the zeros are outside the unit circle, and all the poles are inside the unit circle in the z-plane.

A minimum-phase system means that it has a minimum delay function, while a maximum-phase system implies that it has a maximum delay characteristic. In addition, if we are given a magnitude-square function of $H(z)$ and $H(z)$ is a minimum-phase system as follows:

$$\begin{aligned} C(z) &= |H(z)|^2 \\ &= H(z)H^*(1/z), \end{aligned} \quad (A.63)$$

we can then determine the z-transform transfer function $H(z)$ uniquely.

A.4.3 Decomposition of Phase Systems

The z-transform of any rational transfer function system $H(z)$ can be decomposited as follows:

$$H(z) = H_{min}(z)H_{ap}(z), \quad (A.64)$$

where $H_{min}(z)$ is a minimum-phase system and $H_{ap}(z)$ is an all-phase system. This is to say that any nonminimum-phase system can be formed into the minimum-phase system $H_{min}(z)$ by reflecting one or more zeros within the unit circle into their conjugate reciprocal locations outside the unit circle by using (A.64). On the other hand, a minimum-phase system can be created from a nonminimum-phase system by reflecting all the zeros lying outside the unit circle into their conjugate reciprocal locations inside the unit circle. However, the frequency response transfer function system in (A.64) will have the same magnitudes in the case of both the minimum-phase and nonminimum-phase systems.

A.4.4 Compensation Systems

Consider a transmitting signal over a communication channel that has an undersirable frequency response $H_d(z)$. Let $x[n]$ denote the input signals and $y[n]$ denote the output signals of the communication channel. We want to develop a perfect transfer function system $H_c(z)$ that is able to compensate the communication channel distortion, that is, $H_c(z)$ is the inverse of $H_d(z)$ and $y[n] = x[n]$.

Assume that the transfer function system $H_d(z)$ of the communication channel is stable and causal and is a rational system. Thus, by using (A.64), we can decomposite $H_d(z)$ as follows:

$$H_d(z) = H_{min}(z)H_{ap}(z), \qquad (A.65)$$

where $H_{min}(z)$ is the minimum-phase system and $H_{ap}(z)$ is the all-phase system for the communication channel. Now, if we choose the compensation system $H_c(z)$ to have the following form

$$H_c(z) = \frac{1}{H_{min}(z)}, \qquad (A.66)$$

then the overall transfer function system $H(z)$ such that $Y(z) = H(z)X(z)$ is given by

$$H(z) = H_d(z)H_c(z) = H_{ap}(z). \qquad (A.67)$$

It is interesting to see that (A.67) corresponds to an all-pass system. Therefore, the magnitude of frequency response is exactly compensated, but the phase response is changed to $H_{ap}(e^{j\omega})$.

A.4.5 FIR Systems to Minimum-Phase Systems

The z-transform of any FIR linear phase system can be factored in terms of a mininum-phase system $H_{min}(z)$, a maximum-phase system $H_{max}(z)$, and a $H_{uc}(z)$ that only contains zeros in the unit circle given by Oppenheim and Schafer [3] as follows:

$$H(z) = H_{min}(z)H_{uc}(z)H_{max}(z), \qquad (A.68)$$

where

$$H_{max}(z) = z^{-M}H_{min}(1/z), \qquad (A.69)$$

where M is the number of zeros in the minimum-phase system $H_{min}(z)$. Note that the minimum-phase system $H_{min}(z)$ in (A.68) has all M zeros inside the unit circle, $H_{uc}(z)$ has all N zeros on the unit circle, and $H_{max}(z)$ has all M zeros outside the unit circle. In fact, all the zeros of the $H_{max}(z)$ are the reciprocals of the zeros of the minimum-phase system $H_{min}(z)$. Thus, the order of the z-transform of any FIR linear phase system $H(z)$ is equal to $2M + N$.

For further reading, we recommend to the interested reader the books by Miao and Clements [1], Oppenheim and Schafer [3], and Proakis and Manolakis [6].

References

[1] Miao, G. J., and M. A. Clements, *Digital Signal Processing and Statistical Classification*, Artech House, Norwood, Massachusetts, 2002.

[2] Oppenheim, A. V., A. S. Willsky, and I. T. Young, *Signal and Systems*, Prentice Hall, Englewood Cliffs, New Jersey, 1983.

[3] Oppenheim, A. V., and R. W. Schafer, *Discrete-Time Signal Processing*, Prentice Hall, Englewood Cliffs, New Jersey, 1989.

[4] D'Azzo, J. J., and C. H. Houpis, *Linear Control System Analysis and Design: Conventional and Modern*, 3rd ed., McGraw-Hill Publishing Company, New York, 1988.

[5] Churchill, R. V., and J. W. Brown, *Introduction to Complex Variables and Applications*, McGraw-Hill Book Company, New York, 1984.

[6] Proakis, J. G., and D. G. Manolakis, *Digital Signal Processing: Principles, Algorithms, and Applications*, 3rd ed., Prentice Hall, Upper Saddle River, New Jersey, 1996.

Appendix B: Matrix Theory

B.1 Introduction

In this appendix, we provide a summary of fundamental definitions and results in a matrix theory. The goal here is to review those results from the matrix theory that have a direct relevance to this book. Thus, we present the necessary definitions and results of the matrix theory along with some of the proofs in detail.

B.2 Vector Definitions

An array of real-valued numbers or complex-valued numbers, which are denoted by the lowercase bold letter **x**, is called a vector. The vector **x** is assumed to be a row vector given by

$$\mathbf{x} = [x_1, x_2, x_3], \tag{B.1}$$

or a column vector given by

$$\mathbf{x} = \begin{bmatrix} x_1 \\ x_2 \\ x_3 \end{bmatrix}. \tag{B.2}$$

If the vector **x** contains real values, the vector **x** is then said to be a *real vector*. If the vector **x** has complex values, the vector **x** is then said to be a *complex vector*. If the vector **x** includes N values, the vector **x** is then called an N-dimensional vector defined by

$$\mathbf{x} = \begin{bmatrix} x_1 \\ x_2 \\ \vdots \\ x_N \end{bmatrix}, \quad (B.3)$$

or in another form denoted by $\mathbf{x}^T$,

$$\mathbf{x}^T = [x_1, x_2, ..., x_N], \quad (B.4)$$

where $\mathbf{x}^T$ is known as the *transpose of the vector* $\mathbf{x}$. In general, we often write a column vector as the transpose of a row vector.

A complex conjugate of the transpose of the vector $\mathbf{x}^T$, denoted by $\mathbf{x}^H$, is defined by

$$\begin{aligned} \mathbf{x}^H &= \{\mathbf{x}^T\}^* \\ &= [x_1^*, x_2^*, ..., x_N^*], \end{aligned} \quad (B.5)$$

where $\mathbf{x}^H$ is called the *Hermitian transpose*.

In many vector operations, we are interested in finding the magnitude of a vector $\mathbf{x}$. The magnitude of the vector $\mathbf{x}$ is defined as

$$\|\mathbf{x}\| = \left[\sum_{j=1}^{N} x_j^2\right]^{1/2}. \quad (B.6)$$

Equation (B.6) is also referred to as the *Euclidean norm*.

Normalizing a vector $\mathbf{x}$ by dividing its Euclidean norm is to have a unit magnitude given by

$$\mathbf{y}_x = \frac{\mathbf{x}}{\|\mathbf{x}\|}. \quad (B.7)$$

Equation (B.7) is called a *unit norm vector*.

Consider two real vectors $\mathbf{a} = [a_1, a_2, \cdots, a_N]^T$ and $\mathbf{b} = [b_1, b_2, \cdots, b_N]^T$. The *inner product* of the two vectors is defined by

$$\begin{aligned} <\mathbf{a}, \mathbf{b}> &= \mathbf{a}^T \mathbf{b} \\ &= \sum_{j=1}^{N} a_j b_j. \end{aligned} \quad (B.8)$$

If (B.8) is equal to zero, then the two vectors **a** and **b** are said to be *orthogonal*. Furthermore, if the **a** and **b** are orthogonal and have unit norm, then the **a** and **b** are said to be *orthonormal*.

B.3 Matrix Definitions

A rectangular array of numbers is referred to as a *matrix*. We denote it by **A**. If the matrix **A** has n rows and p columns, then the matrix **A** is said to be $(n \times p)$ matrix defined by

$$\mathbf{A} = \{a_{ij}\} = \begin{bmatrix} a_{11} & a_{12} & \cdots & a_{1p} \\ a_{21} & a_{22} & \cdots & a_{2p} \\ \vdots & \vdots & \vdots & \vdots \\ a_{n1} & a_{n2} & \cdots & a_{np} \end{bmatrix}, \qquad (B.9)$$

where a_{ij} is the value in ith row and jth column of the matrix **A**, $i = 1, 2, ..., n$, and $j = 1, 2, ..., p$. The transpose of the matrix **A** is produced by interchanging the rows and columns defined by

$$\mathbf{A} = \begin{bmatrix} a_{11} & a_{21} & \cdots & a_{n1} \\ a_{12} & a_{22} & \cdots & a_{n2} \\ \vdots & \vdots & \vdots & \vdots \\ a_{1p} & a_{2p} & \cdots & a_{np} \end{bmatrix}. \qquad (B.10)$$

A matrix $(n \times n)$ **A** is called a *square matrix* if its number of rows and number of columns are equal. If a square matrix **A** has all of its elements equal to zero except those elements along the main diagonal, this matrix **A** is referred to as a *diagonal matrix*. The diagonal matrix **A** is defined by

$$\mathbf{A} = \begin{bmatrix} a_{11} & 0 & \cdots & 0 \\ 0 & a_{22} & \cdots & 0 \\ \vdots & \vdots & \vdots & \vdots \\ 0 & 0 & \cdots & a_{nn} \end{bmatrix}. \qquad (B.11)$$

Furthermore, a diagonal matrix with the diagonal elements equal to 1 is referred to as an *identity matrix* given by

$$\mathbf{I} = \begin{bmatrix} 1 & 0 & \cdots & 0 \\ 0 & 1 & \cdots & 0 \\ \vdots & \vdots & \vdots & \vdots \\ 0 & 0 & \cdots & 1 \end{bmatrix}. \tag{B.12}$$

Note that the identity matrix is the multiplicative identity for a square matrix $\mathbf{A}$, that is,

$$\mathbf{IA} = \mathbf{AI} = \mathbf{A}. \tag{B.13}$$

A square matrix $\mathbf{A}$ is said to be a *Toeplitz matrix* if all the elements along with each of diagonals have the same value. For instance, we show a (4×4) Toeplitz matrix as follows:

$$\mathbf{A} = \begin{bmatrix} 1 & 2 & 9 & 10 \\ 3 & 1 & 2 & 9 \\ 5 & 3 & 1 & 2 \\ 15 & 5 & 3 & 1 \end{bmatrix}. \tag{B.14}$$

Moreover, if a Toeplitz matrix is symmetric, then all of the elements in the matrix $\mathbf{A}$ can be determined by using either the first column or the first row of the matrix. To see this, we show a (4×4) Toeplitz matrix with symmetric elements as follows:

$$\mathbf{A} = \begin{bmatrix} 1 & 4 & 7 & 9 \\ 4 & 1 & 4 & 7 \\ 7 & 4 & 1 & 4 \\ 9 & 7 & 4 & 1 \end{bmatrix}. \tag{B.15}$$

A matrix $\mathbf{A}$ that has nonzero values on and below its main diagonal is called a *lower triangular matrix* defined as

$$\mathbf{A} = \begin{bmatrix} a_{11} & 0 & 0 & 0 \\ a_{21} & a_{22} & 0 & 0 \\ a_{31} & a_{32} & a_{33} & 0 \\ a_{41} & a_{42} & a_{43} & a_{44} \end{bmatrix}. \tag{B.16}$$

An *upper triangular matrix* can be defined in a similar way.

B.4 Orthogonal Matrices

Consider a square matrix with a dimension $(p \times p)$ denoted by $\mathbf{A}$. If the square matrix $\mathbf{A}\mathbf{A}^T = \mathbf{I}$, the matrix $\mathbf{A}$ is then called the orthonormal. The orthogonal matrix $\mathbf{A}$ has many properties as follows:

1. $\mathbf{A}^{-1} = \mathbf{A}^T$.
2. $|\mathbf{A}| = \pm 1$.
3. $\mathbf{a}_i^T \mathbf{a}_j = \begin{cases} 1, & i = j \\ 0, & i \neq j, \end{cases}$ where $\mathbf{a}_i$ is the columns of matrix $\mathbf{A}$.
4. If $\mathbf{A}$ and $\mathbf{B}$ are orthogonal matrices, respectively, then, $\mathbf{C} = \mathbf{A}\mathbf{B}$ is also orthogonal.

B.5 Trace

The trace of the square matrix $\mathbf{A}$ with dimension $(p \times p)$ is defined as

$$\text{tr}(\mathbf{A}) = \sum_{i=1}^{p} a_{ii}. \tag{B.17}$$

Equation (B.17) also satisfies for the operation of two square matrices $\mathbf{A}$ and $\mathbf{B}$ and the scalar β with the following properties:

$$\text{tr}(\beta) = \beta, \tag{B.18}$$

$$\text{tr}(\mathbf{A} \pm \mathbf{B}) = \text{tr}(\mathbf{A}) \pm \text{tr}(\mathbf{B}), \tag{B.19}$$

$$\text{tr}(\beta \mathbf{A}) = \beta \text{tr}(\mathbf{A}), \tag{B.20}$$

$$\text{tr} \sum \mathbf{x}_i^T \mathbf{A} \mathbf{x}_i = \text{tr}(\mathbf{A} \sum \mathbf{x}_i \mathbf{x}_i^T), \tag{B.21}$$

and

$$\text{tr}(\mathbf{A}\mathbf{B}) = \text{tr}(\mathbf{B}\mathbf{A}) = \sum_i \sum_j a_{ij} b_{ji}. \tag{B.22}$$

In a special case of (B.22) when the square matrix $\mathbf{B} = \mathbf{A}^T$, we obtain

$$\text{tr}(\mathbf{A}\mathbf{A}^T) = \text{tr}(\mathbf{A}^T \mathbf{A}) = \sum a_{ij}^2. \tag{B.23}$$

If the square matrix $\mathbf{A}$ is orthogonal, then (B.23) is given by

$$\mathrm{tr}(\mathbf{A}\mathbf{A}^T) = \mathrm{tr}(\mathbf{I}) = \sum_{i=1}^{p} I_{ii} = p. \qquad (\text{B.24})$$

In addition, for matrices $\mathbf{C}$ and $\mathbf{D}$ with $n \times p$ and $p \times n$, respectively, the trace function can be expressed as

$$\mathrm{tr}(\mathbf{C}\mathbf{D}) = \mathrm{tr}(\mathbf{D}\mathbf{C}) = \sum_i \sum_j c_{ij} d_{ji} = \sum_j \sum_i d_{ji} c_{ij}. \qquad (\text{B.25})$$

B.6 Matrix Differentiation

Consider an $n \times p$ matrix $\mathbf{X}$ and a function of the matrix $\mathbf{X}$, denoted by $f(\mathbf{X})$. We then take the derivative of matrix function $f(\mathbf{X})$ with respect to the matrix $\mathbf{X}$ and obtain

$$\frac{\partial f(\mathbf{X})}{\partial \mathbf{X}} = \frac{\partial f(\mathbf{X})}{\partial x_{ij}}, \qquad (\text{B.26})$$

where $i = 1, 2, ..., n$ and $j = 1, 2, ..., p$. The derivative result is also an $n \times p$ matrix. Some useful results of the matrix differentiation are listed as follows:

1. $\frac{\partial \mathbf{b}^T \mathbf{x}}{\partial \mathbf{x}} = \mathbf{b}$.

2. $\frac{\partial \mathbf{x}^T \mathbf{x}}{\partial \mathbf{x}} = 2\mathbf{x}$.

3. $\frac{\partial \mathbf{x}^T \mathbf{B} \mathbf{x}}{\partial \mathbf{x}} = \mathbf{B}\mathbf{x} + \mathbf{x}^T \mathbf{B}$.

4. $\frac{\partial \mathbf{x}^T \mathbf{B} \mathbf{y}}{\partial \mathbf{x}} = \mathbf{B}\mathbf{y}$.

B.7 Determinants

The determinant of an $(n \times n)$ square matrix $\mathbf{A}$ is denoted by $\det(\mathbf{A})$ and can be solved recursively in terms of the determinants of $(n-1) \times (n-1)$ matrices given by

$$\det(\mathbf{A}) = \sum_{i=1}^{n} (-1)^{i+j} a_{ij} \det(\mathbf{A}_{ij}), \qquad (\text{B.27})$$

where $\mathbf{A}_{ij}$ is the $(n-1) \times (n-1)$ matrix that is obtained by deleting the ith row and the jth column of the matrix $\mathbf{A}$.

If the matrix $\mathbf{A}$ is singular, its determinant will be equal to zero. In this case, there does not exist an inverse of the matrix $\mathbf{A}$. On the other hand, a square matrix $\mathbf{A}$ is said to be singular if $\det(\mathbf{A}) = 0$, and nonsingular if $\det(\mathbf{A}) \neq 0$.

Assume that $\mathbf{A}$ and $\mathbf{B}$ are $n \times n$ matrices. Then we can list the properties of the determinants as follows:

1. $\det(\mathbf{AB}) = \det(\mathbf{A})\det(\mathbf{B})$.
2. $\det(\mathbf{A}^T) = \det(\mathbf{A})$.
3. $\det(\alpha \mathbf{A}) = \alpha^n \det(\mathbf{A})$, where α is a constant.
4. $\det(\mathbf{A}^{-1}) = \frac{1}{\det(\mathbf{A})}$, if $\mathbf{A}$ is invertible.
5. The determinant of a diagonal matrix is the product of its diagonal elements. It is also true for lower or upper triangular square matrices.

Example B.1

Let a square matrix

$$\mathbf{A} = \begin{bmatrix} a & b \\ c & d \end{bmatrix}. \tag{B.28}$$

Then the determinant of the square matrix $\mathbf{A}$ using (B.27) is obtained by

$$\det(\mathbf{A}) = ad - bc. \tag{B.29}$$

Now, let a square matrix

$$\mathbf{A} = \begin{bmatrix} 1 & 2 & 3 \\ 2 & 3 & 2 \\ 3 & 4 & 3 \end{bmatrix}. \tag{B.30}$$

Then the determinant of the square matrix $\mathbf{A}$ using (B.27) is obtained by

$$\det(\mathbf{A}) = 1 \times \begin{vmatrix} 3 & 2 \\ 4 & 3 \end{vmatrix} - 2 \times \begin{vmatrix} 2 & 3 \\ 4 & 3 \end{vmatrix} + 3 \times \begin{vmatrix} 2 & 3 \\ 3 & 2 \end{vmatrix}$$
$$= -2. \tag{B.31}$$

B.8 Matrix Inversion

If a square matrix $\mathbf{A}$ is a nonsingular, $\det(\mathbf{A}) \neq 0$, then there is a closed form expression for a matrix inversion given by

$$\mathbf{A}^{-1} = \frac{\text{adj}(\mathbf{A})}{\det(\mathbf{A})}, \tag{B.32}$$

where $\text{adj}(\mathbf{A})$ is the *adjugate* of the square matrix $\mathbf{A}$ defined by

$$\text{adj}(\mathbf{A}) = \text{cofactor of } A_{ji}, \tag{B.33}$$

where the (i,j) element of the adjugate is equal to the cofactor of the (j,i) element of the square matrix $\mathbf{A}$.

Another useful formula for inverse matrix is known as the matrix inversion lemma given by

$$(\mathbf{A} + \mathbf{BCD})^{-1} = \mathbf{A}^{-1} - \mathbf{A}^{-1}\mathbf{B}(\mathbf{C}^{-1} + \mathbf{DA}^{-1}\mathbf{B})^{-1}\mathbf{DA}^{-1}, \tag{B.34}$$

where $\mathbf{A}$ is $n \times n$, $\mathbf{B}$ is $n \times m$, $\mathbf{C}$ is $m \times m$, and $\mathbf{D}$ is $m \times n$, and $\mathbf{A}$ and $\mathbf{C}$ are nonsingular matrices.

In addition, if matrices $\mathbf{A}$ and $\mathbf{B}$ are invertible matrices, we have a property of $(\mathbf{AB})^{-1} = \mathbf{B}^{-1}\mathbf{A}^{-1}$.

B.9 Eigenanalysis

Eigenanalysis is one of the most powerful concepts in the matrix operation. This is because the eigenanalysis, including eigenvalues and eigenvectors, provides useful and important information about a matrix. It also contributes to an important representation for matrices known as the *eigenvalue decomposition*.

Consider any square matrix $\mathbf{A}$ with dimension $n \times n$. We establish a matrix function as follows:

$$E(\lambda) = \mid \mathbf{A} - \lambda \mathbf{I} \mid, \tag{B.35}$$

which is an nth order polynomial in λ, where the n roots of $E(\lambda)$, $\lambda_1, \lambda_2, ..., \lambda_n$ are called eigenvalues of the square matrix $\mathbf{A}$.

For each of eigenvalues, λ_i, $i = 1, 2, ..., n$, the matrix function $(\mathbf{A} - \lambda \mathbf{I})$ is singular. There is at least one nonzero vector, $\mathbf{v}_i$, such that

$$\mathbf{A}\mathbf{v}_i = \lambda_i \mathbf{v}_i. \tag{B.36}$$

These vectors, $\mathbf{v}_i$, are called the eigenvectors of the square matrix $\mathbf{A}$. In addition, an eigenvector $\mathbf{v}_i$ is called normalized if it has unit norm, $\| \mathbf{v}_i \| = \mathbf{v}^T \mathbf{v} = 1$.

Another theory of the eigenanalysis is that if a matrix $\mathbf{B}$ is a nonsingular square matrix, then matrices $\mathbf{A}$ and $\mathbf{BAB}^{-1}$ have the same eigenvalues, λ_i. Furthermore, the matrix $\mathbf{BAB}^{-1}$ has the eigenvector $\mathbf{u}_i = \mathbf{B}\mathbf{v}_i$. The theory can be proved as follows:

Note that the matrix function in (B.35) can be rewritten as

$$\begin{aligned} E(\lambda) &= |\mathbf{A} - \lambda \mathbf{I}| \\ &= |\mathbf{A} - \lambda \mathbf{B}^{-1}\mathbf{B}| \\ &= |\mathbf{B} \| \mathbf{A} - \lambda \mathbf{B}^{-1}\mathbf{B} \| \mathbf{B}^{-1}| \\ &= |\mathbf{BAB}^{-1} - \lambda \mathbf{I}|. \end{aligned} \tag{B.37}$$

Now, it is clear to see from (B.37) that the matrix $\mathbf{A}$ and the matrix $\mathbf{BAB}^{-1}$ have the same eigenvalues. Since $\mathbf{v}_i$ is an eigenvector of the matrix $\mathbf{A}$ for eigenvalue λ_i, then there exists a nonzero vector $\mathbf{u}$ satisfying the condition as follows:

$$\mathbf{BAB}^{-1}(\mathbf{B}\mathbf{v}_i) = \lambda_i (\mathbf{B}\mathbf{v}_i). \tag{B.38}$$

Therefore, the eigenvector $\mathbf{u}_i = \mathbf{B}\mathbf{v}_i$ is an eigenvector of the matrix $\mathbf{BAB}^{-1}$ for the eigenvalue λ_i.

B.10 Spectral Decomposition Theorem

Consider any symmetric matrix $\mathbf{A}$ with dimension $(p \times p)$. Then the symmetric matrix $\mathbf{A}$ can be written as

$$\begin{aligned} \mathbf{A} &= \mathbf{R}\Lambda\mathbf{R}^T \\ &= \sum_{i=1}^{p} \lambda_i \mathbf{r}_i \mathbf{r}_i^T, \end{aligned} \tag{B.39}$$

where Λ is a diagonal matrix consisting of the eigenvalues of $\mathbf{A}$, $\mathbf{R}$ is an orthogonal matrix whose columns are normalized eigenvectors, and $\mathbf{r}_i$ is the corresponding eigenvectors. Equation (B.39) is known as the *spectral decomposition theorem* or sometimes referred to as the *Jordan decomposition theorem*. This theorem can be approved as follows.

Assume that there are orthonormal eigenvectors $\mathbf{r}_1, \mathbf{r}_2, \cdots, \mathbf{r}_p$ such that $\mathbf{A}\mathbf{r}_i = \lambda_i \mathbf{r}_i$ for some numbers λ_i. Then we obtain

$$\begin{aligned} \mathbf{r}_i^T \mathbf{A} \mathbf{r}_j &= \lambda_i \mathbf{r}_i^T \mathbf{r}_j \\ &= \begin{cases} \lambda_i, & \text{if } i = j, \\ 0, & \text{otherwise.} \end{cases} \end{aligned} \qquad (B.40)$$

In a matrix form, we can write (B.40) as follows:

$$\mathbf{R}^T \mathbf{A} \mathbf{R} = \Lambda. \qquad (B.41)$$

Multiplying by using $\mathbf{R}$ and $\mathbf{R}^T$ on both sides of (B.41), we obtain

$$\mathbf{R}\mathbf{R}^T \mathbf{A} \mathbf{R}\mathbf{R}^T = \mathbf{R}\Lambda\mathbf{R}^T. \qquad (B.42)$$

Note that $\mathbf{R}\mathbf{R}^T = \mathbf{I}$. Thus, we can rewrite (B.42) as

$$\mathbf{A} = \mathbf{R}\Lambda\mathbf{R}^T. \qquad (B.43)$$

Compared with (B.37), it is interesting to see that the matrix $\mathbf{A}$ and the matrix $\mathbf{R}\Lambda\mathbf{R}$ have the same eigenvalues. Therefore, the elements of Λ are exactly the eigenvalues of symmetric matrix $\mathbf{A}$ with the same multiplicity.

Furthermore, if $\mathbf{A}$ is a nonsingular symmetric matrix, then for any integer n, we can have as follows:

$$\Lambda^n = \text{diag}(\lambda_i^n), \qquad (B.44)$$

or in a matrix form,

$$\mathbf{A}^n = \mathbf{R}\Lambda^n\mathbf{R}^T. \qquad (B.45)$$

In addition, if all the eigenvalues of the symmetric matrix $\mathbf{A}$ are positive, we can even define the rational powers as follows:

$$\Lambda^{p/q} = \mathrm{diag}(\lambda_i^{p/q}), \tag{B.46}$$

or in a matrix form

$$\mathbf{A}^{p/q} = \mathbf{R}\Lambda^{p/q}\mathbf{R}^T. \tag{B.47}$$

for integers q and p, where $q > 0$. If some of the eigenvalues of matrix $\mathbf{A}$ are zero, then (B.45) and (B.47) still hold if the exponents p/q are restricted to be positive. This expanded theory can be approved as follows.

In order to prove the theory in (B.45), using the spectral decomposition theorem in (B.39), we obtain

$$\mathbf{A} = \mathbf{R}\Lambda\mathbf{R}^T. \tag{B.48}$$

If we use the power of 2 on both sides of (B.48), then we obtain

$$\begin{aligned} \mathbf{A}^2 &= (\mathbf{R}\Lambda\mathbf{R}^T)^2 \\ &= \mathbf{R}\Lambda\mathbf{R}^T\mathbf{R}\Lambda\mathbf{R}^T \\ &= \mathbf{R}\Lambda^2\mathbf{R}^T. \end{aligned} \tag{B.49}$$

Because of $\mathbf{R}^T\mathbf{R} = \mathbf{I}$, then we obtain,

$$\mathbf{A}^{-1} = \mathbf{R}\Lambda^{-1}\mathbf{R}^T, \tag{B.50}$$

where $\Lambda^{-1} = \mathrm{diag}(\lambda_i^{-1})$. Therefore, by using induction, the theory of (B.45) can be proved completely.

In order to prove the theory in (B.47), we use the notation operation as follows:

$$\begin{aligned} \mathbf{A}^p &= (\mathbf{A}^{p/q})^q \\ &= \mathbf{R}\Lambda^{p/q}\mathbf{R}^T \cdots \mathbf{R}\Lambda^{p/q}\mathbf{R}^T \\ &= \mathbf{R}\Lambda^{p/q}\mathbf{R}^T. \end{aligned} \tag{B.51}$$

There is an important case in (B.47) when $p = 1$ and $q = 2$. In this case, we have

$$\mathbf{A}^{1/2} = \mathbf{R}\Lambda^{1/2}\mathbf{R}^T, \tag{B.52}$$

where $\Lambda^{1/2} = \text{diag}(\lambda_i^{1/2})$ and $\lambda_i \geq 0$ for all i, and

$$\mathbf{A}^{-1/2} = \mathbf{R}\Lambda^{-1/2}\mathbf{R}^T, \tag{B.53}$$

where $\Lambda^{-1/2} = \text{diag}(\lambda_i^{-1/2})$ and $\lambda_i > 0$ for all i. Equations (B.52) and (B.53) are called the *symmetric square root decomposition* of the matrix $\mathbf{A}$.

Another theory states that for any symmetric matrix $\mathbf{A}$, there exists an orthogonal transformation given by

$$\mathbf{y} = \mathbf{R}^T\mathbf{x}, \tag{B.54}$$

such that

$$\mathbf{x}^T\mathbf{A}\mathbf{x} = \sum_i \lambda_i y_i^2. \tag{B.55}$$

To prove the theory of (B.54) and (B.55), we again consider using the spectral decomposition theorem of $\mathbf{A} = \mathbf{R}\Lambda\mathbf{R}^T$. Thus, we can express it as follows:

$$\begin{aligned} \mathbf{x}^T\mathbf{A}\mathbf{x} &= \mathbf{x}^T\mathbf{R}\Lambda\mathbf{R}^T\mathbf{x} \\ &= \mathbf{y}^T\mathbf{R}^T\mathbf{R}\Lambda\mathbf{R}^T\mathbf{R}\mathbf{y} \\ &= \mathbf{y}^T\Lambda\mathbf{y} \\ &= \sum_i \lambda_i y_i^2, \end{aligned} \tag{B.56}$$

where the $\mathbf{R}$ is the columns of the eigenvectors for the matrix $\mathbf{A}$ and $\lambda_1, ..., \lambda_p$ are the corresponding eigenvalues.

B.11 Singular Value Decomposition

If $\mathbf{S}$ is an $(n \times p)$ matrix with rank r, then $\mathbf{S}$ can be written as

$$\mathbf{S} = \mathbf{U}\mathbf{D}\mathbf{V}^T, \tag{B.57}$$

where $\mathbf{U}$ and $\mathbf{V}$ are orthonormal matrices of dimensions $(n \times r)$ and $(p \times r)$, respectively, and $\mathbf{D}$ is a diagonal matrix with positive elements. Equation (B.57) is known as the *singular value decomposition theorem*.

To prove the singular value decomposition theorem in (B.57), we note that $\mathbf{U}$ and $\mathbf{V}$ are orthonormal matrices of dimensions $(n \times r)$ and $(p \times r)$, respectively. Thus, we immediately obtain as follows:

$$\mathbf{U}^T\mathbf{U} = \mathbf{I}_r, \tag{B.58}$$

and

$$\mathbf{V}^T\mathbf{V} = \mathbf{I}_r. \tag{B.59}$$

Now, letting $\mathbf{Y} = \mathbf{S}^T\mathbf{S}$, we note that $\mathbf{Y}$ is a symmetric matrix with a rank r. We then rewrite the matrix $\mathbf{Y}$ by using the spectral decomposition theorem

$$\begin{aligned}\mathbf{Y} &= \mathbf{S}^T\mathbf{S} \\ &= \mathbf{V}\Lambda\mathbf{V}^T,\end{aligned} \tag{B.60}$$

where $\mathbf{V}$ is a column orthonormal matrix of eigenvectors of $\mathbf{Y}$ and $\Lambda = \operatorname{diag}(\lambda_1, \lambda_2, \cdots, \lambda_r)$ with $\lambda_i > 0$ since

$$\begin{aligned}\lambda_i &= \mathbf{v}_i^T\mathbf{S}^T\mathbf{S}\mathbf{v}_i \\ &= \|\mathbf{S}\mathbf{v}_i\| \\ &> 0.\end{aligned} \tag{B.61}$$

Further let $d_i = \lambda_i^{1/2}$, for $i = 1, 2, \cdots, r$. This implies that

$$\mathbf{D} = \operatorname{diag}[d_1, d_2, \cdots, d_r]. \tag{B.62}$$

Also, the $\mathbf{U}$ matrix with dimension $(n \times r)$ is defined by

$$\mathbf{u}_i = d_i^{-1}\mathbf{S}\mathbf{v}_i, \quad i = 1, 2, \cdots, r. \tag{B.63}$$

Then, we have the results as follows:

$$\begin{aligned}\mathbf{u}_i^T\mathbf{u}_j &= d_i^{-1}d_j^{-1}\mathbf{v}_i^T\mathbf{S}^T\mathbf{S}\mathbf{v}_i \\ &= \lambda_i d_i^{-1}d_j^{-1}\mathbf{v}_i\mathbf{v}_j \\ &= \begin{cases} 1, & \text{if } i = j, \\ 0, & \text{otherwise}. \end{cases}\end{aligned} \tag{B.64}$$

Therefore, it is interesting to see that the matrix $\mathbf{U}$ is also an orthonormal matrix.

Assume that $\mathbf{x}$ is any $p \times 1$ vector; it can then be written as

$$\mathbf{x} = \sum_i \alpha_i \mathbf{v}_i. \tag{B.65}$$

Also assume that $\mathbf{e}_i$ is an $r \times 1$ vector that has 1 in the ith position and 0 elsewhere. Using (B.63), we have the results as follows:

$$\begin{aligned}\mathbf{UDV}^T\mathbf{x} &= \sum_i \alpha_i \mathbf{UDe}_i \\ &= \sum_i \alpha_i d_i \mathbf{u}_i \\ &= \sum_i \alpha_i \mathbf{Sv}_i \\ &= \mathbf{Sx}.\end{aligned} \tag{B.66}$$

Therefore, the singular value decomposition theorem of $\mathbf{S} = \mathbf{UDV}^T$ has been approved since the formula in (B.66) holds for any matrix $\mathbf{x}$.

B.12 Quadratic Forms

The quadratic form of a symmetric matrix $\mathbf{A}$ is defined by

$$\mathcal{Q}(\mathbf{x}) = \mathbf{x}^T \mathbf{A} \mathbf{x} = \sum_{i=1}^n \sum_{j=1}^n a_{ij} x_i x_j, \tag{B.67}$$

where $\mathbf{x} = [x_1, x_2, ..., x_n]$ is a vector of n variables.

If $\mathcal{Q}(\mathbf{x}) > 0$ for all $\mathbf{x} \neq \mathbf{0}$, then $\mathcal{Q}(\mathbf{x})$ is called the *positive definite quadratic form*. If $\mathcal{Q}(\mathbf{x}) \geq 0$ for all $\mathbf{x} \neq \mathbf{0}$, then $\mathcal{Q}(\mathbf{x})$ is said to be the *positive semidefinite quadratic form*. A symmetric matrix $\mathbf{A}$ is said to be positive definite or positive semidefinite if $\mathcal{Q}(\mathbf{x})$ is a positive definite quadratic form or positive semidefinite quadratic form, respectively.

An important quantity in matrix theory to manipulate in terms of eigenvalues and eigenvectors is to use the ratio of two quadratic

forms given by
$$F = \frac{\mathbf{x}^T \mathbf{A} \mathbf{x}}{\mathbf{x}^T \mathbf{B} \mathbf{x}}, \tag{B.68}$$
where $\mathbf{B}$ must be positive definite. Equation (B.68) is called the *F-ratio quotient* or sometimes referred to as the *Rayleigh quotient*. The F-ratio quotient is bounded by the minimum and maximum eigenvalues
$$\lambda_{min} \leq F \leq \lambda_{max}, \tag{B.69}$$
where λ_{min} and λ_{max} are the minimum and maximum eigenvalues, respectively.

B.13 Maximization and Minimization Analysis

Consider two symmetric matrices $\mathbf{A}$ and $\mathbf{B}$, and $\mathbf{B} > 0$. A maximum or minimum analysis of
$$\mathbf{x}^T \mathbf{A} \mathbf{x} \quad \text{subject to } \mathbf{x}^T \mathbf{B} \mathbf{x} = 1 \tag{B.70}$$
is to solve when $\mathbf{x}$ is the eigenvector of $\mathbf{B}^{-1}\mathbf{A}$ corresponding to the largest or the smallest eigenvalue of $\mathbf{B}^{-1}\mathbf{A}$, respectively.

Let λ_{max} and λ_{min} be the largest and the smallest eigenvalues of $\mathbf{B}^{-1}\mathbf{A}$, respectively. Subject to the constraint (B.70), optimization solutions provide the results as follows:
$$\max_x (\mathbf{x}^T \mathbf{A} \mathbf{x}) = \lambda_{max}, \tag{B.71}$$
and
$$\min_x (\mathbf{x}^T \mathbf{A} \mathbf{x}) = \lambda_{min}. \tag{B.72}$$

To prove this theory of the maximization and the minimization analysis, we let $\mathbf{B}^{1/2}$ denote the symmetric square root of symmetric matrix $\mathbf{B}$, and let a linear transform be given by
$$\mathbf{y} = \mathbf{B}^{1/2} \mathbf{x}. \tag{B.73}$$
Then the theory in (B.70) can be rewritten as
$$\mathbf{y}^T \mathbf{B}^{-1/2} \mathbf{A} \mathbf{B}^{-1/2} \mathbf{y} \quad \text{subject to } \mathbf{y}^T \mathbf{y} = 1. \tag{B.74}$$

Now, by using the spectral decomposition for the symmetric matrix, the matrix $\mathbf{B}^{-1/2}\mathbf{A}\mathbf{B}^{-1/2}$ can be written in the following form:

$$\mathbf{B}^{-1/2}\mathbf{A}\mathbf{B}^{-1/2} = \mathbf{G}\Lambda\mathbf{G}, \tag{B.75}$$

where $\mathbf{G}$ is an orthogonal matrix and Λ is a diagonal matrix of the eigenvalues of $\mathbf{B}^{-1/2}\mathbf{A}\mathbf{B}^{-1/2}$ with $\lambda_{max} \geq \lambda_2 \geq, ..., \geq \lambda_{min} \geq 0$.
Further let $\mathbf{z} = \mathbf{G}^T\mathbf{y}$. Then we obtain

$$\begin{aligned} \mathbf{z}^T\mathbf{z} &= \mathbf{y}^T\mathbf{G}\mathbf{G}^T\mathbf{y} \\ &= \mathbf{y}^T\mathbf{y}. \end{aligned} \tag{B.76}$$

Thus, by using (B.75) and (B.76), the theory in (B.74) can be rewritten as follows:

$$\max_z(\mathbf{z}^T\Lambda\mathbf{z}) = \max_z\left(\sum \lambda_i z_i^2\right) \quad \text{subject to } \mathbf{z}^T\mathbf{z} = 1. \tag{B.77}$$

If the eigenvalues are in descending order, then (B.77) satisfies the following:

$$\max_z\left(\sum \lambda_i z_i^2\right) \leq \lambda_{max}\max_z\left(\sum z_i^2\right) = \lambda_{max}. \tag{B.78}$$

Therefore, the maximization theorem is proved for the maximization result.

For the minimization theory, the theory in (B.74) can be rewritten in the following form by using (B.75) and (B.76) as follows:

$$\min_z(\mathbf{z}^T\Lambda\mathbf{z}) = \min_z\left(\sum \lambda_i z_i^2\right) \quad \text{subject to } \mathbf{z}^T\mathbf{z} = 1. \tag{B.79}$$

If the eigenvalues are in descending order, then (B.79) satisfies the following:

$$\min_z\left(\sum \lambda_i z_i^2\right) \geq \lambda_{min}\min_z\left(\sum z_i^2\right) = \lambda_{min}. \tag{B.80}$$

Again, this minimization theorem is proved for the minimization result. This, in turn, proves the result in (B.69) for F-ratio quotient.

In this appendix, we have summarized the important developments of the matrix theory, which are particularly useful for understanding the theory development of signal processing for digital communications. There are many excellent references on the matrix theory. To research the relevant topic further, we recommend the interested reader to Mardia, Kent, and Bibby [1]. For understanding the matrix theory in applications, the interested reader is referred to Kailath [2], Miao [3], Miao and Clements [4], Vaidyanathan [5], Johnson and Dudgeon [6], Hayes [7], and Neter et al. [8] to name a few.

References

[1] Mardia, K. V., J. T. Kent, and J. M. Bibby, *Multivariate Analysis*, 3rd ed., Academic Press Inc., New York, 1982.

[2] Kailath, T., *Linear Systems*, Prentice-Hall, Englewood Cliffs, New Jersey, 1980.

[3] Miao, J., *Component Feature-Based Digital Waveform Analysis and Classification*, UMI, A Bell & Howell Information Company, Ann Arbor, Michigan, 1996.

[4] Miao, G. J., and M. A. Clements, *Digital Signal Processing and Statistical Classification*, Artech House, Norwood, Massachusetts, 2002.

[5] Vaidyanathan, P. P., *Multirate Systems and Filter Banks*, Prentice Hall, Englewood Cliffs, New Jersey, 1993.

[6] Johnson, D. H., and D. E. Dudgeon, *Array Signal Processing*, Prentice Hall, Englewood Cliffs, New Jersey, 1993.

[7] Hayes, M. H., *Statistical Digital Signal Processing and Modeling*, John Wiley & Sons, New York, 1996.

[8] Neter, J., W. Wasserman, and M. H. Kutner, *Applied Linear Regression Models*, Richard D. Irwin, Inc., Homewood, Illinois, 1983.

Appendix C: The Discrete Fourier Transform

C.1 Introduction

For a finite-duration discrete-time signal, it is possible to develop a Fourier series representation. When the Fourier series is used to represent the finite-duration discrete-time signal, we refer to it as the *discrete Fourier transform* (DFT). The DFT is also a discrete sequence that corresponds to samples of the Fourier transform of the signal, equally spaced in frequency. The DFT plays a central role in the implementation of a variety of digital signal processing (DSP) and communication algorithms, such as orthogonal frequency division multiplexing (OFDM) and discrete multitone (DMT) modulations.

C.2 DFT Operation

The DFT can be defined as an operation for an N-point discrete-time input signal $x[n]$ given by [1–3]

$$X[k] = \sum_{n=0}^{N-1} x[n] W_N^{nk}, \ k = 0, 1, \cdots, N-1, \qquad (C.1)$$

where W_N is the complex quantity expressed as

$$\begin{aligned} W_N &= e^{-j2\pi/N} \\ &= \cos\left(\frac{2\pi}{N}\right) - j\sin\left(\frac{2\pi}{N}\right). \end{aligned} \qquad (C.2)$$

Equation (C.1) is a transformation from an N-point discrete-time input signal $x[n]$ to an N-point set of frequency-domain samples $X[k]$. Note that (C.1) can be considered as two nested loops of a scalar operation containing a complex multiplication and a complex addition. It can also be viewed as a single loop that computes each point of frequency-domain samples $X[k]$ by an inner product of an N-point discrete-time input signal $x[n]$ and an N-point basis factor W_N^{nk} for $n = 0, 1, \cdots, N-1$ when k is fixed.

The DFT can also be interpreted as a "frequency sampling" of the discrete-time Fourier transform (DTFT) given by [2]. It provides the frequency spectrum of the discrete-time input signal $x[n]$,

$$X(f) = \sum_{n=-\infty}^{\infty} x[n] e^{\frac{-j2\pi f n}{F_s}}, \tag{C.3}$$

where F_s is called the sampling rate and f is a frequency value, both in hertz (Hz). Since the discrete-time input signal $x[n]$ has N data points, the signal frequency range, from 0 to F_s, is divided into N discrete points with equally spaced frequencies, $f = \frac{kF_s}{N}$. The DFT can then be expressed as a discrete variable k for frequency,

$$\begin{aligned} X[k] &= \sum_{n=0}^{N-1} x[n] e^{\frac{-j2\pi kn}{N}} \\ &= \sum_{n=0}^{N-1} x[n] W_N^{kn}, \quad k = 0, 1, \cdots, N-1, \end{aligned} \tag{C.4}$$

where $W_N = e^{-j(2\pi/N)}$. The region around each of the N points is referred to as a *frequency bin* or a "tone." Each tone has a bandwidth of $\frac{1}{NT_s}$ or $\frac{F_s}{N}$ Hz.

C.3   IDFT Operation

The corresponding inverse discrete Fourier transform (IDFT) is given by

$$x[n] = \frac{1}{N} \sum_{k=0}^{N-1} X[k] W_N^{-kn}, \quad n = 0, 1, ..., N-1. \tag{C.5}$$

Appendix C: The Discrete Fourier Transform

Note that both the discrete-time input signal $x[n]$ in (C.5) and the frequency-domain samples $X[k]$ in (C.1) may be complex valued.

C.4 DFT Matrix Operation

In a matrix form, we can consider the N-point DFT as a linear transformation of the discrete-time complex input signal vector

$$\mathbf{x} = \left\{ \begin{array}{cccc} x[0] & x[1] & \cdots & x[N-1] \end{array} \right\}^T \quad \text{(C.6)}$$

to another complex vector in the frequency domain

$$\mathbf{X} = \left\{ \begin{array}{cccc} X[0] & X[1] & \cdots & X[N-1] \end{array} \right\}^T. \quad \text{(C.7)}$$

Thus, the DFT operation in the matrix form becomes

$$\mathbf{X} = \mathbf{W}\mathbf{x}, \quad \text{(C.8)}$$

where the matrix $\mathbf{W}$ contains the complex exponentials

$$\mathbf{W} = \begin{bmatrix} 1 & 1 & 1 & \cdots & 1 \\ 1 & W & W^2 & \cdots & W^{N-1} \\ 1 & W^2 & W^4 & \cdots & W^{2(N-1)} \\ \cdot & \cdot & \cdot & \cdots & \cdot \\ \cdot & \cdot & \cdot & \cdots & \cdot \\ \cdot & \cdot & \cdot & \cdots & \cdot \\ 1 & W^{N-1} & W^{2(N-1)} & \cdots & W^{(N-1)^2} \end{bmatrix} \quad \text{(C.9)}$$

where $W = e^{\frac{-j2\pi}{N}}$. The inverse DFT matrix can be obtained by using the inverse matrix $\mathbf{W}^{-1}$ as follows:

$$\mathbf{x} = \mathbf{W}^{-1}\mathbf{X}. \quad \text{(C.10)}$$

It should be pointed out that the column vectors $\mathbf{w}_i$ of matrix $\mathbf{W}$ are *orthonormal* because the inner product $\mathbf{w}_i^T \mathbf{w}_j = 0$ when $i \neq j$ and the inner product $\mathbf{w}_i^T \mathbf{w}_i = 1$ when $i = j$. This, in turn, implies that

$$\mathbf{W}\mathbf{W}^* = N\mathbf{I}, \quad \text{(C.11)}$$

where $\mathbf{W}^*$ denotes the complex conjugate of the matrix $\mathbf{W}$, and $\mathbf{I}$ is an $(N \times N)$ identity matrix. This is to say that the inverse matrix $\mathbf{W}^{-1}$ can be obtained as follows:

$$\mathbf{W}^{-1} = \frac{1}{N}\mathbf{W}^*. \tag{C.12}$$

As a result, the inverse DFT matrix in (C.10) can be expressed as

$$\mathbf{x} = \frac{1}{N}\mathbf{W}^*\mathbf{X}. \tag{C.13}$$

The matrix representations of the DFT in (C.8), (C.10), and (C.13) are also useful for analyses of uniform DFT filter bank–based OFDM and DMT modulations.

Note that N complex multiplications and $(N-1)$ complex additions are required to compute each value of the DFT in (C.4) since the discrete-time input signal $x[n]$ may be complex. Therefore, in order to compute all N values directly, we require a total of N^2 complex multiplications and $N(N-1)$ complex additions. In other words, we need N^2 complex multiplication and addition operations. For a real discrete-time input signal $x[n]$, the number of computations in (C.4) can be halved. This is because the DFT will be conjugate-symmetric with respect to the frequency value $\frac{N}{2}$.

References

[1] Oppenheim, A. V., and R. W. Schafer, *Discrete-Time Signal Processing*, Prentice Hall, Englewood Cliffs, New Jersey, 1989.

[2] Miao, G. J., and M. A. Clements, *Digital Signal Processing and Statistical Classification*, Artech House, Norwood, Massachusetts, 2002.

[3] Chen, C. H., (ed.), *Signal Processing Handbook*, Marcel Dekker, New York, 1988.

Appendix D: The Fast Fourier Transform

D.1 Introduction

In Appendix C, we discussed the DFT and IDFT operations as well as their matrix forms. A direct computation for the DFT operation is expensive. However, by exploiting both the symmetry and periodicity properties of W_N^{kn}, the DFT computational complexity can be reduced. The method of greatly reduced computation complexity was popularized when Cooley and Tukey [1] published a method of an efficient algorithm for the computation of the DFT in 1965. The efficient algorithm is called the fast Fourier transform (FFT). In fact, FFT algorithms are developed based on a method of decomposing a DFT into successively smaller DFTs. This method leads to a variety of different algorithms, all with comparable improvements in computational complexity.

D.2 FFT Methods

In this appendix, we present two basic types of the FFT algorithms with *radix-2*, which have been extensively used for DMT and OFDM modulations, including *decimation-in-time* FFT and *decimation-in-frequency* FFT. In the method of decimation in time, the discrete-time input signal $x[n]$ is decomposed into successively smaller subsequences such that the computations are made on smaller transformations. On the other hand, the method of decimation-in-frequency FFT derives its name from the method in which the frequency-domain sequence $X[k]$ is decomposed into smaller subsequences.

D.2.1 Decimation-in-Time FFT Algorithm

The operation of the decimation-in-time FFT algorithm is based on decomposing the discrete-time input signal $x[n]$ into successively smaller subsequences. To illustrate, we consider a DFT with a length of N given by (C.4) in Appendix C,

$$X[k] = \sum_{n=0}^{N-1} x[n]W_N^{kn}, \quad k = 0, 1, \cdots, N-1. \tag{D.1}$$

By separating the discrete-time input signal $x[n]$ into its even-numbered ($r = 2n$) and odd-numbered ($r = 2n+1$) samples, we obtain

$$\begin{aligned} X[k] &= \sum_{r=0}^{(N/2)-1} x[2r]W_N^{2rk} + \sum_{r=0}^{(N/2)-1} x[2r+1]W_N^{(2r+1)k} \\ &= \sum_{r=0}^{(N/2)-1} x[2r](W_N^2)^{rk} + W_N^k \sum_{r=0}^{(N/2)-1} x[2r+1](W_N^2)^{rk}. \end{aligned} \tag{D.2}$$

Since
$$W_N^2 = e^{-2j(2\pi/N)} = e^{-j2\pi/(N/2)} = W_{N/2}, \tag{D.3}$$

we obtain $W_N^2 = W_{N/2}$. Thus, (D.2) can be expressed as

$$\begin{aligned} X[k] &= \sum_{r=0}^{(N/2)-1} x[2r]W_{N/2}^{rk} + W_N^k \sum_{r=0}^{(N/2)-1} x[2r+1]W_{N/2}^{rk} \\ &= E[k] + W_N^k O[k], \end{aligned} \tag{D.4}$$

where $E[k]$ and $O[k]$ are the $\left(\frac{N}{2}\right)$-point DFTs of $\frac{N}{2}$ length sequences $x[2r]$ and $x[2r+1]$, respectively. Even though a $\frac{N}{2}$-point DFT is only of length $\frac{N}{2}$, for values of k greater than $\frac{N}{2}$, we use the property that $E[k] = E[((k))_{\frac{N}{2}}]$ and $O[k] = O[((k))_{\frac{N}{2}}]$. If $\left(\frac{N}{2}\right)$ is still even, each

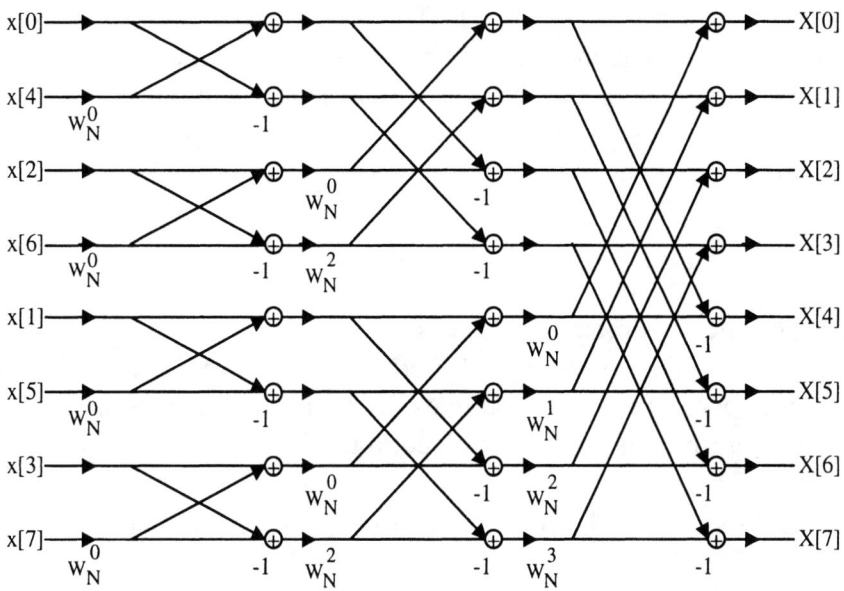

Figure D.1 The flow graph of a radix-2 decimation-in-time for an 8-point FFT algorithm.

of these can be broken into two $\left(\frac{N}{4}\right)$-point DFTs. Furthermore, if N is equal to 2^v, then the decomposition process can be repeated until the DFT can be computed as a combination of $\frac{N}{2}$ 2-point DFTs. This algorithm is known as the *radix-2 decimation-in-time* FFT.

A flow graph for operations of the radix-2 decimation-in-time algorithm with $N = 8$ is shown in Figure D.1. Each stage has 4 complex multiplications and 8 complex additions. Since there are three stages, we have a total of 12 complex multiplications and 24 additions. In general, the number of complex multiplications and additions is equal to $\frac{N}{2} \log_2 N$ and $N \log_2 N$, respectively, because there are $v = \log_2 N$ stages as well as $N/2$ complex multiplications and N complex additions for each stage.

D.2.2 Decimation-in-Frequency FFT Algorithm

The decimation-in-frequency FFT algorithm breaks the frequency-domain output sequence $X[k]$ into smaller and smaller subsequences in the same manner that we did using the decimation-in-time FFT algorithm. We again consider the DFT for an N-point frequency-domain output sequence given by (C.4) in Appendix C,

$$X[k] = \sum_{n=0}^{N-1} x[n] W_N^{nk}, \quad k = 0, 1, \cdots, N-1. \quad \text{(D.5)}$$

The frequency-domain output sequence $X[k]$ in (D.5) can be split into the even-numbered frequency samples $X[2q]$ and the odd-numbered frequency samples $X[2q+1]$. The even-numbered frequency samples $X[2q]$ can be computed by

$$X[2q] = \sum_{n=0}^{N-1} x[n] W_N^{2nq}, \quad q = 0, 1, \cdots, (N/2) - 1. \quad \text{(D.6)}$$

Equation (D.6) can be further expressed as

$$\begin{aligned} X[2q] &= \sum_{n=0}^{\frac{N}{2}-1} x[n] W_N^{2nq} + \sum_{n=N/2}^{N-1} x[n] W_N^{2nq} \\ &= \sum_{n=0}^{\frac{N}{2}-1} x[n] W_N^{2nq} + \sum_{n=0}^{\frac{N}{2}-1} x\left[n + \frac{N}{2}\right] W_N^{2q[n+\frac{N}{2}]}. \end{aligned} \quad \text{(D.7)}$$

Since the factors $W_N^{2q[n+(N/2)]} = W_N^{2qn} W_N^{qN} = W_N^{2qn}$ and $W_N^2 = W_{N/2}$, (D.7) is then equal to

$$\begin{aligned} X[2q] &= \sum_{n=0}^{\frac{N}{2}-1} x[n] W_N^{2nq} + \sum_{n=0}^{\frac{N}{2}-1} x\left[n + \frac{N}{2}\right] W_N^{2nq} \\ &= \sum_{n=0}^{\frac{N}{2}-1} \left(x[n] + x\left[n + \frac{N}{2}\right]\right) W_N^{2nq} \\ &= \sum_{n=0}^{\frac{N}{2}-1} \left(x[n] + x\left[n + \frac{N}{2}\right]\right) W_{N/2}^{nq} \quad \text{(D.8)} \end{aligned}$$

where $q = 0, 1, \cdots, (N/2) - 1$. Therefore, (D.8) is the $(N/2)$-point DFT of the $(N/2)$-point discrete-time input signal $x[n]$ obtained by adding the first and the last half of the discrete-time input signal.

Now we consider the specialization for odd-numbered frequency-domain samples,

$$X[2q+1] = \sum_{n=0}^{N-1} x[n] W_N^{n(2q+1)}, \quad q = 0, 1, \cdots, (N/2) - 1. \quad (D.9)$$

Equation (D.9) can be further broken into

$$\begin{aligned} X[2q+1] &= \sum_{n=0}^{\frac{N}{2}-1} x[n] W_N^{n(2q+1)} + \sum_{n=N/2}^{N-1} x[n] W_N^{n(2q+1)} \\ &= \sum_{n=0}^{\frac{N}{2}-1} x[n] W_N^{n(2q+1)} + \sum_{n=0}^{\frac{N}{2}-1} x\left[n + \frac{N}{2}\right] W_N^{(n+\frac{N}{2})(2q+1)} \\ &= \sum_{n=0}^{\frac{N}{2}-1} x[n] W_N^{n(2q+1)} \\ &\quad + W_N^{(N/2)(2q+1)} \sum_{n=0}^{\frac{N}{2}-1} x\left[n + \frac{N}{2}\right] W_N^{n(2q+1)}. \quad (D.10) \end{aligned}$$

Note that factors $W_N^{(N/2)2q} = 1$, $W_N^{(N/2)} = -1$, and $W_N^2 = W_{N/2}$. Thus, (D.10) can be rewritten as follows:

$$X[2q+1] = \sum_{n=0}^{(N/2)-1} \left(x[n] - x\left[n + \frac{N}{2}\right] \right) W_N^n W_{N/2}^{nq}, \quad (D.11)$$

where $q = 0, 1, \cdots, (N/2) - 1$. Therefore, (D.11) is the $\left(\frac{N}{2}\right)$-point DFT of the discrete-time input sequence $x[n]$ obtained by subtracting the second half of the input sequence from the first half and multiplying the resulting sequence by W_N^n.

Using (D.8) and (D.11), we then define two new $\left(\frac{N}{2}\right)$-point discrete-time sequences $o[n]$ and $e[n]$ as follows:

$$o[n] = x[n] + x\left[n + \frac{N}{2}\right], \quad n = 0, 1, \cdots, (N/2) - 1, \quad (D.12)$$

and

$$e[n] = x[n] - x\left[n + \frac{N}{2}\right], \quad n = 0, 1, \cdots, (N/2) - 1. \quad (D.13)$$

Then, we calculate the $\left(\frac{N}{2}\right)$-point DFT of the discrete-time sequence $o[n]$

$$X[2q] = \sum_{n=0}^{(N/2)-1} o[n] W_{N/2}^{nq}, \quad q = 0, 1, \cdots, (N/2) - 1, \quad (D.14)$$

and the $\left(\frac{N}{2}\right)$-point DFT of the discrete-time sequence $e[n]W_N^n$

$$X[2q+1] = \sum_{n=0}^{(N/2)-1} e[n] W_N^n W_{N/2}^{nq}, \quad q = 0, 1, \cdots, (N/2) - 1.$$
(D.15)

If the length $N = 2^v$, this decomposition process can be continued until the N-point DFT has been completely decomposed into $\frac{N}{2}$ 2-point DFTs. In this case, we refer to the algorithm as the *radix-2 decimation-in-frequency FFT*.

A flow graph for the radix-2 decimation-in-frequency FFT algorithm with $N = 8$ is shown in Figure D.2. In this method, each stage has 4 complex multiplications and 8 complex additions. The branches have transmittances of the form W_8^r. Because there are three stages, we have a total number of 12 complex multiplications and 24 complex additions. In general, the number of complex multiplications and additions is equal to $\frac{N}{2} \log_2 N$ and $N \log_2 N$, respectively; hence, the computational complexity is the same as that of the decimation-in-time FFT algorithm.

D.2.3 Computational Complexity

In general, for both of the radix-2 decimation-in-time and decimation-in-frequency FFT algorithms, a total of $\frac{N}{2} \log_2 N$ complex multiplications and $N \log_2 N$ additions are required. However, note that some of the multiplications with the factors of W_N^0, $W_N^{\frac{N}{2}}$, $W_N^{kN/2}$, and $W_N^{kN/4}$ are equal to 1, -1, $(-1)^k$,

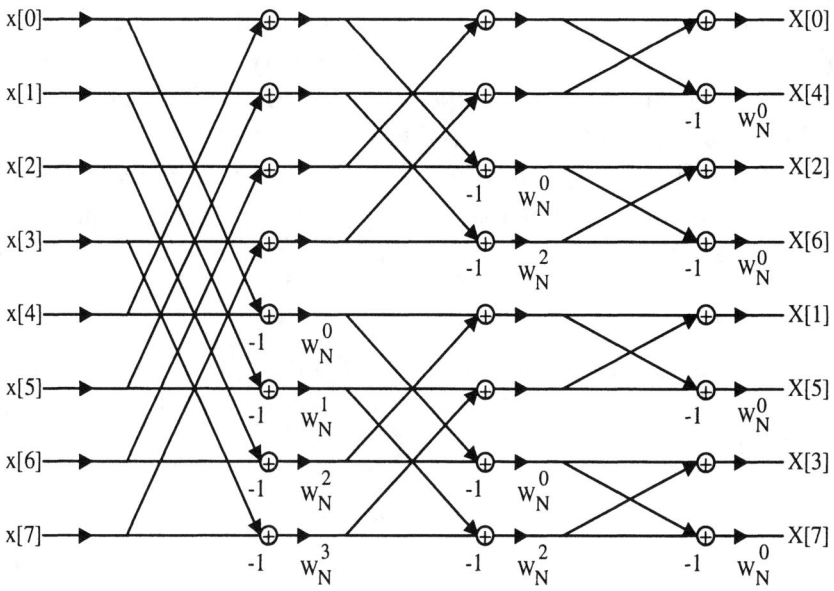

Figure D.2 The flow graph of a radix-2 decimation-in-frequency for an 8-point FFT algorithm.

and $(-j)^k$, respectively. It is clear that these operations are trivial and do not require actual multiplication. In addition, some multiplications by the factor of $W_N^{kN/8}$, which is $[(1-j)/\sqrt{2}]^k$, can be done by two real multiplications and two real additions. Furthermore, general complex multiplications can be computed by three real multiplications and three real additions. Therefore, the total computational complexity of the radix-2 FFT algorithm is reduced to $\left[\frac{N}{2}(3\log_2 N - 10) + 8\right]$ multiplications and $\left[\frac{N}{2}(7\log_2 N - 10) + 8\right]$ additions [2].

D.3 Fixed-Point FFT Algorithm

In DMT and OFDM modulation and demodulation, the FFT and IFFT algorithms are usually implemented by using a fixed-point operation. For the fixed-point operation, the FFT twiddle coefficients and input and output signals need to be quantized. The quantization formats of digital signals usually used a *sign and magnitude*, a *one's complement*, or a *two's complement*. However, the two's complement is most commonly used.

D.3.1 Quantization

Assume that we use a finite number of bits B for the quantization operation. Then, the quantization operation on a real value can be represented as follows:

$$\hat{x}[n] = Q_B\{x[n]\}$$
$$= X_m \hat{x}_B, \qquad (D.16)$$

where X_m is an arbitrary scale factor, and $\hat{x}_B = b_1 \cdots b_B$. The smallest difference between finite numbers is given by

$$\Delta = \frac{2X_m}{2^B}. \qquad (D.17)$$

Considering the effect of quantization, we can define a quantization error given by

$$e[n] = Q_B\{x[n]\} - x[n]. \qquad (D.18)$$

Note that in the case of the two's-complement rounding, we have

$$-\frac{\Delta}{2} < e[n] \leq \frac{\Delta}{2}, \qquad (D.19)$$

and in the case of the two's-complement truncation, we obtain

$$-\Delta < e[n] \leq 0. \qquad (D.20)$$

However, when the real number is greater than X_m, an overflow happens.

In order to analyze the error $e[n]$ in (D.18), let $X_m = 1$ be a full scale. We also assume that the quantization error $e[n]$ has the following properties: (1) the quantization error $e[n]$ is uniformly distributed over a range from $-\frac{1}{2^B}$ to $\frac{1}{2^B}$, and the variance of the quantization error $e[n]$ is given by $\sigma_e^2 = \frac{\Delta^2}{12} = \frac{2^{-2B}}{3}$; and (2) the quantization error $e[n]$ is uncorrelated with each other and also uncorrelated with input signal and output signal. This assumption will be used for analyzing the fixed-point FFT implementation.

D.3.2 Fixed-Point Overflow, Scaling, and SNR

The quantization form for the DFT (or FFT) for the direct computation can be defined as

$$X[k] = \sum_{n=0}^{N-1} Q_B\{x[n]W_N^{kn}\}, \quad k = 0, 1, \cdots, N-1. \quad \text{(D.21)}$$

Note that the term of $x[n]W_N^{kn}$ in (D.21) is a complex product. With the fixed-point operation, the complex product is expressed as follows [3]:

$$\begin{aligned}
Q_B\{x[n]W_N^{kn}\} &= \text{Re}\{x[n]\}\cos\left(\frac{2\pi kn}{N}\right) + e_1[n,k] \\
&\quad + \text{Im}\{x[n]\}\sin\left(\frac{2\pi kn}{N}\right) + e_2[n,k] \\
&\quad + j\left[\text{Im}\{x[n]\}\cos\left(\frac{2\pi kn}{N}\right) + e_3[n,k]\right] \\
&\quad - j\left[\text{Re}\{x[n]\}\sin\left(\frac{2\pi kn}{N}\right) + e_4[n,k]\right],
\end{aligned}$$
(D.22)

where $e_i[n,k]$, $i = 1, 2, 3, 4$, is the quantization errors. Thus, each of the real multiplications contributes a round-off complex error known as $e[n, k]$, and the total of complex errors, $F[k]$, to the output at the kth value is obtained by

$$F[k] = \sum_{n=0}^{N-1} e[n,k]. \quad \text{(D.23)}$$

The squared magnitude of the complex error $e[n, k]$ is given by
$$|e[n, k]|^2 = (e_1[n, k] + e_2[n, k])^2 + (e_3[n, k] + e_4[n, k])^2. \quad (D.24)$$
The corresponding variance of $e[n, k]$ is obtained by
$$E\{|e[n, k]|^2\} = 4\left(\frac{2^{-2B}}{3}\right) = \frac{2^{-2B+1}}{3}. \quad (D.25)$$
Therefore, the variance of the total of complex errors $F[k]$ is obtained by
$$E\{|F[k]|^2\} = \sum_{n=0}^{N-1} E\{|e[n, k]|^2\} = \frac{2^{-2B+1}N}{3}. \quad (D.26)$$
Equation (D.26) indicates that the output noise is proportional to the FFT length N.

In order to avoid overflow, we require
$$|X[k]| < 1, \quad k = 0, 1, \cdots, N - 1. \quad (D.27)$$
This requires that the constraint condition as a bound on the input sequence is given by
$$|x[n]| < \frac{1}{N}, \quad n = 0, 1, \cdots, N - 1. \quad (D.28)$$
Equation (D.28) is sufficient to guarantee no overflow for all stages of the DFT or FFT algorithm. This method is referred to as *input scaling*.

Further assume that the real and imaginary parts of the input sequence $x[n]$ are uncorrelated, each with a uniform density between $-\frac{1}{N}$ and $\frac{1}{N}$. The variance of the complex input signal is then obtained by
$$E\{|x[n]|^2\} = \frac{(1/N + 1/N)^2}{12} = \frac{1}{3N^2}. \quad (D.29)$$
Using (D.5), the corresponding output variance of the DFT (or FFT) is obtained by
$$E\{|X[k]|^2\} = \sum_{n=0}^{N-1} E\{|x[n]|^2\}|W_N^{kn}|^2 = \frac{1}{3N}. \quad (D.30)$$

Combining (D.30) and (D.26) yields the SNR$_{DFT}$ of the fixed-point DFT or FFT algorithm as follows:

$$\begin{aligned} \text{SNR}_{DFT} &= 10\log_{10}\left[\frac{E\{|X[k]|^2\}}{E\{|F[k]|^2\}}\right] \\ &= 10\log_{10}\left[\frac{1/(3N)}{2^{-2B+1}N/3}\right] \\ &= 10\log_{10}\left(\frac{2^{2B-1}}{N^2}\right) \text{ (dB)}. \end{aligned} \qquad \text{(D.31)}$$

Therefore, with quantization by a fixed number of bits B, the SNR$_{DFT}$ of the fixed-point DFT or FFT algorithm is inversely proportional to N^2 and decreases as the length of N increases.

D.3.3 Quantization Analysis of the FFT Algorithm

In Section D.3.2, we have discussed the quantization analysis of the direct implementation for the DFT or FFT algorithm. The quantization effects depend on the specific FFT algorithms used for the OFDM modulation. In this section, we focus on the quantization analysis for the radix-2 FFT algorithm since it has been the most commonly used for the OFDM modulation.

To prevent overflow, we can require the condition of $|X[n]| < \frac{1}{N}$. However, there is an alternative scaling procedure that incorporates an attenuation of $\frac{1}{2}$ at the input to each stage. This is because the maximum magnitude increases by no more than a factor of 2 from stage to stage in a radix-2 butterfly. Figure D.3 shows the radix-2 butterfly with scaling multipliers and associating fixed-point roundoff noise. With scaling by $\frac{1}{2}$ introduced at the input of each of the butterflies, two noise sources are included with each butterfly. We assume that the real and imaginary parts of these noise sources are uncorrelated with each other and also uncorrelated with the other noise sources. In addition, the real and imaginary parts of these noise sources are uniformly distributed between $-\frac{1}{2^B}$ and $\frac{1}{2^B}$. Then we have the variance of $e[m,p]$, the same as (D.25),

$$E\{|e[m,p]|^2\} = E\{|e[m,q]|^2\} = \frac{2^{-2B+1}}{3}. \qquad \text{(D.32)}$$

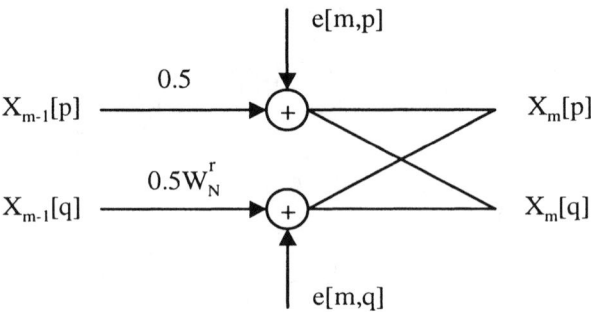

Figure D.3 Butterfly with scaling and corresponding fixed-point roundoff-noise.

In the radix-2 butterfly structure, a noise source originating at the mth array will propagate to the output, multiplying by a complex constant of $\left(\frac{1}{2}\right)^{r-m-1}$, where r is the number of stages. For the general case, Oppenheim and Schafer [3] indicate that each output node connects to 2^{r-m-1} butterflies and to 2^{r-m} noise sources that orginate at the mth array. Therefore, at each output node, the mean-square magnitude of the total noise is obtained by

$$\begin{aligned} E\{|F[k]|^2\} &= \sum_{m=0}^{r-1} \left(2^{r-m}\right) \left[\left(\frac{1}{2}\right)^{2r-2m-2}\right] \left(\frac{2^{-2B+1}}{3}\right) \\ &= \sum_{m=0}^{r-1} \left[\left(\frac{1}{2}\right)^{r-m-2}\right] \left(\frac{2^{-2B+1}}{3}\right) \\ &= 2\sum_{n=0}^{r-1} \left[\left(\frac{1}{2}\right)^n\right] \left(\frac{2^{-2B+1}}{3}\right). \end{aligned} \qquad \text{(D.33)}$$

In order to solve (D.33), we use a closed-form formula given in Appendix E

$$\sum_{n=0}^{N} \alpha^n = \frac{1-\alpha^{N+1}}{1-\alpha}. \qquad \text{(D.34)}$$

Using (D.34), (D.33) can be further simplified as follows:

$$E\{|F[k]|^2\} = 2\left(\frac{2^{-2B+1}}{3}\right) \left[\frac{1-(1/2)^r}{1-(1/2)}\right]$$

$$= 4\left(\frac{2^{-2B+1}}{3}\right)\left[1-\left(\frac{1}{2}\right)^r\right]. \qquad (D.35)$$

If the length of N is large, the number of stages r becomes larger as well. Thus, the term $\left(\frac{1}{2}\right)^r$ is negligible. Then we approximately obtain

$$E\{|F[k]|^2\} \approx 4\left(\frac{2^{-2B+1}}{3}\right). \qquad (D.36)$$

Combining (D.30) and (D.36), we obtain SNR_{FFT} by using the method of *stage-by-stage scaling* as follows:

$$\begin{aligned}
\mathrm{SNR}_{FFT} &= 10\log_{10}\left[\frac{E\{|X[k]|^2\}}{E\{|F[k]|^2\}}\right] \\
&= 10\log_{10}\left[\frac{1/(3N)}{2^{-2B+1}4/3}\right] \\
&= 10\log_{10}\left(\frac{2^{2B-1}}{4N}\right) \text{ (dB)}. \qquad (D.37)
\end{aligned}$$

With the quantization of a fixed number of bits B, the SNR_{FFT} of the fixed-point FFT algorithm using the method of stage-by-stage is inversely proportional to the length of N rather than to the length of N^2. The SNR_{FFT} in (D.37) decreases as the length of N increases. However, as we can see, the SNR_{FFT} in (D.37) obtained by using the method of stage-by-stage scaling is much greater than the SNR_{FFT} in (D.31) obtained by using the method of input scaling.

Figure D.4 shows the SNR of the fixed-point FFT with methods of input scaling and stage-by-stage scaling at the fixed FFT length of $N = 256$ when the number of quantization bits B gradually increases. The method of stage-by-stage scaling clearly has a more superior performance than the method of input scaling. For example, at the quantization bit $B = 15$, the method of stage-by-stage scaling has SNR = 135 dB, while the method of input scaling only has SNR = 95 dB.

Figure D.5 shows the SNR of the fixed-point FFT algorithm with methods of input scaling and stage-by-stage scaling when the

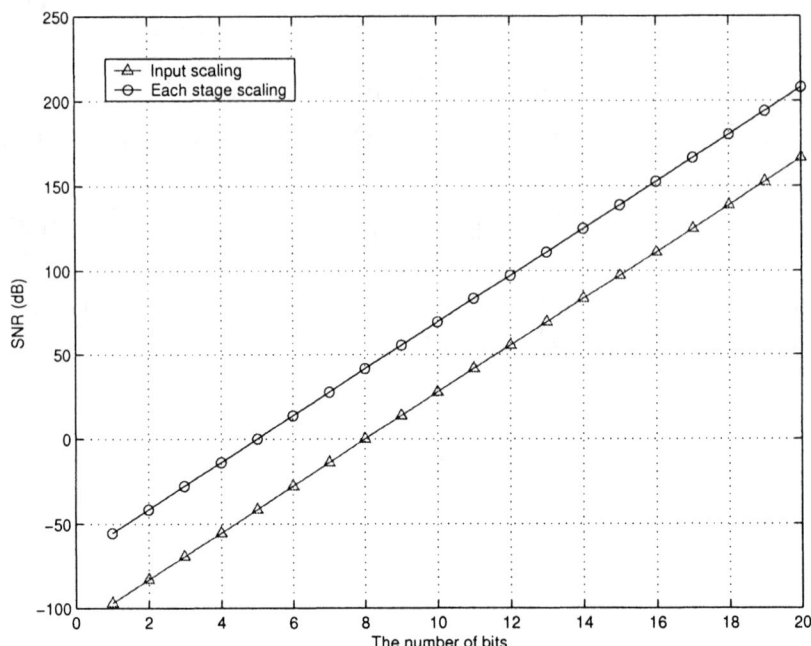

Figure D.4 The SNR of the fixed-point FFT with varying quantization bits B and the fixed length of FFT, $N = 256$.

FFT length of N increases at the fixed quantization bits $B = 15$. The SNR of the fixed-point FFT decreases for both methods of input scaling and stage-by-stage scaling when the FFT length of N increases. However, it is also clear that the method of stage-by-stage scaling is much better than the method of input scaling for the fixed-point FFT implementation. Therefore, we recommend using the attenuators of $\frac{1}{2}$ at each butterfly stage rather than using a large attenuation for the input scaling in the fixed-point FFT implementation for OFDM modulation.

Much research has been done on the FFT algorithms over the past three decades. For reference, we cite the books by Oppenheim and Schafer [3], Miao and Clements [2], Chen [4], and Higgins [5].

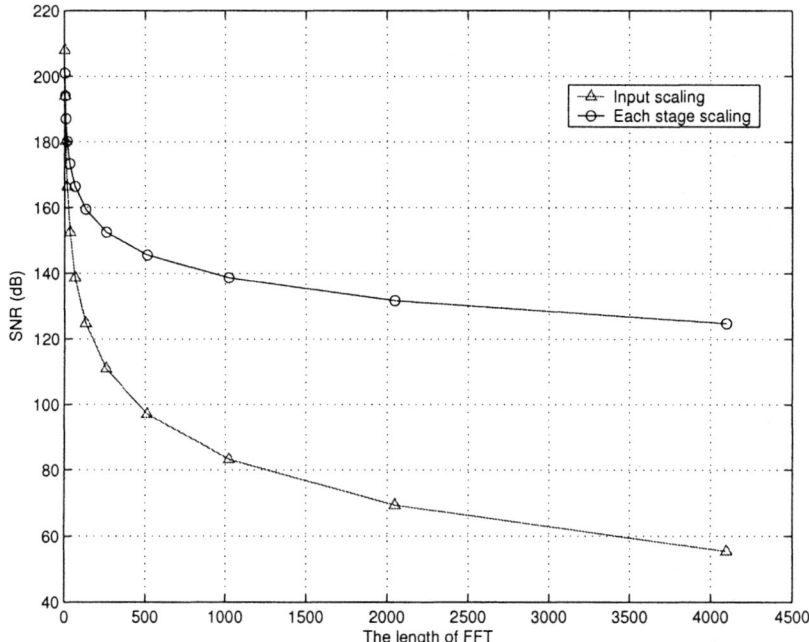

Figure D.5 The SNR of the fixed-point FFT with a varying FFT length N and the fixed quantization bits B.

References

[1] Cooley, J. W., and J. W. Tukey, "An Algorithm for the Machine Computation of Complex Fourier Series," *Mathematics of Computation*, Vol. 19, pp. 297–301, April 1965.

[2] Miao, G. J., and M. A. Clements, *Digital Signal Processing and Statistical Classification*, Artech House, Norwood, Massachusetts, 2002.

[3] Oppenheim, A. V., and R. W. Schafer, *Discrete-Time Signal Processing*, Prentice Hall, Englewood Cliffs, New Jersey, 1989.

[4] Chen, C. H., (ed.), *Signal Processing Handbook*, Marcel Dekker, New York, 1988.

[5] Higgins, R. J., *Digital Signal Processing in VLSI*, Prentice Hall, Englewood Cliffs, New Jersey, 1989.

Appendix E: Discrete Mathematical Formulas

E.1 Complex Exponential Formulas

The discrete complex exponential formulas are listed as follows:

$$e^{j\omega n} = \cos(\omega n) + j\sin(\omega n). \tag{E.1}$$

$$\cos(\omega n) = \frac{1}{2}(e^{j\omega n} + e^{-j\omega n}). \tag{E.2}$$

$$\sin(\omega n) = \frac{1}{2j}(e^{j\omega n} - e^{-j\omega n}). \tag{E.3}$$

E.2 Discrete Closed-Form Formulas

The discrete closed-form formulas are listed as follows:

$$\sum_{k=0}^{N} k = \frac{1}{2}N(N+1). \tag{E.4}$$

$$\sum_{k=0}^{N} k^2 = \frac{1}{6}N(N+1)(2N+1). \tag{E.5}$$

$$\sum_{k=0}^{N} \alpha^k = \frac{1-\alpha^{N+1}}{1-\alpha}. \tag{E.6}$$

$$\sum_{k=N_1}^{N_2} \alpha^k = \frac{\alpha^{N_1} - \alpha^{N_2+1}}{1-\alpha}. \tag{E.7}$$

$$\sum_{k=0}^{N} k\alpha^k = \frac{\alpha}{(1-\alpha)^2}\left[(1-\alpha^N) - (1-\alpha)N\alpha^N\right]. \quad \text{(E.8)}$$

$$\sum_{k=0}^{N} k^2\alpha^k = \frac{\alpha}{(1-\alpha)^3}\left[(1+\alpha)(1-\alpha^N)\right] - \quad \text{(E.9)}$$

$$-\frac{\alpha}{(1-\alpha)^2}\left[2N\alpha^N + (1-\alpha)N^2\alpha^N\right], \quad \text{(E.10)}$$

where $|\alpha| < 1$ and $|\alpha| \neq 0$, and $N_2 \geq N_1$.

$$\sum_{k=0}^{N-1}(a+kd)q^k = \frac{a - [a + (N-1)d]a^N}{1-q} + \frac{dq(1-q^{N-1})}{(1-q)^2}, \quad N \geq 1. \quad \text{(E.11)}$$

$$\sum_{k=0}^{\infty}(a+kd)q^k = \frac{a}{1-q} + \frac{dq}{(1-q)^2}, \quad |q| < 1. \quad \text{(E.12)}$$

$$\sum_{k=1}^{N} \frac{k}{(1+k)!} = 1 - \frac{1}{(1+k)!}. \quad \text{(E.13)}$$

$$\sum_{k=0}^{N} \frac{1}{k!(N-k)!} = \frac{2^N}{N!}. \quad \text{(E.14)}$$

E.3 Approximation Formulas

The approximation formulas are listed as follows:

$$n! = \sqrt{2\pi n}\left(\frac{n}{e}\right)^n e^{\frac{\theta}{12n}}, \quad 0 < \theta < 1. \quad \text{(E.15)}$$

$$n! \approx \sqrt{2\pi n}\left(\frac{n}{e}\right)^n, \quad \text{for a large value of } n. \quad \text{(E.16)}$$

$$\sqrt{2\pi n}\left(\frac{n}{e}\right)^n < n! < \sqrt{2\pi n}\left(\frac{n}{e}\right)^n\left(1 + \frac{1}{12n-1}\right). \quad \text{(E.17)}$$

$$\frac{1}{1+x} \approx 1 - x, \quad x \ll 1. \tag{E.18}$$

$$e^x \approx 1 + x, \quad x \ll 1. \tag{E.19}$$

$$(1+x)^n \approx 1 + nx, \quad x \ll 1 \text{ and } n \geq 1. \tag{E.20}$$

$$\ln(1+x) \approx x, \quad x \ll 1. \tag{E.21}$$

$$\sin(x) \approx x, \quad x \ll 1. \tag{E.22}$$

$$\cos(x) \approx 1 - \frac{x^2}{2}, \quad x \ll 1. \tag{E.23}$$

$$\tan(x) \approx x, \quad x \ll 1. \tag{E.24}$$

E.4 Logarithmic Formulas

Assume that $a > 0$. The logarithmic formulas are listed as follows:

$$\log_a a = 1. \tag{E.25}$$
$$\log_a 1 = 0. \tag{E.26}$$
$$\log_a x^\alpha = \alpha \log_a x. \tag{E.27}$$
$$\log_a y = \frac{\log_b y}{\log_b a}. \tag{E.28}$$
$$a^{\log_a y} = y. \tag{E.29}$$
$$\log_a b \log_b a = 1. \tag{E.30}$$
$$\log_a \frac{x}{y} = \log_a x - \log_a y. \tag{E.31}$$
$$\log_a xy = \log_a x + \log_a y. \tag{E.32}$$
$$\ln = \log_e x, \quad e = 2.718281828459\cdots. \tag{E.33}$$
$$\log_{10} y = M \ln y, \quad M = 0.434294481903\cdots. \tag{E.34}$$
$$\ln y = \frac{1}{M} \log_{10} y. \tag{E.35}$$

About the Author

George J. Miao received a B.Eng. from Shanghai University of Science and Technology, an M.S. from Columbia University, and a Ph.D. in electrical engineering from the Georgia Institute of Technology. He is the president of Sageco Technologies, LLC. He specializes in digital signal processing and wireless and wireline digital communications. His extensive professional experience includes work for several Fortune 500 companies. He is the coauthor of the book *Digital Signal Processing and Statistical Classification*, for which he received the IEEE New Jersey Coast Section Technical Accomplishment Award. He is the author of the book *Signal Processing in Digital Communications*. He has received a number of awards, including the IEEE Section Technical Accomplishment Award for wireless communication patents, the IEEE Chapter Distinguished Service Award, and the IEEE Region-1 Award "for innovative work in digital signal processing leading to applications in wireless communications." He holds a number of U.S. patents and patents pending in wireless, wireline, and UWB communications.

Dr. Miao is a Senior Member of the IEEE, a chairman of the IEEE New Jersey Coast Chapter of Signal Processing/Circuits and Systems, and a board director of the Chinese Association for Science and Technology (Greater New York, United States). He has been named to the *International Who's Who of Information Technology*, *Marquis Who's Who in Science and Engineering*, *Who's Who in the Americas and World*, and *Who's Who in Finance and Business*.

Index

A

A/D converter, 4, 7, 8, 14, 25, 91, 93, 346, 360, 361, 365, 369, 402, 406
   clipping probability, 11
   dynamic range, 9
   filter-band, 10
   high-speed, 4, 12
   minimum number of bits, 12
   multirate, 5
   noise shaping, 10
   oversampling, 10
   resolution, 10
   sigma-delta, 10
Adaptive algorithm, 270, 274, 277
Adaptive channel equalizations, 258
Adaptive channel estimation, 208
Adaptive equalization, 257
Adaptive linear equalizer, 272
Adaptive Rake receiver, 330
Additive white Gaussian noise (*see* AWGN), 83
Adjacent channels, 142
ADSL, 351
Aliasing, 91, 99
All-pass system, 453
All-phase system, 453
Amplitude response, 390

Analog-to-digital converter (*see* A/D converter), 4
Antenna array, 4, 125
Antialiasing analog filter, 10
Antialiasing bandpass filter, 104, 108
Antialiasing filter, 99, 103
Approximation formulas, 496
Asynchronous configuration, 412
Autocorrelation, 71
Autocorrelation function, 134, 204
Autocovariance, 71
AWGN, 83, 87, 264, 359, 366
   channel, 351
   channels, 352
   subchannels, 350, 351
Axiomatic probability, 27

B

Bandedge timing recovery, 406
Bandlimited channel, 112, 114
Bandpass antialiasing filter, 8
Bandpass filter (*see* BPF), 7
Bandpass sampling, 100
Bandpass signal, 100
Baseband sampling, 97
Baseband signal, 14

502 Index

Bayes' minimum risk classifers, 85
Bayes' theorem, 30
Beamformer algorithms, 133
Beamformer gain, 135
Beamformers, 126
    adaptive space-time, 130
    delay-and-sum, 128
    filter-and-sum, 130
    frequency-domain, 133
    interpolation, 132
    space-time, 130
    switched, 126
Beamforming shading, 129
BER, 17, 118, 233
Best linear unbiased estimator, 199, 202
Binary phase-shift keying (*see* BPSK), 299
Biorthogonality property, 375
Bit energy-to-noise ratio, 328
Bit error rate (*see* BER), 17
Bit loading algorithm, 352
Blind channel identification, 238, 239, 241, 242, 283
Blind equalization, 278
Blind equalizer, 259
Blind identification, 189
Blind training, 277
Block adaptive filter, 272
BPF, 7, 13, 429
BPSK, 299
Bridge taps, 167

C

Canonical correlation function, 136
Canonical discriminant function, 136
Capacity, 351
Carrier phase error, 426
Carrier recovery, 428
    open-loop, 428
Carrier-to-noise ratio (*see* CNR), 18
Cascade filter, 181
Cascaded integrator comb (*see* CIC), 15
Cauchy residue theorem, 449
CCI, 189, 315, 323, 341
    cancellation, 318
CDMA, 3, 325
Central limit theorem, 69, 88
Central moment, 36
Central office (*see* CO), 163
Chain matrix, 164
Channel capacity, 110, 115, 116, 119
Channel distortion, 184
Channel equalizer, 262
Channel estimation, 189, 194
    blind, 189
    nonblind, 189
Channel estimator, 194
Channel matrix, 120
Channel mediums, 17
Channel partitioning, 345
Channel shortening, 370
Chebyshev inequality, 67
Chi-square density, 60
    degrees of freedom, 60
    gamma function, 60
Chi-square distribution function, 61

Index

CIC, 15
Clipping, 365, 366
CM, 295
    cost function, 296
CMA, 259, 293, 312
CNR, 18
CO, 163
Cochannel interference
    (*see* CCI), 189
Code division multiple access
    (*see* CDMA), 3
COFDM, 346
Coherence bandwidth,
    153, 156, 257
Coherence time, 155
    maximum Doppler shift, 155
Coherent demodulation, 387
Communication channel(s),
    109, 141
Communication system(s), 142
Compensation system, 455
Complementary error function,
    54, 351
Complex conjugate, 458
Complex exponential formulas,
    495
Complex random variable, 32
Computational complexity, 276
Conditional probability, 29
Constant modulus (*see* CM), 295
Constant modulus algorithm
    (*see* CMA), 259
Conventional Rake receiver, 334
Convergence rate, 275
Correlation coefficient, 37
Correlation receiver, 86
Covariance, 200

Covariance matrix, 119
CPE, 163, 170
Cross-correlation function,
    79, 134, 204
Crosstalk, 168, 186
Cumulative distribution function,
    32
Customer premises equipment
    (*see* CPE), 163
Cycle slip, 397
Cyclic prefix, 361, 369
Cyclostationary,
    231, 238, 242, 246

D

D/A converter, 4, 360
    high-speed, 4
    multirate, 5
Decision feedback equalization,
    258
Decision feedback equalizer, 260
Decision feedback equalizer
    (*see* DFE), 299
Decision-directed method, 277
Decision-directed
    timing recovery, 407
Delay spread, 156
Density function, 33
Detection theory, 83
DFE, 258, 301
DFT, 133, 359, 375, 475, 479, 482
Diagonal matrix, 472
Digital channelization, 14, 15, 21
Digital conversion, 100
Digital downconversion,
    14–16, 21
Digital downconverter, 4

Digital front end, 8
Digital lowpass filters, 15
Digital signal processing
    (see DSP), 8
Digital subscriber loop
    ADSL, 142, 171
    DSL, 142, 171
    HDSL, 142
    LDSL, 163
    long-range extended DSL, 163
    SHDSL, 171
    VDSL, 142, 171
Digital subscriber loop
    (see DSL), 142
Digital upconverter, 4
Digital-to-analog converter
    (see D/A converter), 4
Direct conversion, 12
Discrete binomial distribution, 46
Discrete closed-form formulas, 495
Discrete Fourier transform
    (see DFT), 133, 347
Discrete Poisson distribution, 48
Discrete uniform distribution, 45
Discrete uniform random variable, 45
Discrete wavelet multitone
    (see DWMT), 347
Discrete-time random process, 71
Discrete-time synchronization, 388, 390
Distortion, 171
    amplitude distortion, 172
    delay distortion, 172
Diversity, 258, 323

Diversity combiner, 326
Diversity equalizer, 260, 324
DMT, 190, 345, 359, 369, 383, 418
DMT modulation, 349
DMT transmission, 347, 353, 355
Doppler frequency shift, 156
Doppler shift, 425
Doppler spread, 154
Downconversion, 8
Downsampling, 14, 16
DSL, 163
DSP, 8, 388, 415, 475
Duality function, 156
DWMT, 347
Dynamic range, 9, 10, 103

E

Early-late gate synchronizer, 403
Eigenanalysis, 464
Eigenvalue decomposition, 464
Eigenvalues, 464
    maximum, 471
Eigenvectors, 464
Electromagnetic (see EM), 6
EM, 6
    radiation spectrum, 6
EMSE, 217, 296
End-to-end loop, 168, 169
Energy minimization, 356
Energy-to-noise ratio
    (see ENR), 18
ENR, 18
Equalized Rake receiver, 260, 334, 336, 338
Ergodic, 74
Error function, 54, 367
Error propagation, 312

Error vector magnitude
 (see EVM), 177
Euclidean distance, 87
Euclidean norm, 458
EVM, 177
Examination method, 449
Excess bandwidth, 181
Excess delay, 151
 maximum excess delay, 151
 mean excess delay, 151
 rms excess delay, 151
Excess mean-squared error
 (see EMSE), 217
Expected value, 35
Eye diagram, 176, 186
Eye-closed, 174
Eye-open, 174

F

$\mathcal{F}$ distribution, 61
 mean, 62
 variance, 62
 variance ratio, 61
F-ratio quotient, 471
Fading, 156, 185
 channel impairments, 258
 flat, 161
 frequency-selective, 161
 large-scale, 157
 small-scale, 157, 159
Far-end crosstalk
 (see FEXT), 169
Fast fading, 162
 channel, 162
Fast Fourier transform
 (see FFT), 346
FBF, 300, 302

FDM, 373
Feedback filter (see FBF), 300
Feedforward filter (see FFF), 300
FEQ, 371, 372
FEXT, 169, 171, 346, 383, 418
FFF, 300, 307
FFT, 359, 363, 372, 384, 418, 479
 decimation-in-frequency,
  479, 482
 decimation-in-time, 479
 fixed-point, 489, 491
 flow graph, 484
 radix-2, 489
 radix-2 decimation in
  frequency, 484
 radix-2 decimation in
  time, 481
Filter impulse response, 16
Finite impulse response
 (see FIR), 15
FIR, 15, 130, 132, 210, 229, 236
 filter, 260, 262, 295
First Nyquist zone, 104
Fixed-point digital filter, 25
Flat surface-reflected path, 146
Fractionally spaced equalizer
 (see FSE), 260, 280, 282, 286
Free-space propagation,
  143, 146, 147
Frequency dispersion, 162
Frequency diversity, 117
Frequency division multiplexed
 (see FDM), 373
Frequency error, 420
Frequency offset,
  393, 395, 424, 432, 433
Frequency rotation, 420

Frequency sampling, 476
Frequency synthesizer, 400
Frequency-domain equalization
 (*see* FEQ), 371
Frequency-domain interpolation
 formula, 99
FSE, 260, 280, 290, 291
FSE-CM, 297
 cost function, 297
 criterion, 297
 local minima, 299
 MSE performance, 299
 noisy cost function, 298
FSE-DFE, 312
Full-band digitization, 12
Full-band frequency range, 13

G
Gaussian, 51
Gaussian channel, 110
Gaussian density function, 51, 53
Gaussian distribution function,
 51, 53
Gaussian shaped pulse, 182
Generalized least squares
 estimator, 201
Geometric signal-to-noise ratio,
 352
Global system mobile
 (*see* GSM), 3
Godard algorithm, 293
GSM, 3
Guard time, 369

H
Half-band filter, 15
Hard limiting, 365, 366

Harmonic sampling, 100
Hermitian transpose, 119, 458
Higher-order statistics
 (*see* HOS), 226
HOS, 226, 228
Hypothesis, 85

I
IC, 6
ICI, 361, 369, 383
Ideal discrete-time phase
 detector, 397
Ideal sampled signal, 94
Identifiability condition, 283
IDFT, 360, 476, 479
IDFT in matrix, 477
IF, 13, 100
 center frequency, 14
 sampling, 13
 signal, 14
IF sampling, 100
IFFT, 360, 362
IIR, 453
Impulse radio, 2
Impulse response, 76
In-line loop segment, 166
In-phase component, 427
Independent, 31
Independent and identically
 distributed, 41
Infinite impulse response
 (*see* IIR), 453
Information theory, 2
Inline loop segment
 ABCD matrix, 167
 RLGC parameters, 167
Input scaling, 488, 491

Instantaneous sampling, 93
Integrated circuits (*see* IC), 6
Intercarrier interference
    (*see* ICI), 361
Interference suppression, 321
Intermediate frequency
    (*see* IF), 13
Interpolation filter, 415
Intersymbol interference
    (*see* ISI), 141
Intraband interference, 375
Intuitive probability, 27
Inverse z-transform, 437, 448
Inverse autocorrelation matrix, 221
Inverse channel filter, 263
Inverse discrete Fourier
    transform (*see* IDFT), 360
Inverse fast Fourier transform
    (*see* IFFT), 360
Inverse filtering approach, 228
Inverse orthonormal transform, 349
ISI, 141, 142, 161, 174,
    176–180, 183–185, 257,
    258, 260, 263, 265, 272,
    300, 315, 320, 341, 383,
    418, 432
ISI cancellation, 187
ISI distortion, 162
ISI effects, 184
ISI free, 375
Iterative water filling
    algorithm, 353, 356

J
Jacobian transformation, 43, 65
Johnson noise, 116
Johnson-Nyquist noise, 2
Joint cumulative distribution
    function, 34
Jordan decomposition theorem, 466

K
Kalman filter, 25
Kalman filtering, 275
Karhunen-Loève transform, 84

L
Lag, 73
Lagrange multiplier, 356
Lattice equalizer structure, 258
Law of large numbers, 69
Least mean squares
    (*see* LMS), 210
Least squares estimator,
    198, 199, 201
Likelihood function, 195
Likelihood ratio test, 85
Limited-band digitization, 13
Line of sight (*see* LOS), 143
Linear, 258
Linear channel equalizer, 274
Linear equalizations, 258
Linear equalizer,
    228, 229, 260, 268
Linear predictor, 309
Linear shift-invariant systems, 26

Linear time-invariant, 281
Linearly independent, 37
Link budget, 17
Link margin, 17
LMS, 210, 252, 259, 275
    adaptation, 216
    algorithm, 210, 211, 213, 214, 223, 225, 293, 301
    algorithms, 210
    filter, 219
    misadjustment, 218
    steepest descent, 211
    trade-off, 219
LNA, 8
Log-normal distribution, 158
Logarithmic formulas, 497
Loop filter, 387
LOS, 143, 185, 257
Low-noise amplifier (*see* LNA), 8
LTI difference equation, 447
LTI system, 76–79, 88, 447
    discrete-time, 447, 448

M

M-PSK, 425, 427
MAI, 235
Matched filter, 184, 193
Matched filter receiver, 86
Matched filtering, 142
Matched filters, 387
Matrix, 459
    determinant, 462
    diagonal, 459
    differentiation, 462
    equalizer, 321
    identity, 460
    inversion, 464
    inversion lemma, 464
    lower triangular, 460
    nonsingular, 464
    orthogonal, 461
    singular, 463
    square, 461
    Toeplitz, 460
    trace, 461
    upper triangular, 460
Matrix inversion lemma, 221
Matrix theory, 457, 473
Maximal combining, 326
Maximization theorem, 138, 472
Maximum a posteriori rule, 85
Maximum allowable path loss, 21
Maximum analysis, 471
Maximum likelihood detector, 88
Maximum likelihood estimator, 194, 197
Maximum likelihood sequence estimation equalizer (*see* MLSEE), 258
Maximum likelihood symbol detection (*see* MLSD), 258
Maximum path loss, 21
Maximum-phase system, 453
MCM, 345, 347
Mean, 35
Mean path loss, 158
Mean square error (*see* MSE), 81
Method of least squares, 198
Method of maximum likelihood, 194, 196
MIMO, 22, 110, 117, 226, 253
    channel, 118, 120, 233, 235, 236, 238, 246

channel capacity, 120
channel filtering matrix,
  248, 251
channel matrix, 237
channel model, 121
channels, 234
  system, 120
  systems, 117, 125
Minimization theorem, 138
Minimum analysis, 471
Minimum mean square error
  (see MMSE), 81, 260
Minimum phase, 193
Minimum receiver sensitivity,
  20, 21
Minimum sampling rate, 102
Minimum variance beamformer,
  137
Minimum-phase system, 453
Misadjustment, 275
MISO, 110
  channel, 122
  channel model, 121
  equalizer, 236
  system, 122
Mixed-phase system, 453
MLSD, 258
MLSEE, 258
MMSE, 81, 82, 88, 133, 202,
  204, 209–211, 214, 217,
  218, 260, 269, 270, 291
  DFE, 301, 302
  equalized Rake receiver,
  338
  linear equalizer,
  274, 292, 303, 335
  optimal sense, 287

performance, 290
  Rake receiver, 335
MMSE estimator, 202, 204–207
MMSE timing recovery, 407
Moment generation function,
  37, 38
  binomial, 46
  poisson, 49
  uniform, 46
MSE, 81, 82, 84, 88, 210,
  211, 259, 273, 373
  criterion, 266, 269
Multicarrier modulation
  (see MCM), 345
Multichannel, 4
Multichannel filtering matrix, 231
Multichannel modulation, 345
Multilevel quantizer, 261
Multimode, 4
Multipath, 141
Multipath diversity receiver, 335
Multipath propagation, 149
  diffraction, 149, 150
  reflection, 149
  scattering, 149, 150
Multipath-intensity profile, 150
Multiphase phase-shift keying
  (see M-PSK), 425
Multiple access interference
  (see MAI), 235
Multiple-input multiple-output
  (see MIMO), 22
Multiple-input single-output
  (see MISO), 110
Multirate timing recovery,
  410, 435
Multivariate Gaussian density

function, 58
Mutually exclusive, 28

N
N-dimensional vector, 457
NCO, 389, 400, 413
Near-end crosstalk
 (*see* NEXT), 168
NEE, 258
Neural networks based equalizer
 (*see* NNE), 258
NEVM, 178
NEXT, 168, 171, 186, 346, 383, 418
NLOS, 143, 185, 257
NMSE, 242
 cost function, 289
Noise figure, 18, 19, 21
Noise floor, 20
Noise margin, 174
Noise predictor, 307
Noise PSD, 354
Noise-shaping filter, 10
Noise-to-signal ratio
 (*see* NSR), 355
Nonblind channel estimation, 189
Noncoherent demodulation, 387
Nonline of sight (*see* NLOS), 143
Nonlinear, 258
Nonlinear equalization, 258
Nonminimum-phase, 242
Nonminimum-phase system, 454
Nonuniform filter banks, 382
Nonuniformly spaced samples, 382
Normal, 51
Normal equation, 135

Normalized EVM
 (*see* NEVM), 178
Normalized MSE
 (*see* NMSE), 242
NSR, 355
Numerically controlled oscillator
 (*see* NCO), 389
Nyquist criteria, 104
Nyquist criterion, 179, 180
Nyquist filter, 180
Nyquist frequency, 181
Nyquist sampling frequency, 95, 99
Nyquist sampling rate, 280
Nyquist zones, 96, 100
 third Nyquist zone, 101
Nyquist-Shannon interpolation formula, 98
Nyquist-Shannon sampling theorem, 2, 91, 94, 113

O
OFDM, 22, 189, 345, 359, 369, 383, 418
OFDM demodulation, 363, 372, 379, 382
OFDM modulation, 380
OFDM system performance, 364
Open-circuit transfer admittance, 164
Open-circuit transfer function, 164
Optimal matched filter, 183, 184, 187
Optimal property, 199, 201
Optimality criterion, 83
Optimization array gain, 136
Optimum detector, 85

Optimum receiver, 86
Optimum sampling, 174
Orthogonal, 37, 459
Orthogonal frequency division multiplexing (*see* OFDM), 22
Orthogonal matrix, 461, 472
Orthonormal, 84, 375, 459
Orthonormal basis functions, 84
Orthonormal transform, 349
Overflow, 486, 488
Overlap, 99

P

PAM, 94, 172, 176, 282, 298, 405, 425, 426, 428
PAR, 365
Partial fraction expansion, 450
Partial-band digitization, 13
Path loss, 18, 20, 144
Peak, 174
Peak-to-average ratio (*see* PAR), 365
Perfect equalization, 236
Perfect reconstruction, 379
Perfect source recovery, 289
Phase detector, 387, 397
    modulo-2, 397
Phase error, 390
Phase jitter, 425, 432, 433
Phase locked loop (*see* PLL), 388
Phase offset, 393
Phase reference, 390
Phase rotation, 420
Phase shift, 419
Phase-shift keying (*see* PSK), 176

PLL, 388
    discrete-time, 389, 412, 430
    first-order, 392
    second-order, 394
    transient behavior, 423
Polyphase decomposition, 377, 378
Polyphase filter bank, 16
Polyphase filter bank timing recovery, 414
Polyphase filters, 16
Polyphase-based filter bank DFE, 312
Posteriori, 30
Power spectral density (*see* PSD), 74
Power spectrum, 74, 75, 78–80, 88, 105, 303
Precoding, 313
Predictive DFE, 307, 310
Priori, 30
Probability density function, 33
Probability distribution function, 32
Probability of error, 87
Probability theory, 26, 27
Prototype filter, 377
PSD, 74, 87, 349, 350
Pseudorandom, 327
PSK, 176, 177, 282, 405
Pulse amplitude modulation (*see* PAM), 94
Pulse-shaping filter, 172

Q

Q-function, 56, 87
QAM, 176, 177, 282, 298, 368,

425, 427
QMF, 379, 383
QPSK, 176, 177
Quadratic form, 470
Quadrature amplitude
 modulation (*see* QAM), 176
Quadrature mirror filter
 (*see* QMF), 379
Quadrature phase-shift keying
 (*see* QPSK), 176
Quantization error, 487
Quantization noise, 10
Quantization noise power, 12
Quantization operation, 486

R
Radio frequency (*see* RF), 6
Raised-cosine frequency
 response, 181
Rake correlators, 326
Rake receiver, 260, 325
 MMSE, 333
Random process, 25
Random signal, 25
Random variable, 26, 31
Random vector, 32
Rational powers, 467
Rayleigh distributed, 64, 329
Rayleigh faded component, 160
Rayleigh fading, 159, 185, 330
Rayleigh fading distribution,
 159
Rayleigh quotient, 471
Rayleigh-Ritz ratio, 136
Received power, 150
 large-scale, 150
 small-scale, 150

Receiver, 17
Receiver sensitivity, 19
Rectangular filter, 180
Recursive least squares
 (*see* RLS), 210
Region of convergence
 (*see* ROC), 438, 439, 442
Resampling operation, 14
RF, 6, 91, 171
 components, 12
 front end, 8
 image-reject bandpass filter,
  13
 regulations, 13
 signal propagation, 9
 system, 19
 system design, 6
 transceiver, 7
Ricean factor, 161
Ricean fading, 160, 185
Ricean fading distribution, 161
Ricean distributed, 66
RLS, 210, 252, 259, 275
 algorithm, 219, 223
 convergence analysis, 223
 convergent in the mean value,
  224
 exponential weighting
  factor, 219
 gain vector, 221
 misadjustment, 224
 normal equations, 220
 priori estimation error, 222
 Riccati, 221
RMS, 366
ROC, 438, 439, 441, 448
Rolloff factor, 181

Root mean square
  (*see* RMS), 360, 366
Rotation error, 423

S
Sample mean, 74
Sample rate conversion, 8
Sample space, 27
Sample variance, 74
Sampler, 191
Sampling, 94
Sampling expansion, 106
Sampling frequency, 92
Sampling instant shifts, 419
Sampling interval, 92
Sampling period, 92
Sampling process, 91
Sampling rate, 15, 92, 99,
  102, 103, 476
Sampling theorem, 91, 95,
  105, 106
SDR, 2, 22
Short-circuit current ratio, 164
Short-circuit transfer
  impendance, 164
Signal-to-interference-and-noise
  ratio, 136
Signal-to-noise ratio
  (*see* SNR), 12, 142
Signal-to-quantization noise
  ratio (*see* SQNR), 365
Signals, 92
  analog signals, 92
  digital signals, 92
  discrete-time sampled
    signals, 92
  discrete-time signals, 92

SIMO, 110, 226, 253
  channel, 229, 232, 235,
    238, 242, 246
  channel model, 121
Simulation loop model, 169, 170
Sinc-function, 98
Single-input multiple-output
  (*see* SIMO), 110
Single-input single-output
  (*see* SISO), 120
Singular value decomposition
  (*see* SVD), 120, 240
Singular value decomposition
  theorem, 468
SISO, 120, 226, 239, 253, 316
  channel, 226, 229, 238
Slice, 261
Slow fading, 162
Slow fading channel, 162
Smart antenna, 4, 125
Smart antenna system, 126, 139
Smart antennas, 234, 325
SNR, 12, 17, 19, 143, 162,
  183–185, 227, 235, 270,
  272, 351, 366, 423
  DFE, 305
Software defined radio
  (*see* SDR), 2, 22
Space-only equalizer, 318, 341
Space-time equalizer, 320, 340
Space-time MMSE equalizer,
  320, 323, 341
Space-time modulation, 22
Space-time processing, 315
Space-time signal processing,
  117, 125, 234

Spatial diversity, 117
Spectral decomposition theorem, 466
Spectral efficiency, 22
SQNR, 365
Square root raised cosine (see SRRC), 182
Square-law device, 428
SRRC, 182
SRRC filter, 182
SSS, 73, 88
Stability, 276
Stage-by-stage scaling, 491
Standard deviation, 36, 51
Standard Gaussian, 53
Standard normal distribution, 111
Stationary, 73
Stationary in the strict sense (see SSS), 73
Statistically independent, 37
Steady-state phase error, 392, 394
Step size, 214, 219
Stochastic jitter, 296
Stochastic process, 73
Stochastic sampling theorem, 106
Stochastic signal, 25
Subband signals, 10
Subspace decomposition, 246
Sufficiently large, 27
Super-Nyquist sampling, 100
SVD, 120, 240
Symbol rate, 263
Symbol-based equalizer, 279
System identification, 208

T

TDM, 373
TED, 410, 413
TEQ, 369, 371, 384
The z-plane, 439
The z-transform, 437
    common pairs, 445
    conjugation, 444
    convolution, 443
    differentiation, 445
    final value, 445
    frequency shifting, 443
    initial value, 445
    inverse, 449
    linearity, 442
    multiplication, 444
    properties, 442
    time reversal, 443
    time shifting, 442
Thermal noise, 115, 116
Threshold detector, 261
Time delay, 151, 156
Time dispersion, 161
Time diversity, 117
Time division multiplexed (see TDM), 373
Time domain interpolation formula, 98
Time-domain equalizer (see TEQ), 369
Time-limited signal, 98
Time-only equalizer, 316, 340
Time-selective fading, 162
Time-varying filter, 272
Timing error, 174
Timing error detector (see TED), 410

Index

Timing jitter, 176, 407, 424
Timing offset, 176
Timing recovery, 402, 418, 435
Tomlinson precoder, 313
Tomlinson-Harashima precoder, 313
Total probability theorem, 29
Training method, 277
Transformation function, 43
Transmitter, 17
Transpose of the vector, 458
Transversal filter, 210, 260
Two-port network, 164, 186

U

Ultra wideband (*see* UWB), 2
Unbiased estimate, 199
Undersampling, 100, 101, 107
Uniform DFT receiver
 filter bank, 377
Uniform random variable, 58
Unit norm vector, 458
UWB, 2, 3, 6, 10, 22, 326

V

Variance, 36
VCO, 388, 390
 discrete-time, 400, 404, 412
VDSL, 351
Voltage-controlled oscillator
 (*see* VCO), 388

W

Water-filling algorithm, 352
Water-filling optimization, 356
WCDMA, 3, 334, 341
Weak law of large numbers, 69

White Gaussian noise, 193
Wide-sense stationary
 (*see* WSS), 73, 202
Wideband code division multiple
 access (*see* WCDMA), 3
Wiener filters, 133
Wiener-Hopf equation, 204
 normal equation, 204
Wired communication, 142
Wireless channels, 143
Wireless communication, 141
Wireless communications, 276
Wireless local area network
 (*see* WLAN), 2
Wireline communications, 276
WLAN, 2, 3, 6
WSS, 73, 74, 76, 88

Z

Zero crossing, 174
Zero ISI, 180, 181
Zero-forcing condition, 265, 286
Zero-forcing equalizer, 264–266
Zero-forcing solution, 179

Recent Titles in the Artech House Signal Processing Library

Computer Speech Technology, Robert D. Rodman

Digital Signal Processing and Statistical Classification, George J. Miao and Mark A. Clements

Handbook of Neural Networks for Speech Processing, Shigeru Katagiri, editor

Hilbert Transforms in Signal Processing, Stefan L. Hahn

Phase and Phase-Difference Modulation in Digital Communications, Yuri Okunev

Signal Processing in Digital Communications, George J. Miao

Signal Processing Fundamentals and Applications for Communications and Sensing Systems, John Minkoff

Signals, Oscillations, and Waves: A Modern Approach, David Vakman

Statistical Signal Characterization, Herbert L. Hirsch

Statistical Signal Characterization Algorithms and Analysis Programs, Herbert L. Hirsch

Voice Recognition, Richard L. Klevans and Robert D. Rodman

For further information on these and other Artech House titles, including previously considered out-of-print books now available through our In-Print-Forever® (IPF®) program, contact:

Artech House	Artech House
685 Canton Street	46 Gillingham Street
Norwood, MA 02062	London SW1V 1AH UK
Phone: 781-769-9750	Phone: +44 (0)20 7596-8750
Fax: 781-769-6334	Fax: +44 (0)20 7630-0166
e-mail: artech@artechhouse.com	e-mail: artech-uk@artechhouse.com

Find us on the World Wide Web at: www.artechhouse.com